Fluid Mechanics and Its Applications

Founding Editor

René Moreau

Volume 137

Series Editor

André Thess, German Aerospace Center, Institute of Engineering Thermodynamics, Stuttgart, Germany

The purpose of this series is to focus on subjects in which fluid mechanics plays a fundamental role. As well as the more traditional applications of aeronautics, hydraulics, heat and mass transfer etc., books will be published dealing with topics, which are currently in a state of rapid development, such as turbulence, suspensions and multiphase fluids, super and hypersonic flows and numerical modelling techniques. It is a widely held view that it is the interdisciplinary subjects that will receive intense scientific attention, bringing them to the forefront of technological advancement. Fluids have the ability to transport matter and its properties as well as transmit force, therefore fluid mechanics is a subject that is particulary open to cross fertilisation with other sciences and disciplines of engineering. The subject of fluid mechanics will be highly relevant in such domains as chemical, metallurgical, biological and ecological engineering. This series is particularly open to such new multidisciplinary domains. The median level of presentation is the first year graduate student. Some texts are monographs defining the current state of a field; others are accessible to final year undergraduates; but essentially the emphasis is on readability and clarity.

Springer and Professor Thess welcome book ideas from authors. Potential authors who wish to submit a book proposal should contact Dr. Mayra Castro, Senior Editor, Springer Heidelberg, e-mail: mayra.castro@springer.com

Indexed by SCOPUS, EBSCO Discovery Service, OCLC, ProQuest Summon, Google Scholar and SpringerLink

José María Montanero

Tip Streaming of Simple and Complex Fluids

Foreword by Alfonso M. Gañán Calvo

Springer

José María Montanero
University of Extremadura
Badajoz, Spain

ISSN 0926-5112 ISSN 2215-0056 (electronic)
Fluid Mechanics and Its Applications
ISBN 978-3-031-52767-8 ISBN 978-3-031-52768-5 (eBook)
https://doi.org/10.1007/978-3-031-52768-5

This Springer imprint is published by the registered company Springer Nature Switzerland AG
The registered company address is: Gewerbestrasse 11, 6330 Cham, Switzerland

Paper in this product is recyclable.

A mis padres, fuente de motivación y confianza a lo largo de mi vida

Foreword

This book reflects the outcome of more than 20 years of collaboration between the author and me, exemplified by our completion of numerous joint research projects back to 2002. The idea to create a book as a joint celebration of the 25th anniversary of the first publication on Flow Focusing (Gañán-Calvo 1998, Phys. Rev. Lett. 80, 285) was born in September 2022. However, it is one thing to suggest the joint draft of a book, and quite another to organize it, put the material in perspective, and write it. In this regard, José María Montanero has been exceptionally effective and focused, without concessions to superfluous considerations and ramifications. Thus, I firstly express my gratitude to José María for compiling the work on the development initiated by my invention of Flow Focusing in 1994, including its impact on the related field of fluid physics or other relevant domains, which was brought to fruition thanks to the collective endeavor of contemporary researchers, leading their own ideas and developments or being inspired by the concept.

Flow Focusing has revolutionized microfluidics since Howard Stone and colleagues envisioned a way to incorporate its basic geometry (Gañán-Calvo, US Pat 6,116,516) into PDMS in a two-dimensional or flat format. Their famous publication (Anna, Bontoux and Stone 2003, Appl. Phys. Lett. 82, 364) established this technique as a standard microfluidic method in the scientific literature, where one can find other configurations derived from Flow Focusing, such as that of Utada et al. 2005, Science 308, 537. This technique allows the continuous production of very uniform microbubbles, microdroplets or double or triple microemulsions with a geometric deviation close to one.

One of the strongest illustrations of the enabling capacity of Flow Focusing is its application in Serial Femtosecond Crystallography (SFX). Flow Focusing is capable of generating aerodynamically focused liquid microjets that exhibit extraordinary stability, travel at very high velocities, and possess perfectly controllable sizes as minute as a micron. Previously, these features could only be achieved with electrospray techniques (also described in this volume). The concept of sequentially exposing particles such as microcrystals to electromagnetic beams using Flow Focusing was patented in 2002 (Gañán-Calvo et al. 2003, Spanish patent file 200301500). In November 2007, I was invited by John Spence and Bruce Doak and toured their laboratory at Arizona State University. The concept for SFX, which merges X-ray molecular imaging (Chapman et al. 2006, Nat. Phys. 2, 839) with Flow Focusing, originated from Henry Chapman, collaborating with Spence and others (Chapman et al. 2011, Nature 470, 73). Thanks to Flow Focusing, now small biological samples such as protein microcrystals can be sequentially probed by ultra-powerful and focused photon beams (X-ray Free Electron Lasers, XFELs) of femtosecond duration. The high repetition rates of the X-ray pulses at such facilities enable thousands of measurements per second to enable the rapid measurement of the 3D molecular structures in high-throughput experiments that follow complex biochemical reactions in time. Flow Focusing preserves the microcrystals in their necessary liquid environment and supplies fresh crystals to each X-ray pulse in a way that is gentle to these fragile samples. All this has established Flow Focusing as a standard sample handling procedure at XFEL facilities.

From a physics standpoint, Flow Focusing significantly advanced the development of capillary micro-focusing (tip streaming) techniques, going beyond proposing alternatives to the Taylor cone-jet electrospray method. In effect, mass production of particles with controlled sizes below 100 microns at a rate greater than 1 kHz per emission point could only be achieved through the application of intense electric fields to fluid interfaces, resulting in the appearance of cone-shaped points (reducible to singularities) on the charged surface, above a certain surface charge. Thanks to the discovery of Flow Focusing, it was evident that the concentration of forces leading to cusps and caustics—concepts thoroughly discussed by Jens Eggers (Eggers and Fontelos 2015, Singularities: formation, structure, and propagation. Cambridge UP)—could be utilized for the consistent generation of very small-sized streams. Mechanical means, such as a current forced through an orifice or an extensional current, thermal techniques like the surface tension gradient produced by thermal concentration, and even chemical gradient procedures, can all be considered. These techniques are apt to generate controlled points of force concentration (tips), which can lead to micro-emissions of fluid from localized areas (streaming) due to the localized concentration of stresses or mass forces against surface tension and viscosity.

The fundamental point of the book authored by my esteemed colleague becomes evident through the first two figures of the initial chapter, taken from one of our papers. When I received the National Research Award from the King of Spain in February 2010 for my invention of Flow Focusing, I explained its use in mass-producing controlled size and structured templates at the micro and nano-scale. This process is critical in several high-end industries, including pharmaceuticals, chemicals, food, and electronics. During the presentation, an audience member asserted that the use of mass-produced micro-templates was not a new concept, as it had been previously researched and implemented. I responded that whatever the age of the templating concept, Flow Focusing facilitates the mass production of particles in varying sizes and shapes, such as drops, bubbles, or microcapsules, with an extreme precision of control and statistical dispersion. No other mechanical method can produce them with such accuracy and high output volume. Now, tip streaming not only generalizes the concept introduced by Flow Focusing but also gathers a plethora of tiny, elusive, yet critical processes to controllably disaggregate immiscible fluids.

In summary, José María Montanero has placed in your hands, dear reader, an extraordinary and up-to-date compendium of all the current knowledge related to the controlled disintegration of one or more immiscible fluids into another (or vacuum), either stationary or non-stationary, by means of singular stress concentration. In this sense, a wide variety of processes leading to singular stress concentrations are considered, obviously beyond Flow Focusing, such as the implosion of surface cavities (bubble bursting). A very detailed discussion is also given on the stability of the generated flows in the case of the consideration of stationary emission flows.

In addition, the author provides a unique insight into the influence of the complex molecular structure and non-Newtonian nature of the fluids involved in the disintegration process. Such a compendium aims to serve as a primary reference for future researchers in the field. As Plutarch put it, "Knowledge is not a vessel to be filled, but a fire to be kindled."

Sevilla, Spain Alfonso M. Gañán-Calvo
November 2023

Preface

Tip streaming is an elusive but fascinating phenomenon. Under particular conditions, some external actuation of hydrodynamic or electrohydrodynamic nature stretches a droplet or film, which emits either tiny drops or an extremely thin fluid thread from its tip. The complexity of some versions of this phenomenon is difficult to find in other fields of mesoscale physics, even when the system adopts a steady or quasi-steady regime. Despite numerous works on tip streaming, there are still open fundamental questions about this phenomenon with significant repercussions at the technological level.

This book comprehensively describes the tip streaming in simple fluids and those containing surfactants and polymeric molecules. It summarizes the theoretical models and approximations commonly adopted to analyze this phenomenon. It provides relevant experimental results and presents the scaling laws for rationalizing those results. The stability of the flows leading to tip streaming is investigated experimentally and theoretically. Attention is paid to the effects of surfactant monolayers and viscoelasticity, including solutocapillarity, interfacial elasticity, surface viscosity, and extensional thickening caused by the polymer coil-stretch transition. This book also offers an overall perspective of the numerous technological applications of the tip streaming phenomenon.

We describe the physical mechanisms responsible for the onset of tip streaming driven by hydrodynamic and electrohydrodynamic forces. This text reviews relevant theoretical and experimental results of the periodic microdripping and continuous microjetting modes of tip streaming produced with microfluidic configurations such as electrospray, Flow Focusing, coflowing, selective withdrawal, and confined selective withdrawal. Similarities and distinctions among these techniques are analyzed. The physical mechanisms responsible for the instability of the microjetting mode are studied in detail.

We discuss the applicability of the above-mentioned configurations for controlled soft matter shaping and fragmentation on the submillimeter scale in several technological fields. Remarkable examples are the production of microemulsions and microencapsulation of active agents for the food and pharmacy industries, the atomization of charged liquids for analytical chemistry, and the ejection of ultra-fast and ultra-thin jets for crystallography. The book collects the scaling laws used to rationalize experimental data and predict the outcome of the above-mentioned microfluidic configurations.

This text combines state-of-the-art experimental results and linear stability analysis to identify the instability mechanisms limiting the applicability of the above-mentioned microfluidic configurations. In this way, the book connects experimental observations with fundamental aspects of tip streaming, bridging the microfluidic and fluid dynamicist communities. The connection between results obtained from the theoretical and empirical approaches will help experimentalists understand the fundamental aspects of their practical problems. This work may constitute a valuable guide for the large community working on hydrodynamic focusing and electrospray. It will help researchers in this fruitful area frame their results in a broad context.

Chapter 1 contextualizes the tip streaming phenomenon in the field of microfluidics. Also, we briefly describe the microfluidic configurations analyzed in this work and the ingredients adding complexity to the problem. Chapter 2 presents the equations governing the flows considered in the book in the presence of electric fields, surfactants, and viscoelasticity. Chapters 3 and 4 are methodological sections that discuss the theoretical and experimental approaches commonly used in the study of tip streaming. We devote Chaps. 5 and 6 to the analysis of the onset (transient) tip streaming caused by electrohydrodynamic and hydrodynamic means, respectively. This phenomenon substantially differs from the periodic and steady modes produced by microfluidic devices that allow the control of the dispersed phase flow rate. Chapter 7 describes in detail those devices and introduces the dimensionless numbers characterizing the flows. Chapters 8 and 9 review relevant works for the specific configurations. We start by discussing some fundamental results of electrospray. Chapter 9 analyzes the tip streaming produced by coflowing and hydrodynamic focusing. We reserve the last two chapters for gaseous Flow Focusing, which enjoys remarkable applications.

This publication contains many results and discussions resulting from the fruitful collaboration with my colleagues at the Universities of Extremadura and Seville over the last 20 years. I want to thank all of them for their essential contribution to this work. Without that contribution, it would have been impossible for me to write the present book. I want to thank Pr. Alfonso Gañán-Calvo for introducing me to this field and helping me with his novel ideas and inspiring suggestions. My very close and daily collaboration with Pr. Miguel A. Herrada has been indispensable for many of the studies presented in this book. Those studies are based on the numerical method he devised and perfected over the years. Also, I would like to acknowledge the helpful discussions with Pr. José M. López-Herrera. Worthy of special mention are my former Ph.D. students and now esteemed colleagues at the University of Extremadura; in alphabetical order, Alberto Ponce-Torres, Alejandro Rubio, Antonio Acero, Beatriz Muñoz, Conrado Ferrera, Emilio Vega, Guadalupe Cabezas, Manuel Rubio, and

Noelia Rebollo. They helped me with their generous effort and continuous support to conduct the work presented in this book.

Badajoz, Spain
November 2023

José María Montanero

Contents

Symbols

α	Womersley number in gaseous Flow Focusing
$\hat{\alpha}$	Fluid volume fraction
α^*, α	Dimensionless conductivity
β	Surfactant elasticity (strength)
$\varUpsilon$	Heat capacity ratio
$\varGamma$	Surfactant surface concentration (surface coverage)
$\varGamma_\infty$	Maximum packing density
$\varGamma_{\mathrm{eq}}$	Surfactant surface density at equilibrium
δ_g	Gaseous boundary layer thickness in Flow Focusing
δ_l	Liquid boundary layer thickness in Flow Focusing
δ_μ	Electrohydrodynamic Reynolds number
δF	Perturbation amplitude of the interface location
$\varDelta p_e$	Effective pressure drop in electrospray
$\varDelta p$	Applied pressure drop in Flow Focusing
$\varDelta p_l$	Pressure drop in the liquid meniscus of Flow Focusing
$\varepsilon^{(j)}$	Electrical permittivity of the phase j
ε^*	Complex dielectric constant
$\varepsilon', \varepsilon$	Relative permittivity
ε''	Dielectric loss
ε_o	Outer dielectric medium permittivity
$\dot{\varepsilon}$	Strain rate
θ_a	Advancing contact angle
θ_r	Receding contact angle
θ_T	Taylor cone semiangle
$\kappa_T^{(j)}$	Thermal conductivity of the phase j
κ_{Ti}	Inner phase thermal conductivity
κ_{To}	Outer phase thermal conductivity
κ^s	Dilatational surface viscosity
κ	Twice the mean curvature of the interface
κ_1	Curvature along the meridian

κ_2	Curvature along the parallel
κ_{tip}	Tip curvature
$\lambda_\mu^{(j)}$	Dilatational coefficient of viscosity of the phase j
$\lambda_{\mu o}$	Dilatational coefficient of viscosity of the outer phase
λ_D	Debye layer thickness
$\lambda_s^{(j)}$	Stress relaxation time evaluated in the phase j
$\lambda_r^{(j)}$	Stress retardation time evaluated in the phase j
λ_e	Extensional relaxation time
$\hat{\lambda}_D$	Thickness of the surfactant boundary layer
$\hat{\lambda}$	Inner to outer viscosity ratio in a extensional flow
Λ	Diameter ratio
$\mu^{(j)}$	Viscosity of the phase j
$\mu_p^{(j)}$	Polymer viscosity of the phase j
$\mu_s^{(j)}$	Solvent viscosity of the phase j
μ^s	Shear surface viscosity
μ	Viscosity, ratio of the outer to the inner viscosity
$\hat{\mu}$	Ratio of the inner to the outer viscosity
μ_0	Viscoelastic solution viscosity
μ_i	Inner phase viscosity
μ_o	Outer phase viscosity
$\rho^{(j)}$	Density of the phase j
ρ_i	Inner phase density
ρ_o	Outer phase density
ρ	Density, ratio of the outer to the inner density
$\hat{\rho}$	Ratio of the inner to outer density
$\rho_e^{(j)}$	Net free charge density in the phase j
σ	Surface tension
σ_0	Surface tension of the clean interface
$\hat{\sigma}$	Surface tension relative to its characteristic value
σ_{eq}	Surface tension of a surfactant-covered interface at equilibrium
σ_e	Surface charge density
$\tau^{(j)}$	Deviatoric stress tensor in the phase j
$\tau_p^{(j)}$	Polymer contribution to the deviatoric stress in the phase j
$\tau_s^{(j)}$	Solvent contribution to the deviatoric stress in the phase j
$\tau_M^{(j)}$	Maxwell stress in the phase j
τ_{Ma}	Marangoni stress
τ_s	Viscous surface stress
τ_{sn}	Normal viscous surface stress
τ_{st}	Tangential viscous surface stress
$\phi^{(i)}$	Electric potential in the inner phase
$\phi^{(o)}$	Electric potential in the outer phase
χ	Electric Bond (Taylor) number
$\omega_{n,m}$	Complex eigenfrequency of the mode (n, m)
a	Mother drop radius

$\mathbf{A}^{(j)}$	Conformation tensor in the phase j
Ai	Airy function
B	Bond number
B_{μ}^{s}	Boussinesq number based on the shear surface viscosity
B_{κ}^{s}	Boussinesq number based on the dilatational surface viscosity
Ca	Capillary number
Ca_G	Capillary number based on the extensional flow intensity
Ca_{tip}	Capillary number based on the tip velocity
Ca_o	Outer phase capillary number
Ca_i	Inner phase capillary number
$\overline{Ca}$	Capillary number based on the interface velocity
$c_v^{(j)}$	Specific heat coefficient of the phase j
$c^{(j)}$	Volumetric monomer concentration in the phase j
$c_s^{(j)}$	Surfactant concentration on the interface side j
c_∞	Reservoir surfactant concentration
c_{cmc}	Critical micelle concentration
d_o	Characteristic (intrinsic) diameter scale of electrospray
d_d	Droplet diameter
d_b	Bubble diameter
d_j	Jet diameter
d_j^{FF}	Jet diameter predicted by the Flow Focusing formula
$\mathbf{D}^{(j)}$	The deformation rate tensor evaluated in the phase j
$\mathscr{D}$	Diffusion coefficient
$\mathscr{D}_c^{(j)}$	Diffusion coefficient for the surfactant as monomers
$\mathscr{D}_m^{(j)}$	Diffusion coefficient for the surfactant as micelles
$\mathscr{D}_s$	Surfactant surface diffusion coefficient
$\mathscr{D}_k$	Diffusion coefficient of the k-species in electrospray
D_o	Focusing orifice diameter
$\mathscr{E}_j$	Dimensionless electric strength
E_o	Characteristic electric field in electrospray
F	Distance of the interface from the z axis
F_0	Distance of the interface from the z axis in the base flow
$\mathscr{F}$	Faraday constant
F_{σ}	Characteristic capillary force in Flow Focusing
F_p	Characteristic pressure force in Flow Focusing
$\mathbf{g}$	Gravitational acceleration
G	Intensity of the uniaxial extensional flow
H	Capillary-to-electrode/focusing orifice distance
$\hat{H}$	Dimensionless capillary-to-focusing orifice distance
H_{opt}	Optimum capillary-to-focusing orifice distance
$\mathbf{I}$	Identity matrix
$\mathbf{I}_s$	Tensor that projects any vector onto the interface
I	Electric current
I_s	Electric current convected by the surface

I_{se}	Electrostatic value of I_s
I_b	Electric current due to conduction across the bulk
I_o	Characteristic (intrinsic) electric current in electrospray
$\mathbf{j}$	Charge flux
$\mathscr{J}_{cm}^{(j)}$	Net rate of formation/breakup of micelles per unit volume
$\mathscr{J}_{\Gamma c}^{(j)}$	Net flux of surfactant from the phase j
k_B	Boltzmann constant
$k_a^{(j)}$	Adsorption constant for the phase j
$k_d^{(j)}$	Desorption constants for the phase j
$\widetilde{k}_a$	Adsorption Biot number
$\widetilde{k}_d$	Desorption Biot number
$\widetilde{k}$	Complex wavenumber
$K^{(j)}$	Electrical conductivity of the phase j
l_b	Jet breakup length
L	Axial characteristic length in electrospray
L_j	Jet length
L^2	Finite extensibility parameter
La	Laplace number
$m^{(j)}$	Volumetric concentration of surfactant as micelles
m	Azimuthal number
Ma	Marangoni number
$\mathbf{n}$	Unit outward normal vector
$n_k^{(j)}$	Number of mols per unit volume of the k-species
Oh	Ohnesorge number
Oh_j	Jet Ohnesorge number
Oh_i	Inner phase Ohnesorge number
Oh_{μ}^{s}	Ohnesorge number based on the shear surface viscosity
Oh_{κ}^{s}	Ohnesorge number based on the dilatational surface viscosity
$P^{(j)}$	Reduced pressure field in the phase j
Pe	Peclet number
Pe_s	Surface Peclet number
$\mathbf{q}^{(j)}$	Heat flux vector
q_R	Rayleigh limit of charge
q_o	Intrinsic charge scale of electrospray
Q	Flow rate, dimensionless flow rate in electrospray
Q_c	Characteristic (intrinsic) flow rate in electrospray
Q_i	Inner phase flow rate
Q_o	Outer phase flow rate
$Q_{i\min}$	Minimum inner phase flow rate
$Q_{\min}$	Minimum flow rate ratio
Q_r	Outer to inner phase flow rate ratio
Q_D	Flow Focusing characteristic flow rate
Q_v	Flow Focusing characteristic flow rate
$Q_{i\mathrm{opt}}$	Optimum flow rate in gaseous Flow Focusing

Q_d	Dragged liquid flow rate in gaseous Flow Focusing
$r_k^{(j)}$	Net production rate of the species k
$\hat{R}_g^{(j)}$	Gas constant of the phase j
R_g	Universal gas constant
R_j	Jet radius
R_i	Feeding capillary radius
R_d	Droplet radius
R_{FF}	Jet radius predicted by the Flow Focusing formula
R_μ	Viscous characteristic length of Flow Focusing
Re_i	Inner phase Reynolds number
Re_j	Jet Reynolds number
$\mathrm{Re}_{\mathrm{FF}}$	Reynolds number based on $\boldsymbol{R}_{\mathrm{FF}}$ and v_j^{FF}
s	Intrinsic (surface) coordinate
$t_{\rho\sigma}$	Inertio-capillary time based on the capillary radius
$\hat{t}_{\rho\sigma}$	Jet inertia-capillary time
$\tilde{t}_{\rho\sigma}$	Inertio-capillary time based on the droplet diameter
t_e	Electric relaxation time
t_o	Characteristic (intrinsic) time scale of electrospray
t_r	Residence time
$\mathbf{T}_s$	Viscous surface stress tensor
u	Radial component of the velocity field
$U_{\mu\sigma}$	Visco-capillary velocity
$U_{\rho\sigma}$	Inertio-capillary velocity based on the capillary radius
v_o	Characteristic (intrinsic) velocity scale of electrospray
v_j	Jet velocity
v_j^{FF}	Jet velocity predicted by the Flow Focusing formula
$\mathbf{v}_s$	Surface velocity
V_i	Inner phase velocity
w	Axial component of the velocity field
w_k	Ion mobility of the k-species
We_j	Jet Weber number
We_i	Inner phase Weber number
$\mathrm{We}_{\mathrm{FF}}$	Weber number based on $\boldsymbol{R}_{\mathrm{FF}}$ and v_j^{FF}
Wi	Weissenberg number
Wi_j	Jet Weissenberg number
z_k	Valence of the k-species
Z	Distance from the ejector orifice

Chapter 1
Introduction

Abstract This book's title requires clarifying two concepts: "tip streaming" and "complex fluid." In essence, tip streaming is the flow producing tiny drops or jets from the tip of a mother droplet, fluid meniscus, or liquid film. We devote the first part of this chapter to motivating and contextualizing the tip streaming phenomenon. We also introduce the microfluidic configurations commonly used to produce this singular flow. These configurations and their specific applications will be analyzed in more detail throughout the text.

The term "complex fluids" has been coined to refer to substances involving the coexistence of several phases, such as foams, suspensions, and emulsions. It also refers to fluids containing colloids and macromolecules, as well as liquid crystals and quantum fluids, among others. Here, we use this term to refer only to liquids containing surfactant molecules or polymer chains, which produce interfacial and bulk viscoelastic effects, respectively. As explained in other chapters, surfactants and polymers substantially alter the tip streaming flow in many technologically relevant applications. In the last part of this chapter, we briefly describe these two elements. We also introduce the effect of externally applied electric fields, which drive the flow in many tip streaming realizations.

Keywords Tip streaming · Complex fluids · Microfluidic configurations · Surfactant · Viscoelasticity

1.1 Matter Fragmentation, Shaping, and Deposition

Tip streaming can be regarded as a microfluidic technique to produce the fragmentation, shaping, and deposition of matter in its fluid phase on the submillimeter scale in a controlled manner. In this section, we briefly describe the advantages of tip streaming over traditional droplet-based microfluidics methods for those purposes. Specific applications of tip streaming will be discussed in other chapters.

J. M. Montanero, *Tip Streaming of Simple and Complex Fluids*, Fluid Mechanics and Its Applications 137, https://doi.org/10.1007/978-3-031-52768-5_1

1.1.1 Matter Fragmentation

The matter fragmentation produced by tip streaming on the submillimeter scale has applications in countless fields, such as industrial and chemical engineering, biotechnology, the food and agricultural industry, and advanced material processing. For instance, electrohydrodynamic tip streaming (electrospray) creates aerosols in mass spectrometry [1]. Pharmacy is a salient example of the tip streaming applicability [2]. In this field, functional products with complex morphologies must be manufactured in a controlled way. For the sake of illustration, we here consider this technological discipline to show the relevance of tip streaming in processing both simple and complex fluids.

Product manufacturing in pharmacy belongs to the technological field of chemistry. Reducing bulk reactants or ingredients to granular matter is a typical preparatory step in this field. Chemistry often makes use of top-down fluid-dynamical processes for this purpose. These processes involve forces and energies much smaller than those necessary to trigger chemical reactions, ensuring that the reactants' chemical nature is not altered [2].

The main features of the produced granular matter are the size, shape, and degree of homogeneity of the grains or particles. The particle size determines the surface-to-volume ratio of the processed bulk material, while the size homogeneity reflects how that ratio is distributed among the grains. Finally, the particle shape affects not only the chemical reaction rate of the product but also other mechanical processes undergone by the grains, such as self-assembly and patterning [3–6].

As mentioned above, fluid-dynamical processes are frequently applied to the bulk material to shape it into a desired morphology (i.e., a specific granular structure). These processes use the liquid phase as the starting state and the surface tension as the *mechanical mold*. If the characteristic lengths of the liquid internal structure are much smaller than the size of the produced fragments, then the shape resulting from the complete fragmentation is a sphere (drop). The spherical shape offers two fundamental advantages: it is reproducible and provides the minimum surface-to-volume ratio for a given volume [7, 8]. The cylindrical shape (the jet) is also reproducible, frequently appearing as an intermediate state during fragmentation. This shape is inherently ephemeral due to the capillary instability [9, 10] unless a sufficiently fast physicochemical process (drying, freezing, curing, …) causes a phase change of the liquid before the breakage.

Fluid-dynamical processes produce controlled or irregular fragmentation depending on the relative strength of the forces involved. If inertial forces are much larger than surface tension, fragmentation is turbulent and dominated by stochastic processes. On the contrary, processes driven by surface tension lead to laminar and smooth fragmentation (Fig. 1.1). As explained below, tip streaming techniques, such as flow focusing, can generate those capillary processes [2]. Figure 1.2 shows scanning electron microscopy photographs of Gem-loaded PLGA particles obtained by turbulent (Flow Blurring®) and laminar (Flow Focusing®) fragmentation followed by solvent extraction [11].

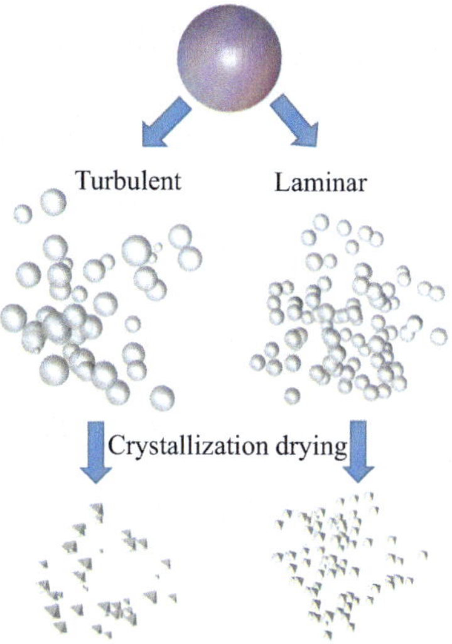

Fig. 1.1 Matter fragmentation: turbulent (flow blurring) versus laminar (flow focusing) fragmentation (reprinted from Ref. [2] with permission of Elsevier)

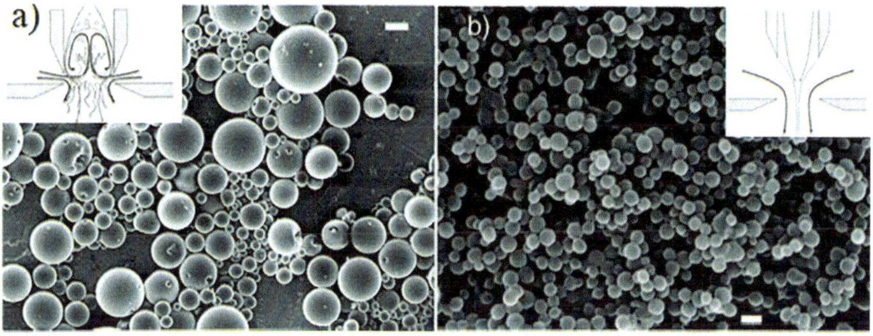

Fig. 1.2 Production of PLGA microparticles: flow blurring (**a**) versus flow focusing (**b**). Bars length: 1 μm (reprinted from Ref. [2] with permission of Elsevier)

1.1.2 Matter Shaping

Some tip streaming methods used for matter fragmentation can also be applied to shape a continuous phase on the submillimeter scale. This process has many applications in very diverse fields. For instance, flow focusing shapes a liquid volume into a fast micrometer jet, as discussed in Chaps 10 and 11. This technique has become the most common method for sample delivery in serial femtosecond crystallography [12].

One of the most remarkable examples of tip streaming for matter shaping is the formation of fibers. Fibers of micro and nanometer sizes possess many relevant

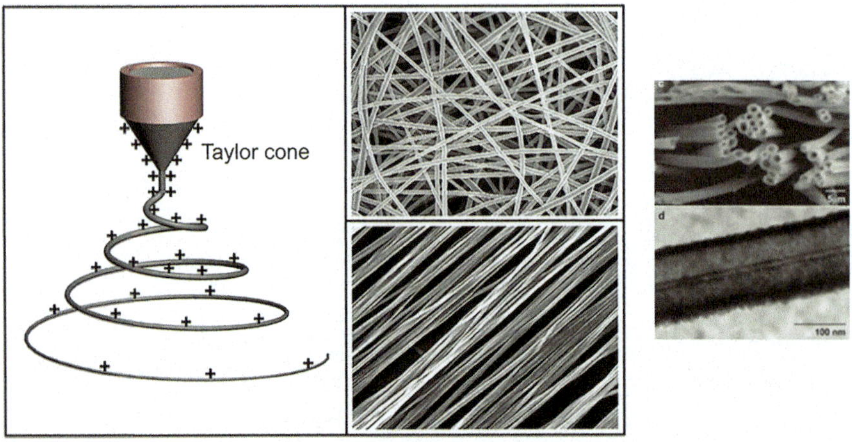

Fig. 1.3 (Left) Electrospinning (reprinted from Ref. [16] with permission of American Chemical Society). (Right) Hollow and coaxial fibers produced with electrospinning (reprinted from Ref. [17] with permission of American Chemical Society)

features, such as high and controllable surface area-to-volume ratios or excellent mechanical properties. Besides, the materials employed to form the fibers exhibit a great variety of physicochemical and biological properties. All these characteristics make micro and nanofibers and their mats particularly attractive for a vast class of applications, including filters, membranes, microelectronics, military, optics, or health and personal care, among many others [13]. Glass and silk fibers [14] have obvious applications in the telecommunications and textile industries, respectively.

Electrospinning has become one of the most popular methods to form micro and nanofibers. It allows the mass production of one-by-one continuous fibers with different morphologies from various polymers (Fig. 1.3). This tip streaming technique and its varied applications have been extensively reviewed over the last years (see, e.g., [13–15]). For this reason, the present work does not intend to describe this method in depth.

Fibers can also be produced from the viscoelastic counterparts of flow blurring [18] and flow focusing [19]. Analogously to what occurs in the Newtonian case, massive production of fibers with a high degree of polydispersity can be extruded with a turbulent gas flow. In contrast, quasi-monodisperse collections of fibers are obtained when the polymer solution is shaped in the laminar mode (Fig. 1.4).

1.1.3 Matter Deposition

Another scope of application of the tip streaming phenomenon is the deposition of a liquid phase on a substrate with submillimeter resolution in a controlled manner. A salient example of this is printed and flexible electronics. Polymers with high elec-

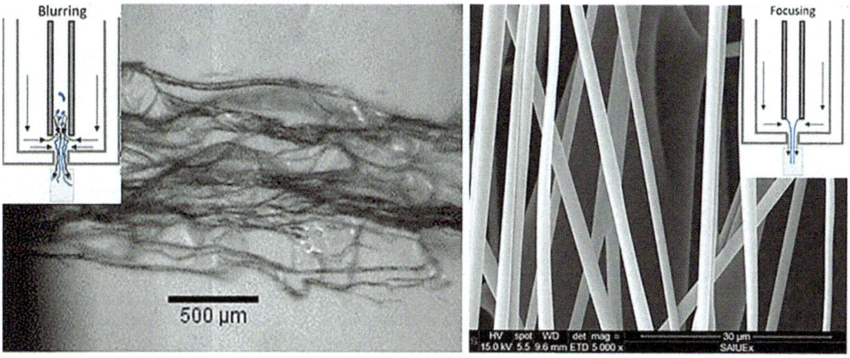

Fig. 1.4 Production of fibers from viscoelastic solutions: flow blurring (left) (reprinted with permission from Ref. [18]) versus flow focusing (right)

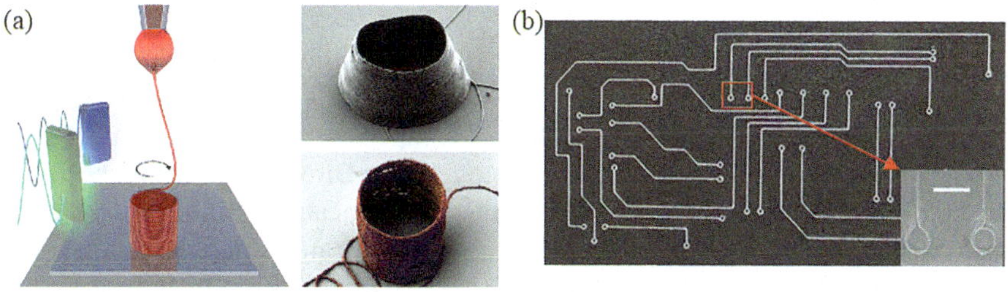

Fig. 1.5 **a** Ultrafast 3D printing of PEO cylindrical microstructures by controlling the voltage applied to two electrodes placed around the jet emitted in electrospinning [26] https://creativecommons.org/licenses/by/4.0/. **b** A circuit pattern fabricated by electrohydrodynamic printing on a glass slide (scale bar 500 μm) (reprinted with permission from Ref. [27])

trical conductivity, transparency, and physical and chemical stability are printed to form films. These films can be produced with inkjet printing [20], which constitutes a simple, flexible, and fast procedure to deposit/pattern a polymer with low cost, waste, and contamination. Inkjet printing has been used in the fabrication of organic solar cells [21, 22], electronic circuitry components [23], LEDs, sensors, and biosensors, among other electronic systems. Near-field electrospinning [24] and electrohydrodynamic printing [25] are tip streaming realizations used to emit and deposit polymeric jets much thinner than all the passages and orifices of the device (Fig. 1.5-left), increasing the spatial resolution of the printing technique (Fig. 1.5-right).

Bioengineering is another technological field that benefits from the smooth and gentle deposition of a liquid phase on a substrate. In extrusion bioplotting (also referred to as Direct Ink Write, DIW) [28–30], threads of "bioinks" are continuously extruded and deposited in spatially controlled filaments by a robotic system to build 3D structures. These bioinks usually consist of aqueous media, thermoreversible polymers, or polymer/hydrogel precursors combined with living cells. Tip streaming, as applied to bioplotting, allows reducing the diameter of the extruded threads and, therefore, of the deposited filaments.

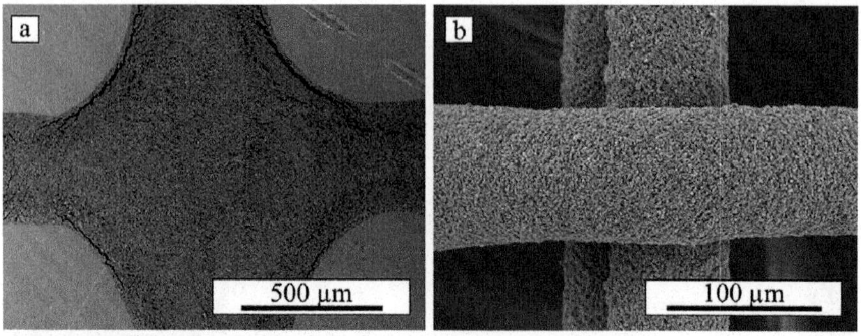

Fig. 1.6 SEM image of the intersection of ceramic (tricalcium phosphate, TCP) fibers produced with near-field electrospinning and deposited orthogonally in different layers under **a** low and **b** high UV exposure (reprinted from Ref. [31]). https://creativecommons.org/licenses/by-nc-nd/4.0/

Thin ceramic fibers are widely used in a range of applications ranging from catalysis to biomedical materials. Due to their high surface-to-volume ratio and superior strength, they are instrumental in fabricating composites with enhanced mechanical properties in terms of strength and fracture toughness. Direct ink writing is becoming an additive manufacturing technique of choice to fabricate scaffolds from bioactive materials like calcium phosphates or bioglasses. As occurs in bioplotting, one of the main limitations of this technique is that clogging of the extrusion nozzle prevents the production of ceramic struts less than 200 μm in diameter. Near-field electrospinning and other tip streaming methods can overcome this limitation (Fig. 1.6).

The applications described in this section are just some relevant examples of the utility of tip streaming. An exhaustive description of all these applications is beyond the scope of this text.

1.2 Droplet-Based Microfluidics

As mentioned above, numerous technological applications require the fragmentation of a continuous fluid phase into droplets/bubbles of different nature and morphology with narrow diameter distributions on the micro and nanometer scales. An ample variety of methods can be used for this purpose. Among them, we can distinguish drop-on-demand techniques from those in which droplets/bubbles are continuously generated [32, 33] (Fig. 1.7).

In a drop-on-demand technique, an external actuator produces the ejection of an individual droplet through the nozzle. Thermal and piezoelectric actuators are frequently used to this end. For instance, a resistor heats the ink until it vaporizes in the thermal inkjet method [34]. The quasi-instantaneous collapse of the resulting bubble drives the ejection of a droplet through the nozzle. In the piezoelectric inkjet method [33, 35], the droplet is ejected by the action of a pressure wave triggered

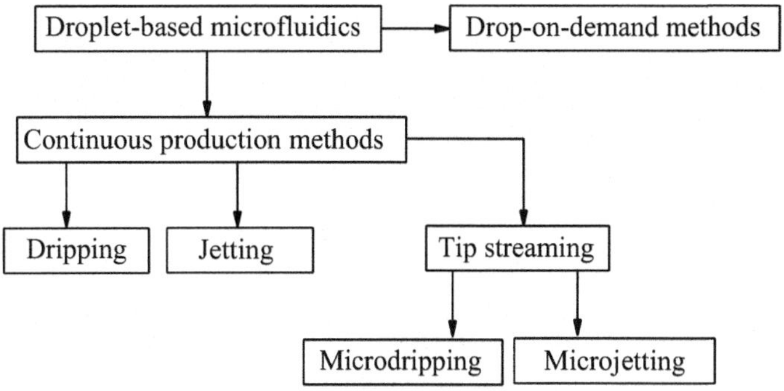

Fig. 1.7 Methods for producing droplets/bubbles on the micrometer and nanometer scales. Tip streaming is a route to microdripping and microjetting

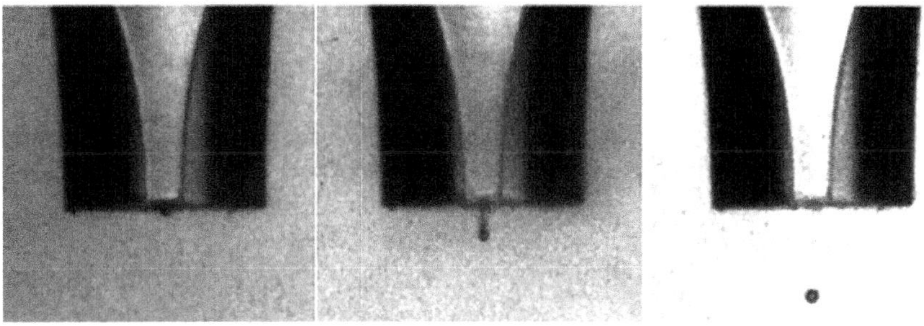

Fig. 1.8 Formation of a drop about 32 μm in diameter upon application of a voltage waveform to a piezoelectric transducer (reprinted from Ref. [36] with permission of AIP Publishing)

by the contraction of a piezoelectric element. Drop-on-demand techniques produce droplets with diameters similar to or even bigger than that of the ejecting nozzle. There are very few exceptions in which the diameter can be significantly reduced (Fig. 1.8) [36].

In a continuous production method, the nozzle ejects the fluid phase steadily. This ejection is powered by an external energy source constantly applied to that phase [37]. The nozzle can operate either in the dripping/bubbling or in the jetting mode depending on the ratio of the fluid kinetic energy to the interfacial one at the ejector orifice (Fig. 1.9). In the dripping/bubbling mode, drops/bubbles are emitted right behind the exit of the nozzle, while a fluid thread long compared with its diameter is formed in the jetting regime. In this case, the surface tension triggers the Plateau-Rayleigh instability [8, 9], which eventually leads to the breakup of the thread into a collection of droplets/bubbles whose diameters are commensurate with that of the jet (nozzle). The dripping/bubbling mode generally leads to very high monodispersity degrees, while the jetting regime yields larger production rates.

The distinction between the dripping/bubbling and jetting regimes is not always clear. In many applications, jetting becomes dripping/bubbling as the precursor fluid

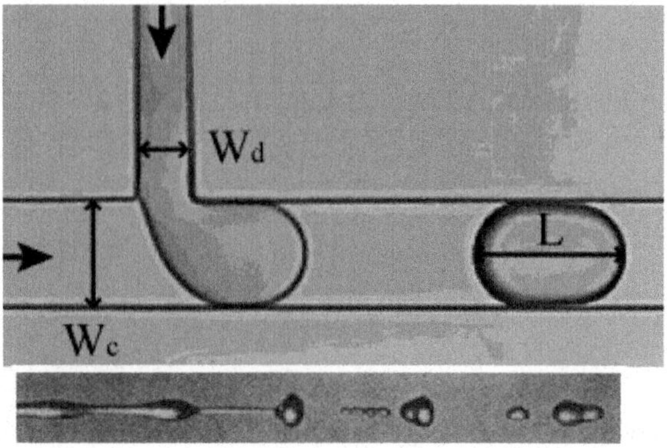

Fig. 1.9 Dripping in a T-junction (upper image) (reprinted from Ref. [38] under Creative Commons CC. BY. 4.0 license) and jetting (lower image) (reprinted from Ref. [39] with permission)

thread shortens. In this case, the free surface pinches off at the ends of the thread (the so-called end-pinching mechanism), also driven by surface tension [40, 41].

Reducing the droplet size and increasing the population's monodispersity degree is very important in many technological applications. However, these two features are somehow antagonistic because the size can be decreased only at the expense of monodispersity, and *vice versa*. Decreasing the size of the droplets requires overcoming the surface tension and viscosity force. This can only be achieved by injecting a significant amount of energy into the process, which generally increases the polydispersity in the usual atomization techniques. High monodispersity degrees can be obtained only if the energy is carefully focused on the droplet tip. When this occurs, the tip streaming phenomenon arises. As explained below, tip streaming is the route to the microdripping and microjetting modes, producing droplets with much smaller sizes than that of the microfluidic device (Fig. 1.7).

1.3 The Tip Streaming Phenomenon

Localized stresses can generate a point of force concentration (tip) that can lead to micro-emissions of fluid (streaming) from diminutive regions, overcoming the resistance of surface tension and viscosity. The stresses driving this tip streaming can have different origins (electrical, mechanical, chemical, thermal, ...). This book will focus on the first two.

More than 100 years ago, Zeleny [42] and subsequently Wilson and Taylor [43] observed that when a pendant droplet or sessile bubble is subjected to a sufficiently strong electric field, a fine jet tapers from the tip of that droplet or bubble. Nolan [44] and Macky [45] observed a similar phenomenon with an electrically neutral levitating droplet. The conical shapes tapering the jets are called Taylor cones [46], and the jet emission is known as electrohydrodynamic tip streaming [47, 48]. More recently, numerous works have contributed to the understanding of the physical mechanism underlying the electrohydrodynamic tip streaming [48–51]. Fenn et al. [1] developed

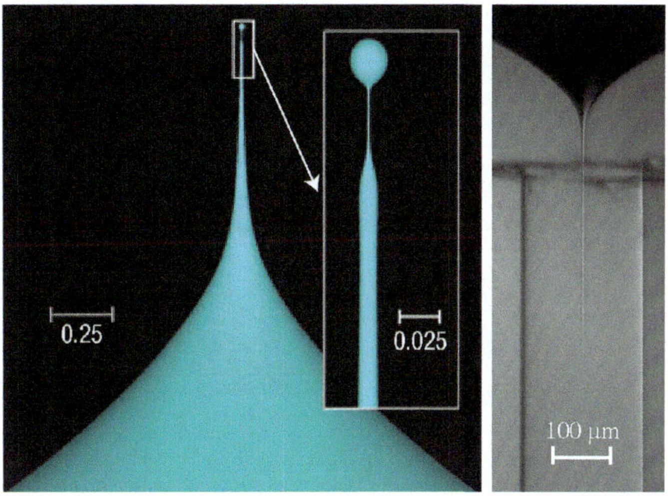

Fig. 1.10 Examples of electrohydrodynamic (left) (reprinted from Ref. [48] with permission) and hydrodynamic (right) (reprinted from Ref. [58] with permission from Elsevier) tip streaming

a novel method of mass spectrometry of large biomolecules based on Zelenny's approach, which resulted in the awarding of the 2002 Nobel Prize in Chemistry to John Fenn.

Tip streaming can also be produced by hydrodynamic means. The antecedents of hydrodynamic tip streaming are more recent. De Bruijn [52] observed this phenomenon in droplets submerged in simple shear flows and covered by surfactants. Gañán-Calvo [53] proposed a method to produce the microjetting mode through the focusing effect caused by a high-speed gas current. The microbubbling mode was obtained when the liquid and gas phases were exchanged [54]. The flow focusing method was popularized by Anna et al. [55] by implementing it in the 2D topology commonly used in microfluidics. Takeuchi et al. [56], Utada et al. [57] demonstrated the utility of this technique for microencapsulation and fabrication of monodisperse double emulsions.

Tip streaming is an elusive but fascinating phenomenon. Under particular conditions, the tip of a deformed droplet or film emits either tiny drops or a very thin fluid thread. As mentioned above, this ejection results from some external actuation of hydrodynamic or electrohydrodynamic nature in the form of interfacial stresses (Fig. 1.10). It is still an open question to determine which factors determine the tiny spatial and temporal scales arising at the onset of tip streaming. In other words, why does the mother droplet decide to break up this way?

In electrohydrodynamic tip streaming, the flow is powered by the electric (Maxwell) stress originated by the accumulation of free electric charge at the interface and the jump of electrical permittivity across this surface [59]. As explained in this text, the tangential electric stress caused by an imperfect screening of the external electric field plays a fundamental role in the onset of tip streaming.

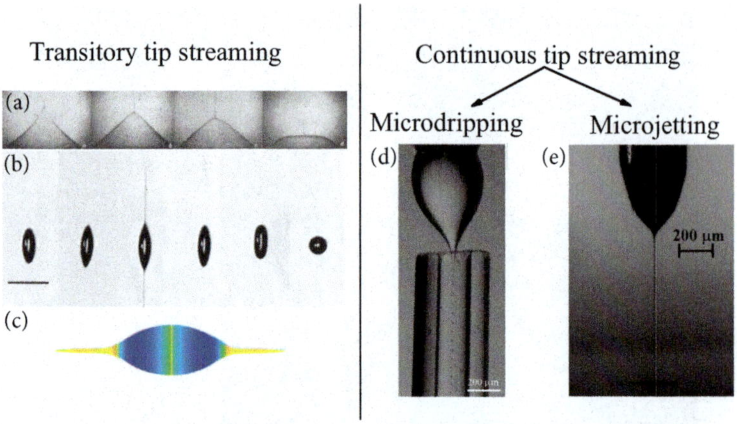

Fig. 1.11 Transitory versus continuous tip streaming. Electrohydrodynamic tip streaming in a soap film (**a**) (reproduced from Ref. [46] with permission) and a suspended droplet (**b**) (reproduced from Ref. [49] with permission). Tip streaming in a droplet loaded with a surfactant in an extensional flow (**c**) (reproduced from Ref. [60]). Microdripping in confined selective withdrawal (**d**) (reproduced from Ref. [61]). Microjetting in electrospray (**e**) (reproduced from Ref. [62] with permission of American Physical Society)

In hydrodynamic tip streaming, the driving stresses are caused by the suction (decrease of hydrostatic pressure) and viscous traction exerted by an outer continuous stream moving faster than the inner dispersed phase. This type of tip streaming is also referred to as hydrodynamic focusing, alluding to the fact that the energy transmitted to the dispersed phase is concentrated in the source tip. This effect is typically achieved by forcing the continuous outer stream to cross a discharge orifice right in front of the fluid source. As explained below, several microfluidic configurations have been proposed to this end.

In principle, the tip streaming phenomenon described above is an intrinsically transient flow that necessarily entails the loss of mass, charge, and surfactant of the precursor fluid shape. This implies that the system eventually ceases to verify the stringent conditions for the phenomenon to occur, and the phenomenon can survive only for a short time. However, tip streaming can keep running indefinitely if the ejected volume is replaced at the appropriate flow rate. In this case, the system finds autonomously a state that allows continuous fluid emission through tip streaming (Fig. 1.11). It must be pointed out that the scaling laws derived for continuous tip streaming cannot be extrapolated to the transitory counterpart because they are fundamentally different phenomena.

One can distinguish two continuous ejection modes via tip streaming: microdripping and microjetting. In microdripping (Fig. 1.12), tiny drops (bubbles) are periodically formed at the tip of the mother droplet (bubble). The collection of drops exhibits a very high degree of monodispersity, and the drop size and the ejection frequency can be controlled by adjusting the intensity of the external actuation and the dispersed phase flow rate. In microjetting (Fig. 1.13), a jet constantly tapers

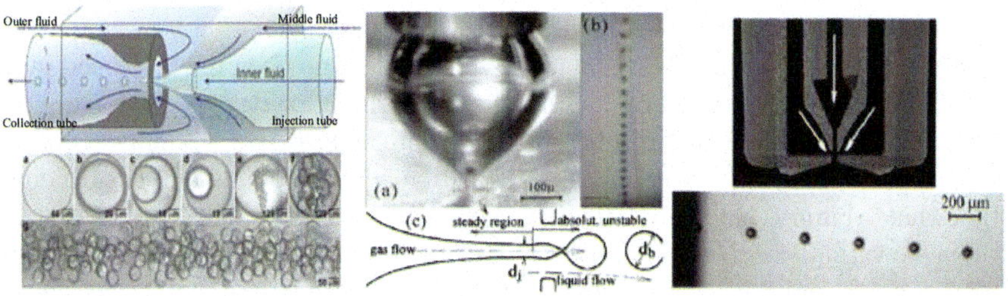

Fig. 1.12 Some examples of microdripping/microbubbling: double emulsions produced with hydrodynamic focusing (left-hand images) [57], bubbling in flow focusing (central images) [54, 63], and dripping in gaseous flow focusing (right-hand images) [64] (reprinted from Ref. [65] with permission of IOP Publishing)

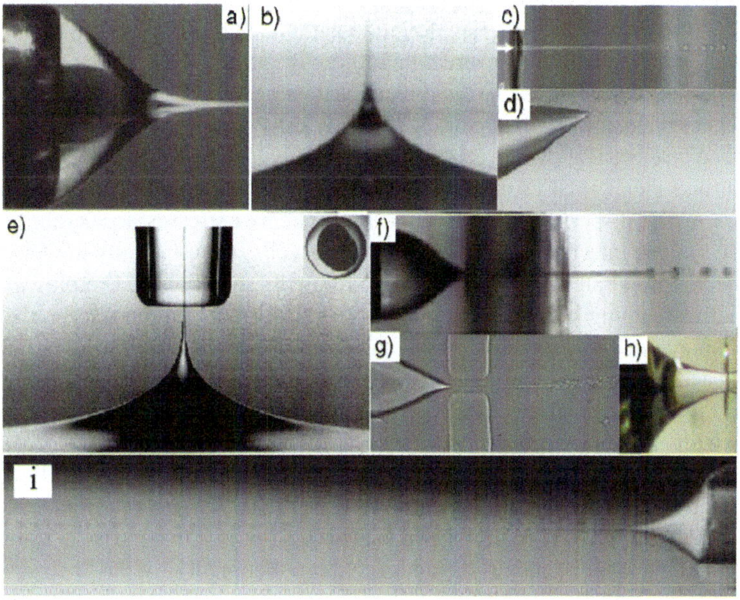

Fig. 1.13 Some examples of microjetting: **a** double electrospray configuration to produce capsules down to the nanometer scale [66], **b** electrified film of a low-conductivity liquid [48], **c** liquid transport due to light scattering [67], **d** tip streaming due to the action of a shear flow in the presence of surfactants [68], **e** selective withdrawal to produce microcapsules [69], **f** flow focusing plate-orifice configuration [70], **g** flow focusing in a 2D microfluidic device [55], **h** double flow focusing arrangement to coat microparticles [71] and **i** liquid-liquid coflowing configuration to produce emulsions [72] (reprinted from Ref. [65] with permission of IOP Publishing)

from the tip and breaks into droplets due to capillary instability. The jet diameter is essentially determined by the dispersed phase flow rate and, to a lesser extent, the strength of the driving force. In both microdripping and microjetting, the size of the resulting droplet is much smaller than any passage of the microfluidic device, which constitutes the fundamental characteristic of these flows.

In many tip streaming realizations, there is a gradual transition between microdripping and microjetting. This situation typically occurs when the dispersed inner phase acceleration is caused by the viscous shear stress exerted by the continuous outer phase, as happens in the coflowing and confined selective withdrawal configurations (see Chap. 9). In this case, the distinction between the two modes becomes somewhat arbitrary. One can define an axial characteristic length ℓ_c as that along which the variation Δd_j of the emitted thread diameter is of the order of its value d_j. If $\ell_c \gg d_j$, the microjetting regime has been established, while the system adopts the microdripping mode otherwise.

Microdripping and microjetting result from a delicate balance between the forces driving and opposing the flow (viscosity and interfacial tension). When that balance is tilted in favor of one of those forces, tip streaming destabilizes, and regular or intermittent dripping is obtained. This explains why tip streaming is sometimes elusive, only found in a small stability island in the parameter space. Despite the important advances in understanding tip streaming, there are still many open questions about the instability mechanisms that limit its appearance.

As mentioned above, tip streaming is gently driven by interfacial stresses, whose origin can be electrical (Maxwell stresses), hydrodynamic (an outer stream), or both. Normal interfacial stresses create favorable hydrostatic pressure gradients pushing the bulk of the disperse phase. In contrast, tangential stresses accelerate the fluid layer next to the interface and feed recirculation cells for sufficiently low viscosities and injected flow rates [73]. This recirculation pattern is believed to play a crucial role in tip streaming stability. The shear stress applied on the interface is the major energy source in many tip streaming realizations due to the large surface-to-volume ratios reached in the critical region. The normal stress caused, for instance, by the electrostatic pressure in electrospraying may play a subdominant role in the final energy budget. However, this stress is necessary for this phenomenon to occur.

Droplets can be produced by regular dripping and jetting, in which the diameter of the ejected fluid shape is commensurate with that of the discharge orifice. However, tip streaming has proved to offer significant advantages for that purpose. In fact, and as mentioned above, it enables the controlled production of droplets with sizes well below that of the feeding channels, thus avoiding clogging effects. It is easy to show from the mass conservation principle that reducing the size of the emitted droplets requires decreasing the injected flow rate in tip streaming. However, the flow rate cannot be decreased indefinitely because the flow becomes unstable at a certain point.

There are still fundamental unresolved questions about the critical minimum flow rate stability limit of the tip streaming microjetting mode. One is whether the instability depends only on the flow in the tapering meniscus tip or is also affected by the flow pattern in the entire dispersed phase. As explained in Chap. 8, the disparity of the spatial scales characterizing the tip and the rest of the meniscus of electrohydrodynamic tip streaming decouples the corresponding flows, which allows one to determine the microjetting stability only in terms of the forces arising in the meniscus tip [74]. On the contrary, it has been hypothesized that the minimum flow rate stability limit of hydrodynamic tip streaming is affected by global flow features, such as the meniscus shape [75] and the size [76] and location [77] of the recirculation cells (see Chaps. 9 and 10). In fact, the injection system can be modified to eliminate recirculating patterns in the liquid source and stabilize hydrodynamic tip streaming [78–81].

1.4 Microfluidic Configurations for Producing Tip Streaming. An Overview

1.4.1 Two-Dimensional Versus Axisymmetric Configurations

Most microfluidic devices present a planar (2D) topology. These devices can be manufactured in one single step, either via photolithography and etching in silicon substrates or through soft lithography in substrates of polymer materials (PDMS) [82, 83]. This characteristic has conferred great popularity on the 2D configuration. However, the planar topology also suffers from relevant disadvantages. These devices can damage fragile particles and cause problems due to the competitive wetting between immiscible liquids when emerging droplets and jets touch the walls of the channel [56]. Besides, PDMS channels swell in strong organic solvents and siloxane-based compounds. They can deform under intense applied pressures due to their high elasticity. The 2D geometry frequently requires specific coatings and dedicated surfactants.

Axisymmetric devices do not present the problems mentioned above. The continuous phase surrounds completely and shields the dispersed one at all flow rates. Axisymmetric devices can be fabricated with glass, which is a resistant, smooth, and transparent material. More importantly, axisymmetry devices considerably increase the droplet production rate with respect to that taking place in the 2D topology. The major drawback of the axisymmetric geometry is probably the fact that microfluidic devices usually consist of several pieces that must be carefully aligned to obtain the desired outcome. This renders the manufacturing techniques art-dependent and not scalable in most cases.

This section briefly describes the axisymmetric microfluidic configurations most commonly used to produce the microdripping and microjetting modes of tip streaming. The main results obtained for these configurations will be described in other chapters.

1.4.2 Coflowing

The coflowing configuration is probably the simplest way of producing tip streaming. In a coflowing device, the inner (dispersed) phase is injected at a constant flow rate through a feeding capillary of radius R_i (Fig. 1.14a). An external fluid stream is coaxially injected across a tube of radius R_o, and coflows parallel to the inner dispersed

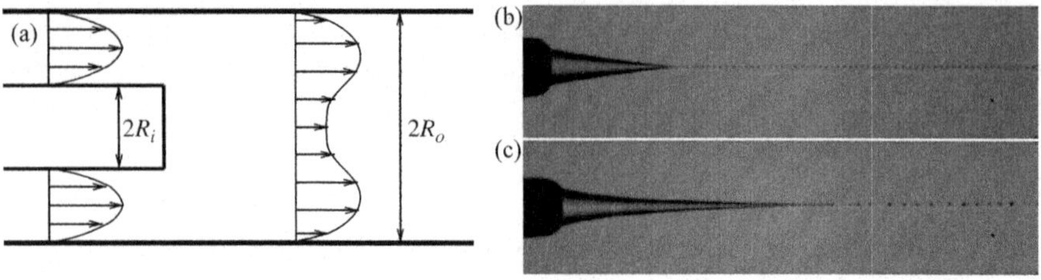

Fig. 1.14 Sketch of a coflowing device (**a**) and images of the microdripping (**b**) and microjetting (**c**) modes. (reproduced with permission, Ref. [84])

Fig. 1.15 Submicrometer fibers spun from polymeric solutions using SBS (reprinted from Ref. [87] with permission)

phase. The tangential viscous stress exerted by the outer stream transfers momentum to the inner fluid, which accelerates and forms a stretched meniscus attached to the inner capillary. The meniscus tip emits either tiny droplets (microdripping) (Fig. 1.14b) or a long and thin fluid thread (microjetting) (Fig. 1.14c), which breaks up downstream due to the capillary instability. The distinction between these two modes is sometimes unclear.

The production of Newtonian liquid jets assisted by an outer gaseous stream in a coflowing configuration is much less popular than its liquid-liquid counterpart due to the limited force exerted by the gas on the liquid jet in this geometry. In addition, the large difference between the gas and liquid speeds triggers the whipping (bending) instability, leading to a chaotic motion of the ejected fluid thread. Submicrometer fibers are spun from polymeric solutions using the Solution Blow Spinning (SBS) (Fig. 1.15) [85–87] or airbrushing technique [85], which can be regarded as an example of gaseous coflowing geometry. Gaseous coflowing must not be considered a proper tip streaming realization because the jet diameter is commensurate with that of the inner orifice, and the gaseous current is turbulent.

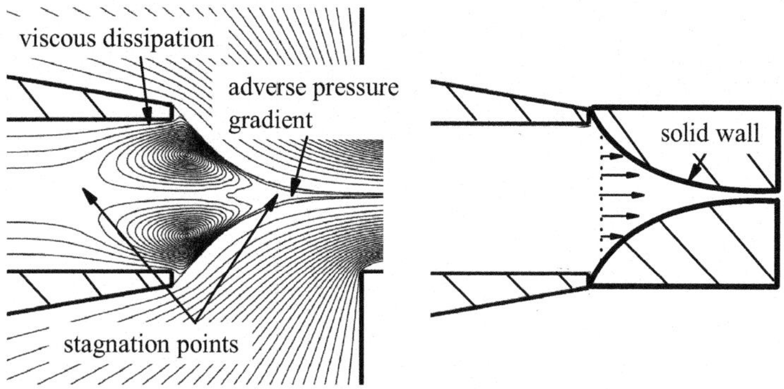

Fig. 1.16 Sketch to illustrate the difference between the flow in the virtual fluid nozzle formed in hydrodynamic focusing (left) and the solid counterpart (right)Sketch to illustrate the difference between the flow in the virtual fluid nozzle formed in hydrodynamic focusing (left) and the solid counterpart (right)

1.4.3 Hydrodynamic Focusing

Tip streaming can be produced by hydrodynamic focusing. This effect occurs when an outer stream moves faster than the inner dispersed phase when crossing a discharge orifice or tube. The drop of hydrostatic pressure in front of the orifice and the viscous traction exerted by the outer stream gently shape a steady tapering meniscus. The outer phase can be seen as a fluid converging "nozzle," which directs the inner fluid toward a virtual orifice whose diameter is much smaller than that of the real one. There is, however, a fundamental difference between this fluid nozzle and that delimited by a solid wall. The velocity would vanish at the surface of a solid nozzle, and a quasi-parabolic velocity profile would develop at any section. On the contrary, the speed of a low-viscosity fluid reaches its maximum value at the surface of the fluid nozzle. This frequently translates into the growth of recirculation cells in the tapering meniscus of hydrodynamic focusing, a flow pattern fundamentally different from that in a solid nozzle (Fig. 1.16).

The critical regions of the above-mentioned flow pattern are the two stagnation points delimiting the recirculation cells, the narrow streamtube formed next to the inner wall of the feeding capillary, where viscous dissipation sharply increases, and the mensicus-jet transition region, where the fluid flows against an adverse pressure gradient. In some liquid-liquid configurations, the interface slightly deforms and an inflection point appears in the interface contour as the passage of the dragged fluid narrows. These critical regions may be responsible for the instability of the microjetting mode as the inner injected flow rate decreases.

Tiny droplets of practically equal size can be directly ejected at the end of the fluid nozzle (microdripping). Under certain conditions, the outer stream sweeps away the capillary waves growing on an emitted fluid thread (convective instability), which enables the formation of a thin and long jet [88]. The jet eventually breaks up due to

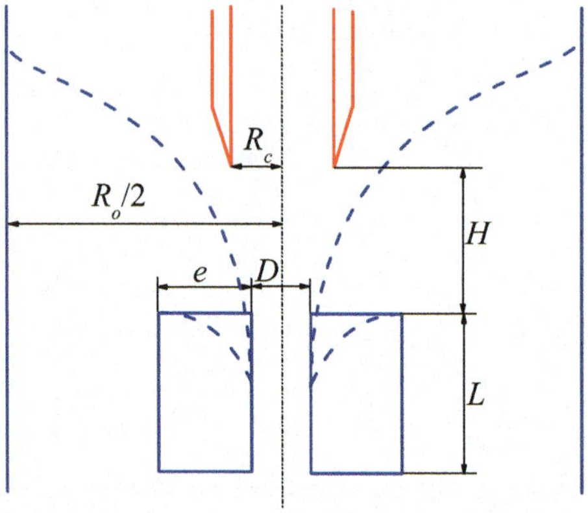

Fig. 1.17 Microfluidic configurations based on the hydrodynamic focusing principle. The dashed lines represent the nozzles used in counter [57] and confined [77] flow focusing

Table 1.1 Parameter relations corresponding to the microfluidic configurations implementing the hydrodynamic focusing principle [61]. The last column indicates the shape of the focusing orifice

Configuration	R_c/D	e/D	R_o/D	L/D	Shape
Flow focusing [89]	~ 1	$\gg 1$	$\gg 1$	~ 1	Orifice
Confined flow focusing [77]	~ 1	–	$\gtrsim 1$	~ 1	Nozzle
Selective withdrawal [69]	$\gg 1$	$\lesssim 1$	$\gg 1$	$\gg 1$	Tube
Confined selective withdrawal [90]	~ 1	$\lesssim 1$	$\gg 1$	$\gg 1$	Tube
Raydrop [91]	$\lesssim 1$	$\lesssim 1$	$\gg 1$	$\gg 1$	Tube
Counter flow focusing [57]	$\gtrsim 1$	$\lesssim 1$	$\gtrsim 1$	~ 1	Nozzle
Opposed flow focusing [92]	~ 1	$\ll 1$	$\gtrsim 1$	$\gg 1$	Tube
Stretched flow focusing [61]	~ 1	$\gg 1$	$\gg 1$	$\gg 1$	–

the Rayleigh capillary instability [9], giving rise to a collection of droplets with an acceptable degree of monodispersity.

Several microfluidic configurations have been proposed to implement the hydrodynamic focusing principle in liquid-liquid systems. Table 1.1 shows the choices for the main geometrical parameters governing the focusing effect (see Fig. 1.17). These choices lead to the different configurations: flow focusing [89], confined flow focusing [77], selective withdrawal [69], confined selective withdrawal [90], Raydrop [91], counter flow focusing [57], and opposed flow focusing [92]. The focusing tube/orifice can be cylindrical or have a converging shape, as indicated in the figure. The distance H between the feeding capillary and the focusing orifice is a control parameter in all the configurations, which affects the flow stability and, therefore, the minimum attainable size of the ejected droplets. The hydrodynamic focusing applied to liquid-liquid systems will be analyzed in Chap. 9.

In gaseous flow focusing (Fig. 1.18) [53], a liquid stream is focused by a high-speed gaseous current. This technique has been implemented in the original plate-orifice configuration [53] and the so-called Gas Dynamic Virtual Nozzle (GDVN) [93]. The droplets are produced in the microjetting regime in practically all the exper-

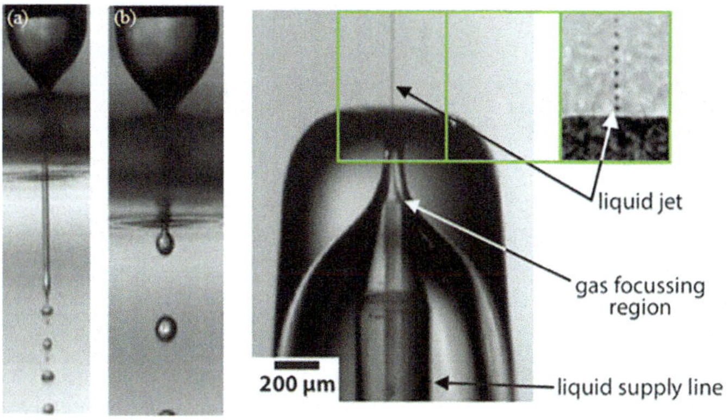

Fig. 1.18 (Left) Gaseous flow focusing. Images of the plate-orifice configuration in the microjetting (**a**) and microdripping (**b**) modes (reproduced with permission, Ref. [70]). (Right) Microjetting mode of GDVN (reproduced with with permission, Ref. [93])

iments. The microdripping mode can be observed only under particular conditions [64, 70]. Chapters. 10 and 11 are entirely devoted to gaseous flow focusing.

1.4.4 Electrospray

The microdripping mode of electrospray (Fig. 1.19-left) [94] is obtained when a liquid is supplied at a constant flow rate through a feeding capillary while a DC voltage is applied between the capillary and a counter-electrode. For specific intervals of voltage and flow rate, the electric stress at the liquid surface stretches the meniscus. A fluid thread develops and eventually detaches, forming a single droplet. This process repeats itself periodically. If the flow rate and applied voltage are appropriately selected, a steady quasi-conical meniscus continuously ejects a very thin jet (Fig. 1.19 right). This jet breaks up downstream owing to the capillary instability, giving rise to charged droplets. This microjetting regime is the popular steady cone-jet mode of electrospray, which has enormous applications in very diverse fields. We will devote Chap. 8 to this tip streaming configuration.

1.5 Electric Fields, Surfactants, and Viscoelasticity

1.5.1 Electric Fields

Many microfluidic applications take advantage of the inherently clean action of electric fields. Low-conductivity liquid solvents contain free ions. The interfaces between immiscible fluid phases prevent the continuous diffusion of these ions under applied electric fields, which causes the accumulation of net free charge onto those

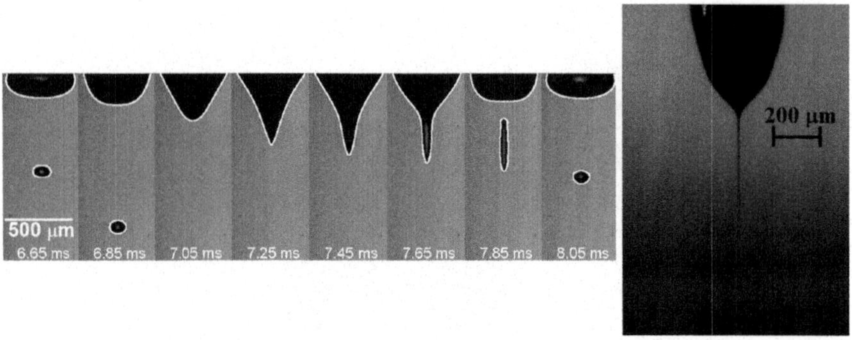

Fig. 1.19 (Left) Electrospray microdripping (reprinted with permission from Ref. [94]. Copyright (2015) by the American Physical Society). (Right) Electrospray microjetting (the steady cone-jet mode) (reprinted with permission from Ref. [62]. Copyright (2012) by the American Physical Society)

surfaces. The electrostatic force exerted on the interface charge and the jump of electrical permittivity in that region considerably modify the surface forces and alter their equilibrium. In a broad sense, *electrohydrodynamics* is the study of the evolution of fluids owing to the action of both surface and volumetric electric forces.

In electrohydrodynamics, transient, periodic, quasi-steady, and steady flows result from the competition between the electric driving force and the hydrodynamic ones arising when the liquid sets in motion. The complexity of electrohydrodynamics is difficult to find in other fields of mesoscale physics, even in simple situations where the system adopts a steady or quasi-steady regime. The tip streaming in electrified droplets constitutes a perfect example of this.

Electrokinetics models describe the production, destruction, and transport of electric charges in a continuous medium. In principle, these models must be coupled to the hydrodynamic equations to describe the evolution of a charged flowing phase. Due to the complexity of the problem, electrokinetics and hydrodynamic studies have proceeded in parallel, turning their backs on each other. The Taylor-Melcher leaky-dielectric approximation [59, 95] gave the definitive impulse to overcome the breach between these two disciplines.

The leaky-dielectric approximation describes the charge transport in the liquid bulk from the Ohmic model with a constant conductivity. In addition, the net free charge is assumed to accumulate at the interface; therefore, the electric force is taken into account only through the Maxwell stresses applied on that surface. Schnitzer and Yariv [96] culminated the work of Baygents and Saville [97] by deriving the leaky-dielectric approximation from an electrokinetic model under proper simplifications and limits. The Taylor-Melcher leaky-dielectric approximation has become so popular that the term *electrohydrodynamics* is frequently reserved solely for studies using this or similar approximations.

This work reviews experimental and theoretical results of the electrohydrodynamic tip streaming arising when a droplet, bubble, liquid meniscus, or film is subject

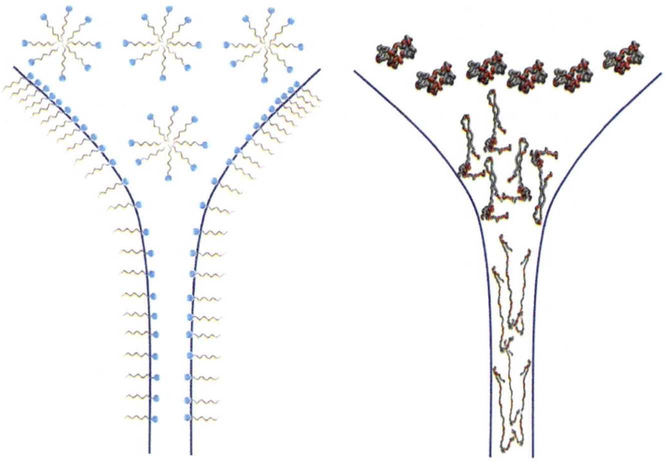

Fig. 1.20 (Left) Micelles and surfactant molecules adsorbed at the interface in the tapering meniscus tip of flow focusing. (Right) Coil-stretch transition in the tapering meniscus tip of flow focusing or electrospinning

to a strong electric (see Chaps. 5 and 8). The theoretical results are obtained from the leaky-dielectric model and a charge conservation approximation similar to that model. We describe these theoretical approaches in Chap. 2.

1.5.2 Surfactants

The term "complex fluids" has been coined in microfluidics and in this text to refer to liquids containing complex molecules either dissolved in the hydrodynamic bulk or adsorbed onto the interface with an immiscible fluid. In most cases, these molecules are surfactants and polymer chains (Fig. 1.20), which reduce the interfacial tension and confer viscoelasticity on the Newtonian solvent, respectively. As shown throughout this text, these complex molecules fundamentally change the flow in tip streaming realizations. We devote this section to describing the effects of surfactants very briefly. These molecules have been called the "hidden variables controlling fluid flows" [98], stressing their complicated but important role in a multitude of natural and industrial processes. Although surfactants alter cohesion forces on a molecular level, they can change macroscopic flow patterns.

Surfactants are amphiphilic molecules, i.e., they consist of a hydrophilic (polar) head and a hydrophobic tail (Fig. 1.20-left). For this reason, their solubility in water and organic liquids as oils is quite limited. In most cases, they get readily adsorbed at liquid-fluid interfaces forming monolayers. These monolayers substantially alter the balance between the stresses applied on the interface. The thickness of a surfactant monolayer is much smaller than any hydrodynamic length; therefore, it can be treated as an infinitely thin layer (surface). Ionic surfactants have either negatively or positively charged heads, while anionic surfactants are electrically neutral. This distinction may be necessary for electrohydrodynamic tip streaming.

Soluble surfactants form aggregates in the bulk. In many cases, these aggregates are micelles whose cores are formed by the hydrophobic tails, while the hydrophilic heads remain in contact with the surrounding liquid (Fig. 1.20-left). Micelles are formed for concentrations above the so-called *critical micelle concentration* (CMC). This surfactant volume concentration is smaller than that leading to the monolayer saturation, which occurs when the adsorbed surfactant molecules reach their maximum packing density.

The surfactant molecules dissolved in the liquid are convected and diffused across the hydrodynamic bulk. They accumulate at the sublayer next to the interface, which constitutes a source/sink of surfactant molecules for the interface during the system evolution. The net adsorption/desorption flux can be obtained from a kinetic model, which typically involves the adsorption and desorption constants and the maximum packing density. The surfactant molecules are also convected and diffused over the interface. The surfactant surface density at any point is affected by the adsorption/desorption kinetics, the interface surfactant transport, and the dilatation/compression of the interface element (Fig. 1.20-left).

The process described above results in an inhomogeneous surfactant distribution at the interface. The surfactant concentration gradient produces two effects not present in a clean interface: local soluto-capillarity and surface elasticity (Marangoni stresses). The term local soluto-capillarity has been coined to refer to the local reduction of the interfacial tension due to the surfactant accumulation at that point. One implicitly assumes that the chemical equilibrium is restored instantaneously at any interface point because this process is much faster than any other characteristic time. For this reason, the interfacial tension is calculated from the *equilibrium* equation of state for the corresponding value of the surfactant surface concentration.

The interface region with high surface tension pulls on the surrounding liquid more strongly than that with lower values of that quantity. The Marangoni stress generated by the imbalances of the interfacial tension causes the liquid to flow from the latter to the former. This is a surface stress that becomes more relevant as the fluid surface-to-volume ratio increases, as occurs in tip streaming. For this reason, the Marangoni stress considerably affects the tip streaming phenomenon. The equations governing the surfactant dynamics and its effects will be presented in Chap. 2.

There is an interesting parallelism between the role played by surfactant molecules and electric charges at the onset of tip streaming. Consider a droplet (meniscus or film) immersed in a focusing current, as occurs in flow focusing or selective withdrawal. Suppose that the interface is loaded with a surfactant monolayer (Fig. 1.21-left). The flow driven by the focusing effect drags the surfactant molecules towards the apex of the deformed droplet, reducing the interfacial tension there. If the focusing effect is sufficiently intense, the apex curvature increases, resulting in a sharp tip. When the droplet tip becomes unstable, it accelerates and stretches the interface. This process entails the production of an interface empty of surfactant molecules. Typically, the stretching rate is sufficiently high for surfactant molecules not to adsorb from the

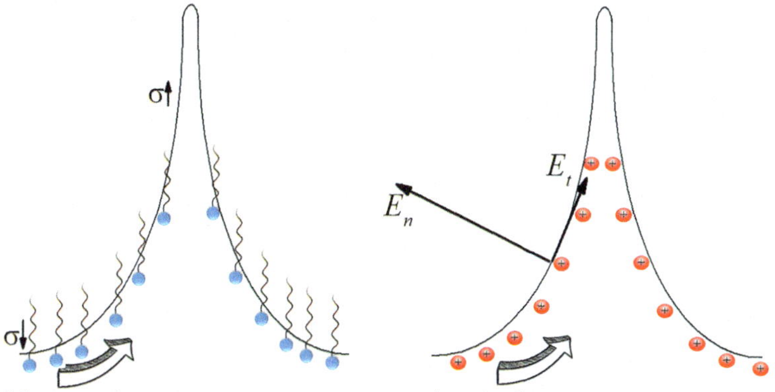

Fig. 1.21 Parallelism between the role played by surfactant molecules and electric charges at the onset of tip streaming. Effect of the Marangoni stress (left) and electric shear stress (right) on the onset of tip streaming. A fresh interface is created. Surfactant adsorption and diffusion (electric conduction) do not manage to restore the initial surfactant concentration (electrostatic surface density) at the tip. The shear stress produced by the surface tension gradient (tangential electric field) drives the liquid toward the tip

bulk and fill the created interface. Consequently, the surface density decreases in the ejected thread, and the interfacial tension increases there, giving rise to Marangoni convection towards the thread. This convection contributes to feeding the emerging jet against the adverse capillary pressure gradient caused by the large curvature of the droplet tip.

Consider now a droplet (meniscus or film) of a low-conductivity liquid immersed in a dielectric fluid and subjected to a constant electric field (Fig. 1.21-right). The droplet stretches in the direction of the electric field. Free charges distribute over the interface screening the outer electric field so that the flow in the equipotential droplet vanishes. As the electric field intensity increases, the charges accumulate in the droplet tip. The resulting electric pressure produces an effect similar to the interfacial tension reduction due to the accumulation of surfactant molecules. The apex curvature increases for a sufficiently large electric field, resulting in a sharp tip. As occurs in the phenomenon described above, the interface rapidly stretches at the onset of this electrified tip streaming. In a low-conductivity liquid, the electric relaxation time is not short enough for the equipotential condition to restore *instantaneously*. Consequently, an inner electric field arises, giving rise to electric shear stress. This stress pumps liquid towards the droplet tip favoring the jet ejection, as done by the Marangoni stress in the hydrodynamic tip streaming described above.

Local soluto-capillarity and surface elasticity (Marangoni stress) are the only effects considered in practically all the tip streaming flows involving surfactant monolayers. However, an interface separating two immiscible media offers viscous resistance to surface velocity gradients. This resistance can significantly affect the onset of tip streaming, even for low-viscosity surfactants. This occurs because this phenomenon exhibits considerable surface velocity gradients.

The simplest constitutive equation relating the viscous surface stress with the surface rate of strain is the Boussinesq-Scriven law [99, 100], which assumes that the instantaneous and isotropic surface stress is linear in the surface rate of strain. This constitutive equation can be regarded as the surface analog of the Navier-Poisson (Newton) law for compressible flows. For this reason, the surfactant monolayers described by the Boussinesq-Scriven law are called Newtonian monolayers. There is, however, a significant difference between the Navier-Poisson and Boussinesq-Scriven laws. The shear and dilatational viscosity coefficients in the Boussinesq-Scriven law generally depend not only on the temperature but also on the surfactant monolayer concentration. Non-Newtonian surface rheology arises in surfactant monolayers formed by complex molecules exhibiting, for instance, viscoelasticity [101]. However, this effect has not yet been described in the context of tip streaming.

1.5.3 Viscoelasticity

Many dilute polymer solutions exhibit a practically constant viscosity (negligible shear thinning) over a wide range of shear rates [102] so that the major polymer effects are the increase of the solution shear viscosity and elasticity. These solutions are commonly referred to as *Boger fluids* [102].

In sufficiently dilute solutions, neutral polymer chains adopt a random coil configuration at equilibrium. It has been recognized for a long time that high-velocity gradients can cause the transition from a coil to a nearly completely stretched conformation of those chains [103]. When the coil-stretch transition occurs, the polymeric contribution to the total stress increases dramatically (the extensional thickening effect), and the flow dynamics substantially change. When the velocity gradient triggering the coil-stretch transition disappears, the polymeric chain relaxes to the equilibrium configuration. In an ideal polymer solution, this relaxation consists of multiple processes in an individual polymeric chain. If the relaxation time λ_p characterizing the slowest process is much larger than the rest of the spectrum, the polymer solution elasticity can approximately be quantified by this parameter.

Modeling the relationship between the velocity field and stress tensor is a challenging problem for viscoelastic liquids, even for Boger fluids. As stated by Eggers and Villermaux [104], "The best one can hope for is to build in some simple physical ideas in a consistent fashion." Several constitutive equations, such as the Oldroyd-B and FENE-P models, provide reasonably accurate predictions for extensional flows while keeping certain simplicity [105].

Non-Newtonian effects in an Oldroyd-B liquid are quantified by the stress relaxation time λ_s and the polymer viscosity (equivalently, the retardation time). According to the Oldroyd-B model, the exponential thickening due to the coil-stretch transition occurs in a uniaxial extensional flow if the stretching rate exceeds the threshold $1/(2\lambda_s)$ [104]. This threshold is exceeded in the tip of the tapering meniscus of some tip streaming configurations, such as flow focusing and electrospinning, even for $\lambda_s \sim 10\,\mu s$ [106] (Fig. 1.20-right).

As explained in Chap. 4, the response of the viscoelastic solution to a straining flow is experimentally characterized by the so-called extensional relaxation time λ_r, measured in filament thinning rheometers such as CaBER [107] and FiSER [108]. This parameter is usually linked to the polymer relaxation time λ_p, although it increases with the polymer concentration of a dilute solution due to the hydrodynamic (flow-mediated) interaction among polymers. If the filament thinning in a rheometer is described with the Oldroyd-B model, one concludes that $\lambda_s = \lambda_r$.

Polymer chains can also be surface active agents, forming monolayers on the interfaces that reduce the surface tension. Due to the strong extensional flow characterizing tip streaming, this effect is expected to be much less relevant than bulk viscoelasticity. To our knowledge, these interfacial rheological effects have yet to be considered in tip streaming flows.

References

1. Fenn JB, Mann M, Meng CK, Wong SF, Whitehouse CM (1989) Electrospray ionization for mass spectrometry of large biomolecules. Science 246:64–71
2. Gañán-Calvo AM, Montanero JM, Martín-Banderas L, Flores-Mosquera M (2013) Building functional materials for health care and pharmacy from microfluidic principles and flow focusing. Adv Drug Deliv Rev 65:1447–1469
3. Whitesides GM, Grzybowski B (2002) Self-assembly at all scales. Science 295:2418–2421
4. Geissler M, Xia Y (2004) Patterning: principles and some new developments. Adv Mater 16:1249–1269
5. Burda C, Chen X, Narayanan R, El-Sayed MA (2005) Chemistry and properties of nanocrystals of different shapes. Chem Rev 105:1025–1102
6. Xia Y, Xiong Y, Lim B, Skrabalak SE (2009) Shape-controlled synthesis of metal nanocrystals: simple chemistry meets complex physics? Angew Chemie Intl Ed 48:60–103
7. Plateau J (1849) Sur les figures d'equilibre d'une masse liquide sans pesanteur. Mem Acad R Belg Nouv Ser 23
8. Plateau JAF (1873) Statique expérimentale et théorique des liquides soumis aux seules forces moléculaires. Gauthier-Villars
9. Rayleigh L (1878) On the instability of jets. Proc Lond Math Soc s1-10:4–13
10. Lord JWSR (1879) On the capillary phenomena of jets. Proc R Soc Lond Ser A 29:71–97
11. Martín-Banderas L, Sáez-Fernández E, Holgado M, Durán-Lobato MM, Prados JC, Melguizo C, Arias JL (2013) Biocompatible gemcitabine-based nanomedicine engineered by flow focusing for efficient antitumor activity. Int J Pharmaceut 443:103–109
12. Chapman HN et al (2011) Femtosecond X-ray protein nanocrystallography. Nature 470:73–79
13. Yarin AL, Pourdeyhimi B, Ramakrishna S (2014) Fundamentals and applications of micro- and nanofibers. Cambridge University Press, Cambridge, Great Britain
14. Barrero A, Loscertales IG (2007) Micro and nanoparticles via capillary flows. Annu Rev Fluid Mech 39:89–106
15. Haider A, Haider S, Kang IK (2018) A comprehensive review summarizing the effect of electrospinning parameters and potential applications of nanofibers in biomedical and biotechnology. Arab J Chem 11:1165–1188
16. Xue J, Wu T, Dai Y, Xia Y (2019) Electrospinning and electrospun nanofibers: methods, materials, and applications. Chem Rev 119:5298–5415

17. Loscertales IG, Barrero A, Máquez M, Spretz R, Velarde-Ortiz R, Larsen G (2004) Electrically forced coaxial nanojets for one step hollow nanofiber design. J Am Chem Soc 126:5376–77
18. Modesto-López LB, Pérez-Arjona A, Gañán-Calvo AM (2019) Flow blurring-enabled production of polymer filaments from poly(ethylene oxide) solutions. ACS Omega 4:2693–2701
19. Ponce-Torres A, Ortega E, Rubio M, Rubio A, Vega EJ, Montanero JM (2019) Gaseous flow focusing for spinning micro and nanofibers. Polymer 178(121):623
20. Jung S, Sou A, Banger K, Ko DH, Chow PCY, McNeill CR, Sirringhaus H (2014) All-inkjet-printed, all-air-processed solar cells. Adv Energy Mater 4(1400):432
21. Singh A, Katiyarab M, Garg A (2015) Understanding the formation of PEDOT:PSS films by ink-jet printing for organic solar cell applications. RSC Adv 5:78,677–78,685
22. Sun K, Zhang S, Li P, Xia Y, Zhang X, Du D, Isikgor FH, Ouyang J (2015) Review on application of PEDOTs and PEDOT: PSS in energy conversion and storage devices. J Mater Sci Mater Electron 26:4438–4462
23. Guo Y, Otley MT, Li M, Zhang X, Sinha SK, Treich GM, Sotzing GA (2016) PEDOT:PSS "wires? printed on textile for wearable electronics. ACS Appl Mater Interfaces 8:26,998–27,005
24. Sun D, Chang C, Li S, Lin L (2006) Near-field electrospinning. Nano Lett 6:839–842
25. Park JU et al (2007) High-resolution electrohydrodynamic jet printing. Nat Mater 6:782–789
26. Liashenko I, Rosell-Llompart J, Cabot A (2020) Ultrafast 3D printing with submicrometer features using electrostatic jet deflection. Nat Commun 11:753
27. Han Y, Dong J (2018) Electrohydrodynamic (EHD) printing of molten metal ink for flexible and stretchable conductor with self-healing capability. Adv Mater Technol 3(1700):268
28. Ringeisen BR, Othon CM, Barron JA, Young D, Spargo BJ (2006) Jet-based methods to print living cells. Biotechnol J 1:930–948
29. Wust S, Muller R, Hofmann S (2011) Controlled positioning of cells in biomaterials-approaches towards 3D tissue printing. J Funct Biomater 2:119–154
30. Bajaj P, Schweller RM, Khademhosseini A, West JL, Bashir R (2014) 3D biofabrication strategies for tissue engineering and regenerative medicine. Annu Rev Biomed Eng 16:247–276
31. García-Galán M, Martínez-Vázquez FJ, Rebollo-Muñoz N, Montanero J, Miranda P (2022) Applicability of near-field electrospinning for the development of TCP-based thin fibres and scaffold 3D printing. Bol Soc Esp Ceram Vidr
32. Basaran OA (2002) Small-scale free surface flows with breakup: drop formation and emerging applications. AlChE J 48:1842–1848
33. Wijshoff H (2010) The dynamics of the piezo inkjet printhead operation. Phys Rep 491:77–177
34. Nielsen NJ (1985) History of the inkjet printhead development. Hewlett-Packard J 36:4–10
35. Fromm JE (1984) Numerical calculation of the fluid dynamics of drop-on-demand jets. IBM J Res Dev 28:322–333
36. Chen AU, Basaran OA (2002) A new method for significantly reducing drop radius without reducing nozzle radius in drop-on-demand drop production. Phys Fluids 14:L1–L4
37. Christopher GF, Anna SL (2007) Microfluidic methods for generating continuous droplet streams. J Phys D Appl Phys 40:R319–R336
38. Gu H, Duits MHG, Mugele F (2011) Droplets formation and merging in two-phase flow microfluidics. Int J Mol Sci 12:2572–2597
39. Rayleigh L (1891) Some applications of photography. Nature 44:249–254
40. Batchelor GK (1958) Collected works of G. Cambridge University Press, Cambride, I. Taylor, UK
41. Stone HA, Bentley BJ, Leal LG (1986) An experimental study of transient effects in the breakup of viscous drops. J Fluid Mech 173:131–158
42. Zeleny J (1917) Instability of electrified liquid surfaces. Phys Rev 10:1–6
43. Wilson CTR, Taylor GI (1925) The bursting of soap-bubbles in a uniform electric field. Math Proc Cambridge Philos Soc 22:728–730
44. Nolan JJ (1926) The breaking of water-drops by electric fields. Proc R Ir Acad Sect A 37:28–39

45. Macky WA (1931) Some investigations on the deformation and breaking of water drops in strong electric fields. Proc R Soc Lond A 133:565–587
46. Taylor G (1964) Disintegration of water drops in electric field. Proc R Soc Lond A 280:383–397
47. Cloupeau M, Prunet-Foch B (1989) Electrostatic spraying of liquids in cone-jet mode. J Electrostatics 22:135–159
48. Collins RT, Jones JJ, Harris MT, Basaran OA (2008) Electrohydrodynamic tip streaming and emission of charged drops from liquid cones. Nat Phys 4:149–154
49. Duft D, Achtzehn T, Muller R, Huber BA, Leisner T (2003) Coulomb fission: Rayleigh jets from levitated microdroplets. Nature 421:128
50. Gañán-Calvo AM (2004) On the general scaling theory for electrospraying. J Fluid Mech 507:203–212
51. Fernández de la Mora J (2007) The fluid dynamics of Taylor cones. Annu Rev Fluid Mech 39:217–243
52. De Bruijn RA (1993) Tipstreaming of drops in simple shear flows. Chem Eng Sci 48:277–284
53. Gañán-Calvo AM (1998) Generation of steady liquid microthreads and micron-sized monodisperse sprays in gas streams. Phys Rev Lett 80:285–288
54. Gañán-Calvo AM, Gordillo JM (2001) Perfectly monodisperse microbubbling by capillary flow focusing. Phys Rev Lett 87(274):501
55. Anna SL, Bontoux N, Stone HA (2003) Formation of dispersions using flow focusing in microchannels. Appl Phys Lett 82:364–366
56. Takeuchi S, Garstecki P, Weibel DB, Whitesides GM (2005) An axisymmetric flow-focusing microfluidic device. Adv Mater 17:1067–1072
57. Utada AS, Lorenceau E, Link DR, Kaplan PD, Stone HA, Weitz DA (2005) Monodisperse double emulsions generated from a microcapillary device. Science 308:537–541
58. Rubio M, Montanero JM (2023) Influence of a soluble surfactant on the transition to tip streaming. Exp Therm Fluid Sci 141(110):776
59. Melcher JR, Taylor GI (1969) Electrohydrodynamics: a review of the role of interfacial shear stresses. Annu Rev Fluid Mech 1:111–146
60. Herrada MA, Ponce-Torres A, Rubio M, Eggers J, Montanero JM (2022) Stability and tip streaming of a surfactant-loaded drop in an extensional flow. Influence of surface viscosity. J Fluid Mech 934:A26
61. López M, Cabezas MG, Montanero JM, Herrada MA (2022) On the hydrodynamic focusing for producing microemulsions via tip streaming. J Fluid Mech 934:A47
62. Herrada MA, López-Herrera JM, Gañán-Calvo AM, Vega EJ, Montanero JM, Popinet S, (2012) Numerical simulation of electrospray in the cone-jet mode. Phys Rev E 86(026):305
63. Gañán-Calvo AM (2004) Perfectly monodisperse microbubbling by capillary flow focusing: an alternate physical description and universal scaling. Phys Rev E 69(027):301
64. Cruz-Mazo F, Montanero JM, Gañán-Calvo AM (2016) Monosized dripping mode of axisymmetric flow focusing. Phys Rev E 94(053):122
65. Montanero JM, Gañán-Calvo AM (2020) Dripping, jetting and tip streaming. Rep Prog Phys 83(097):001
66. Loscertales IG, Barrero A, Guerrero I, Cortijo R, Marquez M, Gañán-Calvo AM (2002) Micro/nano encapsulation via electrified coaxial liquid jets. Science 295:1695 1698
67. Schroll RD, Wunenburger R, Casner A, Zhang WW, Delville JP (2007) Liquid transport due to light scattering. Phys Rev Lett 98(133):601
68. Eggers J (1997) Nonlinear dynamics and breakup of free-surface flows. Rev Mod Phys 69:865–929
69. Cohen I, Li H, Hougland JL, Mrksich M, Nagel SR (2001) Using selective withdrawal to coat microparticles. Science 292:265–267
70. Si T, Li F, Yin XY, Yin XZ (2009) Modes in flow focusing and instability of coaxial liquid-gas jets. J Fluid Mech 629:1–23
71. Gañán-Calvo AM, González-Prieto R, Riesco-Chueca P, Herrada MA, Flores-Mosquera M (2007) Focusing capillary jets close to the continuum limit. Nat Phys 3:737–742

72. Castro-Hernández E, Gundabala V, Fernández-Nieves A, Gordillo JM (2009) Scaling the drop size in coflow experiments. New J Phys 11(075):021
73. Gañán-Calvo AM, Montanero JM (2009) Revision of capillary cone-jet physics: electrospray and flow focusing. Phys Rev E 79(066):305
74. Gañán-Calvo AM, Rebollo-Muñoz N, Montanero JM (2013) Physical symmetries and scaling laws for the minimum or natural rate of flow and droplet size ejected by Taylor cone-jets. New J Phys 15(033):035
75. Montanero JM, Rebollo-Muñoz N, Herrada MA, Gañán-Calvo AM (2011) Global stability of the focusing effect of fluid jet flows. Phys Rev E 83(036):309
76. Herrada MA, Gañán-Calvo AM, Ojeda-Monge A, Bluth B, Riesco-Chueca P (2008) Liquid flow focused by a gas: jetting, dripping, and recirculation. Phys Rev E 78(036):323
77. Cabezas MG, Rubio M, Rebollo-Muñoz N, Herrada MA, Montanero JM (2021) Global stability analysis of axisymmetric liquid-liquid flow focusing. J Fluid Mech 909:A10
78. Lozano P, Martínez-Sánchez M (2005) On the dynamic response of externally wetted ionic liquid ion sources. J Phys D 38:2371–2377
79. Whitehouse CM, Dreyer RN, Yamashita M, Fenn JB (1985) Electrospray interface for liquid chromatographs and mass spectrometers. Anal Chem 57:675–679
80. Sen AK, Darabi J, Knapp DR, Liu J (2006) Modeling and characterization of a carbon fiber emitter for electrospray ionization. J Micromech Microeng 16:620–630
81. Acero AJ, Rebollo-Muñoz N, Montanero JM, Gañán-Calvo AM, Vega EJ (2013) A new flow focusing technique to produce very thin jets. J Micromech Microeng 23(065):009
82. Duffy DC, McDonald JC, Schueller OJA, Whitesides GM (1998) Rapid prototyping of microfluidic systems in poly(dimethylsiloxane). Anal Chem 70:4974–4984
83. Unger MA, Chou HP, Thorsen T, Scherer A, Quake SR (2000) Monolithic microfabricated valves and pumps by multilayer soft lithography. Science 288:113–116
84. Gordillo JM, Sevilla A, Campo-Cortés F (2014) Global stability of stretched jets: conditions for the generation of monodisperse micro-emulsions using coflows. J Fluid Mech 738:335–357
85. Oliveira JE, Moraes EA, Costa RGF, Afonso AS, Mattoso LHC, Orts WJ, Medeiros ES (2011) Nano and submicrometric fibers of poly(d, l-lactide) obtained by solution blow spinning: process and solution variables. J Appl Polym Sci 122:3396–3405
86. Daristotle JL, Behrens AM, Sandler AD, Kofinas P (2016) A review of the fundamental principles and applications of solution blow spinning. ACS Appl Mater Interf 8:34,951–34,963
87. Rubio A, Martínez-Vázquez FJ, Cabezas MG, Montanero JM, Vega EJ (2022) Effect of the air coflow temperature and spinneret position on the solution blow spinning technique. Fibers Polym 23:2299–2305
88. Huerre P, Monkewitz PA (1990) Local and global instabilites in spatially developing flows. Annu Rev Fluid Mech 22:473–537
89. Gañán-Calvo AM, Riesco-Chueca P (2006) Jetting-dripping transition of a liquid jet in a lower viscosity co-flowing immiscible liquid: the minimum flow rate in flow focusing. J Fluid Mech 553:75–84
90. Evangelio A, Campo-Cortés F, Gordillo JM (2016) Simple and double microemulsions via the capillary breakup of highly stretched liquid jets. J Fluid Mech 804:550–577
91. Dewandre A, Rivero-Rodriguez J, Vitry Y, Sobac B, Scheid B (2020) Microfluidic droplet generation based on non-embedded co-flow-focusing using 3d printed nozzle. Sci Rep 10(21):616
92. Dong J, Meissner M, Faers MA, Eggers J, Seddon AM, Royall CP (2018) Opposed flow focusing: evidence of a second order jetting transition. Soft Matter 14:8344–8351
93. DePonte DP, Weierstall U, Schmidt K, Warner J, Starodub D, Spence JCH, Doak RB (2008) Gas dynamic virtual nozzle for generation of microscopic droplet streams. J Phys D Appl Phys 41(195):505
94. Hijano AJ, Loscertales IG, Ibáñez SE, Higuera FJ (2015) Periodic emission of droplets from an oscillating electrified meniscus of a low-viscosity, highly conductive liquid. Phys Rev E 91(013):011

95. Saville DA (1997) Electrohydrodynamics: the Taylor-Melcher leaky dielectric model. Annu Rev Fluid Mech 29:27–64
96. Schnitzer O, Yariv E (2015) The Taylor-Melcher leaky dielectric model as a macroscale electrokinetic description. J Fluid Mech 773:1–13
97. Baygents JC, Saville DA (1989) The circulation produced in a drop by an electric field: a high field strength electrokinetic model. In: AIP Conference proceedings, American Institute of Physics, pp 7–17
98. Manikantan H, Squires TM (2020) Surfactant dynamics: hidden variables controlling fluid flows. J Fluid Mech 892:P1
99. Boussinesq J (1913) Existence of a superficial viscosity in the thin transition layer separating one liquid from another contiguous fluid. C R Hehbd Seances Acad Sci 156:983–989
100. Scriven LE (1960) Dynamics of a fluid interface equation of motion for Newtonian surface fluids. Chem Eng Sci 12:98–108
101. Jaensson N, Vermant J (2018) Tensiometry and rheology of complex interfaces. Curr Opin Colloid Interface Sci 37:136–150
102. James DF (2009) Boger fluids. Annu Rev Fluid Mech 41:129–142
103. Peterlin A (1966) Hydrodynamics of linear macromolecules. Pure Appl Chem 12:563–586
104. Eggers J, Villermaux E (2008) Physics of liquid jets. Rep Prog Phys 71(036):601
105. Bazilevskii AV, Entov VM, Rozhkov AN (2001) Breakup of an Oldroyd liquid bridge as a method for testing the rheological properties of polymer solutions. Polym Sci Ser A 42:716–726
106. Rubio A, Galindo F, Vega EJ, Montanero JM, Cabezas MG (2022) Viscoelastic transition in transonic flow focusing. Phys Rev Fluids 7(074):201
107. McKinley GH, Tripathi A (2000) How to extract the Newtonian viscosity from capillary breakup measurements in a filament rheometer. J Rheol 44:653–670
108. Anna SL, McKinley GH (2001) Elasto-capillary thinning and breakup of model elastic liquids. J Rheol 45:115–138

Chapter 2
Governing Equations

Abstract This chapter presents the equations framing the tip streaming phenomenon. These equations provide the theoretical framework to analyze the microfluidic applications studied in other chapters. We consider simple models for describing the effects of viscoelasticity, surfactants, and externally applied electric fields. In this latter case, we introduce the frequently used charge-conservative approximation and leaky-dielectric model, which allow for examining electrohydrodynamic processes in a relatively simple way. The relationship between these approaches and the full electrokinetic model is discussed.

We introduce some rheological models to obtain the polymer stress in dilute viscoelastic liquids. We present the equations describing the transport and conservation of surfactants in the hydrodynamic bulk and interface, simple kinetic models to account for the adsorption-desorption processes, and commonly used equations of state to calculate the interfacial tension as a function of the surfactant surface density. The surface viscous stresses caused by a Newtonian surfactant monolayer are considered as well. The chapter closes with the one-dimensional (1D) approximation for slender configurations, such as jets with relatively small spatial accelerations.

Keywords Electrokinetic · Charge-conservative model · Leaky-dielectric model · FENE-P and Oldroyd-B models · Surfactant equations · The slender-jet approximation

2.1 Basic Equations

2.1.1 Laminar and Turbulent Tip Streaming

Liquid-liquid microfluidic devices operate in the laminar regime because of their submillimeter sizes. In fact, the Reynolds number defined in terms of the microfluidic characteristic length and liquid injection velocity is frequently smaller than unity, and

J. M. Montanero, *Tip Streaming of Simple and Complex Fluids*, Fluid Mechanics and Its Applications 137, https://doi.org/10.1007/978-3-031-52768-5_2

the Stokes approximation can describe the viscous-capillary flow. It must be noted that the governing equations involve nonlinear terms even if the flow is dominated by viscosity due to the inherent non-linearity of the capillary pressure term.

Tip streaming is produced with minimal injection velocities of the dispersed phase, leading to very low values of the Reynolds number characterizing the tapering meniscus (film or droplet). However, the sharp diameter reduction in the meniscus tip considerably increases the speed of the jet emitted in the microjetting mode. In this case, the Reynolds number based on the jet's diameter and velocity can reach values of the order of 10^2, and the liquid inertia must be considered in the analysis of the jet's behavior. When the liquids are very viscous, inertia can also be neglected in the jet dynamics, including most of the jet breakup.

There are gas-liquid configurations in which turbulence may play a relevant role. In particular, the mixing layer between the high-speed gaseous jet and the surrounding ambient in flow focusing [1] becomes unstable and renders the flow turbulent at small distances from the discharge orifice. Turbulent viscosity slows down the gaseous jet, which loses most of its kinetic energy a few nozzle diameters beyond the orifice [2]. This effect influences the amount of energy transferred by the gaseous stream to the liquid through viscous shear stresses beyond the discharge orifice. The spinning [3] and electrospinning [4] of polymeric solutions assisted by a high-speed gas current constitute good examples of partially turbulent microfluidic realizations. Flow focusing transforms into flow blurring [5] when the capillary-to-orifice distance decreases and the applied pressure drop increases. Small turbulence scales are produced in flow blurring, leading to efficient atomization.

2.1.2 Bulk Equations

The next sections present the equations describing the laminar flow of a two-phase system. The extension to other multiphase systems is straightforward. The density $\rho^{(j)}(\mathbf{r}, t)$ and velocity $\mathbf{v}^{(j)}(\mathbf{r}, t)$ fields verify the continuity equation

$$\frac{\partial \rho^{(j)}}{\partial t} + \nabla \cdot (\rho^{(j)} \mathbf{v}^{(j)}) = 0, \tag{2.1}$$

where the superscript j denotes the fluid phase.

The use of the superscript j emphasizes that the quantity or physical property is not a continuous field throughout the fluid domain, but there is a jump in its value across the interface. This does not occur in the Volume of Fluid (VoF) method commonly used to integrate the hydrodynamic equations, where the hydrodynamic fields and the fluid properties continuously vary across the diffused interface. We will come back to this point in the next section.

Compressibility effects must be considered only in specific microjetting applications such as solution blow spinning [3, 6] and the transonic flow focusing used in the Serial Femtosecond Crystallography [7], where the gaseous stream is discharged on high-vacuum conditions.

The incompressibility condition holds in all liquid-liquid configurations and also in liquid ejections assisted with airflow [1, 8, 9] if the pressure drop driving the gas stream is smaller than around 100 mbar. This approximation can safely describe gas-liquid flows in which the outer gaseous stream moves with velocities smaller than, say, 100 m/s. The comparison between numerical simulations and experiments shows that the incompressibility condition provides accurate results even for pressure drops up to 250 mbar [10]. The microfluidic devices that produce bubbles [11–13] generate relatively small pressure variations. For this reason, they also work in the incompressible regime. In this regime, Eq. (2.1) reduces to

$$\nabla \cdot \mathbf{v}^{(j)} = 0. \tag{2.2}$$

The general form of the momentum equation reads

$$\rho^{(j)} \left(\frac{\partial \mathbf{v}^{(j)}}{\partial t} + \mathbf{v}^{(j)} \cdot \nabla \mathbf{v}^{(j)} \right) = -\nabla P^{(j)} + \nabla \cdot \boldsymbol{\tau}^{(j)} + \mathbf{F}^{(j)}, \tag{2.3}$$

where $P^{(j)}(\mathbf{r}, t) = p^{(j)}(\mathbf{r}, t) - \rho^{(j)}(\mathbf{g} \cdot \mathbf{r})$ is the *reduced* pressure field, $p^{(j)}$ the hydrostatic pressure, $\mathbf{g}$ is the gravitational acceleration, $\boldsymbol{\tau}^{(j)}$ is the deviatoric stress tensor, and $\mathbf{F}^{(j)}$ is an additional volumetric force per unit volume. Using $P^{(j)}$ instead of $p^{(j)}$ eliminates the gravitational force from (2.3) and shifts it to the boundary condition of normal stress balance at the interface, as explained below. In the VoF method, the volumetric force accounts for the surface tension and electric forces, as explained below.

The deviatoric stress tensor $\boldsymbol{\tau}^{(j)}$ for a Newtonian fluid is the viscous stress tensor given by the expression

$$\boldsymbol{\tau}^{(j)} = 2\mu^{(j)}\mathsf{D}^{(j)} + \lambda_{\mu}^{(j)}(\nabla \cdot \mathbf{v}^{(j)})\mathbf{I}, \tag{2.4}$$

where $\mu^{(j)}$ is the fluid viscosity, $\mathsf{D}^{(j)} = 1/2[\nabla \mathbf{v}^{(j)} + (\nabla \mathbf{v}^{(j)})^T]$ the deformation rate tensor, $\lambda_{\mu}^{(j)}$ the dilatational coefficient of viscosity, and $\mathbf{I}$ the identity matrix. In viscoelastic liquids, $\boldsymbol{\tau}^{(j)}$ is augmented with the polymeric stress, as explained below.

When compressibility effects are accounted for, the energy equation

$$\rho^{(j)} \frac{D(c_v^{(j)} T^{(j)})}{Dt} = -p^{(j)} \nabla \cdot \mathbf{v}^{(j)} + \boldsymbol{\tau}^{(j)} : \nabla \mathbf{v}^{(j)} - \nabla \cdot \mathbf{q}^{(j)} \tag{2.5}$$

is solved, and the ideal gas law $p^{(j)} = \rho^{(j)} \hat{R}_g^{(j)} T^{(j)}$ is considered in the gaseous phase. Here, D/Dt stands for the material derivative, $c_v^{(j)}$ and $\hat{R}_g^{(j)}$ are the specific heat coefficient and gas constant, respectively, and $T^{(j)}$ the temperature field. In addition, $\mathbf{q}^{(j)} = -\kappa_T^{(j)} \nabla T^{(j)}$ is the heat flux vector and $\kappa_T^{(j)}$ the thermal conductivity. The viscous dissipation function $\boldsymbol{\tau}^{(j)} : \nabla \mathbf{v}^{(j)}$ has a small effect and is usually neglected.

2.1.3 Boundary Conditions

The first approach described here follows along the lines of [14], who assumed that the fluid bulk properties are homogeneous except for the (2D) surface dividing them. The excess of mass or energy not accounted for by the homogeneous bulk phases is assigned to that surface. This description is frequently known as the sharp-interface model. This model may not apply to interfaces whose position cannot be easily defined; for instance, between slightly miscible fluids. In the diffuse-interface approximation, the fluid bulk properties and hydrodynamic fields are assumed to vary continuously across the interface. This model introduces the interface thickness as an additional length scale. As seen throughout this text, both the sharp-interface model and the diffuse-interface approximation have been successfully used to describe tip streaming driven by hydrodynamic and electrohydrodynamic forces.

In the absence of electric fields and surfactants, the balance of stresses on the two sides of the sharp interface yields

$$\mathbf{n}\cdot||\boldsymbol{\tau}^{(j)}|| - \left[||P^{(j)}|| + ||\rho^{(j)}||(\mathbf{g}\cdot\mathbf{r})\right]\mathbf{n} - \sigma(\boldsymbol{\nabla}_s\cdot\mathbf{n})\mathbf{n} = 0, \tag{2.6}$$

where $\mathbf{n}$ is the unit outward normal vector, $||A||$ denotes the difference $A^{(o)} - A^{(i)}$ between the values taken by the quantity A on the outer $(j = o)$ and inner $(j = i)$ sides of the interface, and σ is the the interfacial tension. In addition, $-\boldsymbol{\nabla}_s\cdot\mathbf{n}$ is the interface curvature, $\boldsymbol{\nabla}_s = \mathsf{I}_s\boldsymbol{\nabla}$ is the surface gradient operator, $\mathsf{I}_s = \mathsf{I} - \mathbf{nn}$ is the tensor that projects any vector onto the interface, and I the identity tensor. As shown below, Eq. (2.6) is completed by additional stresses when electric fields and surfactants are present. The balance of stresses on the two sides of the interface reflects the complexity of the problem considered.

Suppose that the interface position $\mathbf{r}_s$ is the solution to the equation $f(\mathbf{r}_s, t) = 0$. Neither of the two phases can cross the interface separating immiscible fluids, which leads to the kinematic compatibility boundary condition

$$\frac{\partial f}{\partial t} + \mathbf{v}^{(j)}\cdot\nabla f = 0\,. \tag{2.7}$$

Alternatively, one can locate the interface by defining the distance $F(\theta, z, t)$ of an interface element from the axis z of a cylindrical coordinate system (r, θ, z). This function obeys the equation

$$\frac{\partial F}{\partial t} - v_r^{(j)} + \frac{v_\theta^{(j)}}{F}\frac{\partial F}{\partial\theta} + v_z^{(j)}\frac{\partial F}{\partial z} = 0\,, \tag{2.8}$$

where $v_r^{(j)}$, $v_\theta^{(j)}$, and $v_z^{(j)}$ stand for the radial, angular, and axial components of the velocity field, respectively. This last formulation allows interface-tracking techniques [15] and boundary-fitted methods [16] to integrate the hydrodynamic equations numerically. It also facilitates imposing the anchorage of triple contact lines in the numerical simulation. The velocity field $\mathbf{v}^{(j)}$ is a continuous function at the interface, i.e., $||\mathbf{v}^{(j)}|| = \mathbf{0}$.

The diffusive-interface approximation solved by the VoF method treats immiscible fluids as a single one with spatially varying properties through the interface,

$$\{p^{(j)}(\mathbf{r}, t), \mathbf{v}^{(j)}(\mathbf{r}, t), \rho^{(j)}, \mu^{(j)}, \ldots\} \longrightarrow \{p(\mathbf{r}, t), \mathbf{v}(\mathbf{r}, t), \rho(\mathbf{r}, t), \mu(\mathbf{r}, t), \ldots\}. \tag{2.9}$$

The problem formulation does not explicitly includes the interface location and the corresponding boundary conditions. The interfacial tension force is added to the momentum equation (2.3) using the continuum surface-force approximation [17]. In this approximation, the interfacial tension force is calculated as

$$\mathbf{F}_\sigma = -\sigma\,(\boldsymbol{\nabla} \cdot \mathbf{n})\, \boldsymbol{\nabla}\hat{\alpha}, \quad \mathbf{n} = \frac{\boldsymbol{\nabla}\hat{\alpha}}{|\boldsymbol{\nabla}\hat{\alpha}|}, \tag{2.10}$$

where $\hat{\alpha}(\mathbf{r}, t)$ is the fluid volume fraction calculated by integrating the advection equation

$$\frac{\partial \hat{\alpha}}{\partial t} + \boldsymbol{\nabla} \cdot \left(\hat{\alpha}\mathbf{v}^{(j)}\right) = 0. \tag{2.11}$$

An artificial compression term is usually added to sharpen the interface.

The problem formulation in the incompressible regime is completed by imposing the no-slip boundary condition at the solid surfaces. Several factors can produce the slip of a Newtonian fluid over a solid surface, such as surface wettability, surface roughness, and nanobubbles or gaseous films [18]. Relevant microfluidic applications involve very thin polymer films experiencing significant slip over solid impermeable substrates. Slip-induced instabilities frequently occur in polymer melt extruders [19]. In all these cases, the slip-length model has been proposed to account for the observed phenomena from continuum mechanics [18]. The study of tip streaming does not require the use of this model. On the other hand, either no temperature jump or zero heat flux is prescribed at the solid surfaces when compressibility effects are considered.

Several tip streaming realizations rely on a tapering meniscus (droplet) whose end is anchored to the sharp edge of a feeding capillary (disk). In this case, the triple contact line must be pinned to that edge. The triple contact line anchorage condition must also be imposed when studying the linear stability of capillary systems interacting with real surfaces, i.e., those exhibiting contact angle hysteresis. As discussed by Dussan [20], the contact angle of the unperturbed state takes a value in the interval delimited by the receding and advancing contact angles (contact angle hysteresis), for which the contact line velocity vanishes. Since linear perturbations produce only infinitesimal variations around that angle, the triple contact line remains fixed during the evolution of those perturbations.

Figure 2.1 illustrates the above idea. The upper contact angle is $\theta = \theta_0$. The growth of an unstable linear perturbation shrinks the meniscus upper part, decreasing the

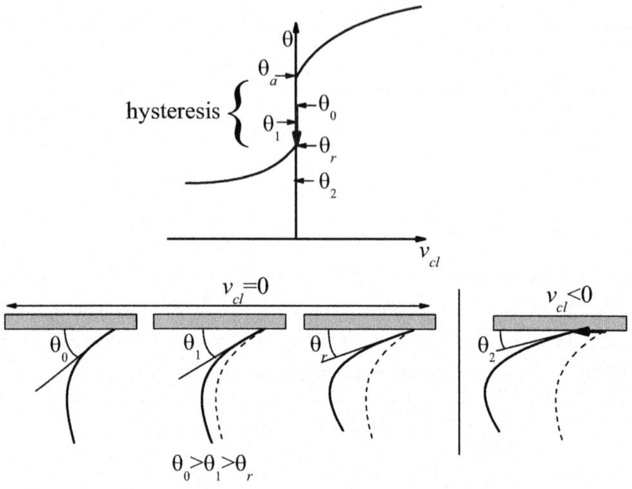

Fig. 2.1 The role of moving contact lines. The upper graph shows schematically the dependence of the dynamic contact angle θ upon the contact line velocity v_{cl} for a partially wetting liquid [20]. The advancing θ_a and receding θ_r contact angles are the largest and smallest angles right before spreading and recoiling, respectively. The sketches in the lower part of the figure illustrate the evolution of the upper part of a meniscus. The dashed and solid lines represent the initial and instantaneous free surface positions, respectively

contact angle ($\theta = \theta_1$). However, the triple contact line remains pinned as long as the contact angle lies in the interval $\theta_r \leq \theta \leq \theta_a$. Therefore, the meniscus dynamics during this interval are exactly the same as that obtained with perfectly sharp edges. The meniscus evolution deviates from that calculated with the idealized boundary condition only for sufficiently large deformations of the interface ($\theta = \theta_2$).

The VoF method does not allow one to prescribe this condition directly because the interface location does not enter into the problem formulation. In this case, the triple contact angle must be selected *ad hoc* to fix the triple contact line approximately at the desired location.

In most tip streaming realizations, the flow is fully developed in the feeding capillary, and the parabolic velocity profile is applied at the numerical domain inlet. Either pressure outlet or outflow boundary conditions are typically applied at the domain outlet.

2.2 Electric Fields

2.2.1 *Electrokinetic Model*

The electric field drives the liquid motion in very popular tip streaming realizations, such as electrospray [21–23] and electrospinning [24–28]. The electric field $\mathbf{E}^{(j)}(\mathbf{r}, t)$

is calculated by integrating the the Maxwell equations

$$\nabla \cdot (\varepsilon^{(j)} \mathbf{E}^{(j)}) = \rho_e^{(j)}, \quad \nabla \times \mathbf{E}^{(j)} = \mathbf{0}, \tag{2.12}$$

with the interface boundary conditions

$$||\varepsilon^{(j)} \mathbf{E}^{(j)}|| \cdot \mathbf{n} = 0, \quad ||\mathbf{E}^{(j)}|| \times \mathbf{n} = \mathbf{0}. \tag{2.13}$$

In these equations, $\varepsilon^{(j)}(\mathbf{r}, t)$ is the electrical permittivity and $\rho_e^{(j)}(\mathbf{r}, t)$ is the net free charge density. In electrospray and electrospinning, a constant value of the electric potential is prescribed at the surface of the perfect conductors (the electrode and counter-electrode). The numerical domain can be closed by imposing an analytical solution for the far-field electric potential [29] on the boundaries without interfaces.

Suppose that the electrical body force exerted on free charge and charge dipoles are transferred directly to the fluid [30]. The Coulomb force due to the free charge present in the liquid equals $\rho_e \mathbf{E}$. The force exerted on a dipole charge Q_d with orientation $\mathbf{d}$ is $(Q_d \mathbf{d}) \cdot \nabla \mathbf{E}$. Therefore, the corresponding force per unit volume is $(N Q_d \mathbf{d}) \cdot \nabla \mathbf{E} = \mathbf{P} \cdot \nabla \mathbf{E}$, where N is the number of dipoles per unit volume and $\mathbf{P}$ the polarization density. Using the Maxwell equations (2.12), the resulting volumetric force $\mathbf{F}_e^{(j)} = \rho_e \mathbf{E}^{(j)} + \mathbf{P}^{(j)} \cdot \nabla \mathbf{E}^{(j)}$ can be obtained as the divergence of the Maxwell stress

$$\boldsymbol{\tau}_M^{(j)} = \varepsilon^{(j)} \left(\mathbf{E}^{(j)} \mathbf{E}^{(j)} - \frac{1}{2} \mathsf{I} E^{(j)2} \right). \tag{2.14}$$

The net free charge density $\rho_e^{(j)}(\mathbf{r}, t)$ does not appear explicitly in (2.14) but through the spatial variation of $\mathbf{E}^{(j)}$ given by the Maxwell equations (2.12).

The dynamical effect of the electric field is simply taken into account by adding the Maxwell stress (2.14) to the viscous stress ($\boldsymbol{\tau}^{(j)} \to \boldsymbol{\tau}^{(j)} + \boldsymbol{\tau}_M^{(j)}$) in both the momentum equation (2.3) and the boundary conditions (2.6) at the interface. Alternatively, one can add the volumetric force [31]

$$\mathbf{F}_e^{(j)} = \nabla \cdot \boldsymbol{\tau}_M^{(j)} = \rho_e^{(j)} \mathbf{E}^{(j)} - \frac{1}{2} E^{(j)2} \nabla \varepsilon^{(j)} + \frac{1}{2} \nabla \left(E^{(j)2} \frac{\partial \varepsilon^{(j)}}{\partial \rho^{(j)}} \rho^{(j)} \right) \tag{2.15}$$

to the momentum equation (2.3). In this case, the Maxwell stress (2.14) is only included in the stress balance at the interface (2.6). The last term of the volumetric force (2.15) is the electrostrictive force, which can be neglected in an incompressible fluid.

As mentioned in the previous section, the VoF method treats immiscible fluids as a single phase with spatially varying properties through the interface. In this case, the permittivity $\varepsilon^{(j)}$ becomes a continuous scalar field with significant variations across the interface, and the term $\frac{1}{2} E^{(j)2} \nabla \varepsilon^{(j)}$ in Eq. (2.15) must be considered, even for fluids with constant permittivities.

In an electrokinetic model, the charge density $\rho_e^{(j)}$ is calculated from the concentrations of ionic species in the liquid as

$$\rho_e^{(j)} = \sum_k \mathcal{F} z_k n_k^{(j)} , \tag{2.16}$$

where $\mathcal{F}$ is the Faraday constant while z_k and $n_k^{(j)}(\mathbf{r}, t)$ are the valence and number of mols per unit volume of the k-species, respectively. The concentration of ions of the species k is calculated throughout the fluid domain by solving the corresponding Nernst-Planck conservation equation [23]

$$\frac{Dn_k^{(j)}}{Dt} = -\nabla \cdot \left[(w_k \, e \, z_k \, n_k^{(j)} \, \mathbf{E}^{(j)}) + \mathcal{D}_k \nabla n_k^{(j)} \right] + r_k^{(j)}, \tag{2.17}$$

where w_k and $\mathcal{D}_k$ are the ion mobility and diffusivity of the k-species, respectively, while e is the elementary charge. The ion diffusivity can be calculated by the Einstein-Smoluchowski relationship $\mathcal{D}_k = w_k k_B T$, where k_B is the Boltzmann constant and T is the temperature. Finally, $r_k^{(j)}(\mathbf{r}, t)$ is the net production rate of the species k.

Several variants of the electrokinetic approximation have been proposed and solved [32]. The electrokinetic model, as applied to the steady cone-jet mode of electrospray, has recently been solved by López-Herrera et al. [33] using a high-accuracy numerical scheme [34]. For strong electrolytes, the ion distribution and, consequently, the conductivity in the bulk can be strongly non-homogeneous, invalidating the leaky-dielectric approximation described below.

2.2.2 A Charge-Conservative Model

A conservation equation for the charge density (2.16) can be readily derived by adding Eq. (2.17) for the different species:

$$\frac{D\rho_e^{(j)}}{Dt} = -\nabla \cdot \left[\left(\sum_k w_k \, \mathcal{F} \, e \, z_k^2 \, n_k^{(j)} \right) \mathbf{E}^{(j)} \right] + \sum_k \mathcal{F} z_k \mathcal{D}_k \nabla^2 n_k^{(j)} + \sum_k \mathcal{F} z_k r_k^{(j)} . \tag{2.18}$$

The production rates $r_k^{(j)}$ depend on the concentrations of the species and the chemical kinetics of the reactions occurring in the system. These reactions can be extremely complex and frequently unknown. The rates $r^{(k)}$ depend on each other because the production/destruction of some species implies the destruction/production of others. Nevertheless, three scenarios can be easily modeled. The first is the unipolar injection, where a single species generated at an upstream electrode migrates through the fluid with zero production rate. The second one is the evolution of strong electrolytes. In this case, neutral species are fully dissociated, and chemical reactions are absent ($r^{(k)} = 0$). The last case is the evolution of a binary z-z electrolyte analyzed by

Saville [30]. Three species are diluted in the fluid: the neutral, the ionic, and the counter-ionic. In the forward reaction, the neutral species dissociate, producing the same amount of ions and counterions. In contrast, in the recombination reaction, the ionic and counter-ionic species disintegrate at the same rate to produce the neutral one.

The source/sink term $\sum_k \mathcal{F} z_k r_k^{(j)}$ in Eq. (2.18) vanishes in the three situations described above. In this case, this equation can be written as

$$\frac{\partial \rho_e^{(j)}}{\partial t} + \mathbf{v}^{(j)} \cdot \nabla \rho_e^{(j)} = -\nabla \cdot \left(K^{(j)} \mathbf{E}^{(j)} \right) + \mathcal{D}_{\rho_e}^{(j)} \nabla^2 \rho_e^{(j)} , \tag{2.19}$$

where the electrical conductivity $K^{(j)}(\mathbf{r}, t)$ and diffusion coefficient $\mathcal{D}_{\rho_e}^{(j)}(\mathbf{r}, t)$ fields are given by the expressions

$$K^{(j)} = \sum_k w_k \mathcal{F} e z_k^2 n_k^{(j)}, \quad \mathcal{D}_{\rho_e}^{(j)} = \frac{\sum_k \mathcal{F} z_k \mathcal{D}_k \nabla^2 n_k^{(j)}}{\nabla^2 \rho_e^{(j)}} = \frac{\sum_k \mathcal{F} z_k \mathcal{D}_k \nabla^2 n_k^{(j)}}{\sum_k \mathcal{F} z_k \nabla^2 n_k^{(j)}} . \tag{2.20}$$

Equation (2.19) can also be postulated as the advection-diffusion conservation equation for the scalar $\rho_e^{(j)}(\mathbf{r}, t)$, where $\mathbf{j} = K^{(j)}(\mathbf{r}, t)\mathbf{E}^{(j)}$ is the charge flux driven by the electric field across a material surface element, and $\mathcal{D}_{\rho_e}^{(j)}(\mathbf{r}, t)$ is an effective diffusion coefficient. This coefficient equals the ionic diffusivity when there is a single ionic species in the liquid or all the species have the same ionic diffusivity [23].

The electrohydrodynamic problems analyzed in this work demand intense electric fields. In this sense, it is sensible to neglect thermal diffusion versus the electric drift over the hydrodynamic bulk. This simplification cannot be applied to the thin Debye layer formed next to the interface, where the accumulation of an ionic species leads to an equilibrium between these two factors. In fact, the Debye layer thickness λ_D can be estimated as follows:

$$\left. \begin{array}{l} \nabla \cdot (K^{(j)} \mathbf{E}^{(j)}) \sim \frac{K^{(j)} \rho_e^{(j)}}{\varepsilon^{(j)}} \\ \\ \mathcal{D}^* \nabla^2 \rho_e^{(j)} \sim \frac{\mathcal{D}^* \rho_e^{(j)}}{\lambda_D^2} \end{array} \right\} \Longrightarrow \lambda_D = \left(\frac{\varepsilon^{(j)} \mathcal{D}^*}{K^{(j)}} \right)^{1/2} , \tag{2.21}$$

where $\mathcal{D}^*$ is the diffusivity of the ionic species whose valence coincides with the polarity of the potential at the interface.

The resolution of the Debye layer structure becomes computationally unaffordable in the numerical simulations of most electrohydrodynamic problems due to the disparity between λ_D and the system size. This problem has been obviated in coarse-grained simulations where thermal diffusion is not considered, and the Debye layer does not need to be resolved [35–37].

If thermal diffusion is neglected, Eq. (2.19) becomes

$$\frac{\partial \rho_e^{(j)}}{\partial t} = -\nabla \cdot (\rho_e^{(j)} \mathbf{v}^{(j)} + K^{(j)} \mathbf{E}^{(j)}) , \tag{2.22}$$

where the incompressibility condition $\nabla \cdot \mathbf{v}^{(j)} = 0$ has been considered. This expression shows that the local variation of the free electrical charge results from the net flux of charge due to both conduction ($K^{(j)} \mathbf{E}^{(j)}$) and the convection caused by the fluid motion ($\rho_e^{(j)} \mathbf{v}^{(j)}$).

It is frequent to assume that the dissolved species are distributed over most of the bulk with a certain degree of uniformity so that the electrical conductivity takes a practically constant value in that region (the so-called Ohmic conduction model) [23]. That value is typically measured in a conductometer from the current between two electrodes under the action of a relatively weak electric field and is taken as an input to feed the electrohydrodynamic model.

If the electrical conductivity and permittivity are constant, then Eq. (2.22) yields

$$\frac{D\rho_e^{(j)}}{Dt} + \frac{K^{(j)}}{\varepsilon^{(j)}} \rho_e^{(j)} = 0. \tag{2.23}$$

This equation shows that the electrical charge transported by a fluid particle relaxes exponentially to its terminal value with the intrinsic characteristic time $t_e = \varepsilon^{(j)}/K^{(j)}$. Electrohydrodynamic phenomena occurring on time scales smaller than t_e can be affected by charge relaxation effects (also called electrokinetic effects). In this case, the Ohmic conduction model fails, and an electrokinetic model must be solved.

In electrified tip streaming, λ_D is typically much smaller than any characteristic length of the problem, and t_e is at most comparable with the characteristic hydrodynamic time. For this reason, the electrokinetic model is practically never required. Instead, the charge-conservative approximation described above is commonly used.

In most studies, these equations are numerically integrated with the VoF method. Therefore, the quantities and physical properties vary across the diffused interface, i.e.,

$$\begin{aligned} &\left\{\mathbf{F}_e^{(j)}(\mathbf{r}, t), \mathbf{E}^{(j)}(\mathbf{r}, t), \rho_e^{(j)}(\mathbf{r}, t), \varepsilon^{(j)}, K^{(j)}\right\} \\ &\quad \rightarrow \{\mathbf{F}_e(\mathbf{r}, t), \mathbf{E}(\mathbf{r}, t), \rho_e(\mathbf{r}, t).\varepsilon(\mathbf{r}, t), K(\mathbf{r}, t)\} . \end{aligned} \tag{2.24}$$

The electrical part of the model solved by the VoF method can be summarized as follows:

$$\mathbf{F}_e = \rho_e \mathbf{E} - \frac{1}{2} E^2 \nabla \varepsilon, \tag{2.25}$$

$$\nabla \cdot (\varepsilon \mathbf{E}) = \rho_e \ , \quad \nabla \times \mathbf{E} = \mathbf{0} \ , \tag{2.26}$$

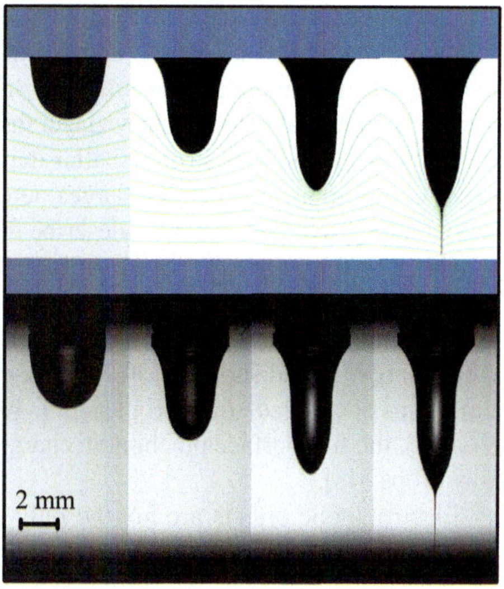

Fig. 2.2 Breakup of a low-conductivity pendant droplet between two electrodes. The droplet stretches until the electric field in front of the apex is intense enough to produce tip streaming. The upper and lower images correspond to the VoF numerical simulation and the experiment, respectively (reprinted from Ref. [35] with permission of AIP publishing)

$$\frac{\partial \rho_e}{\partial t} + \mathbf{v} \cdot \nabla \rho_e = -\nabla \cdot (K\mathbf{E}). \tag{2.27}$$

It must be emphasized that $\varepsilon(\mathbf{r}, t)$ and $K(\mathbf{r}, t)$ are scalar fields due to the interface diffusion in the VoF method, although they are assumed to take constant values throughout each phase in the model[1]. Figure 2.2 illustrates the accuracy of this approximation to describe the breakup of a low-conductivity pendant droplet between two electrodes.

Equations (2.25)–(2.27) can be regarded as an intermediate approximation between a full electrokinetic model, in which the evolution of each ionic species is analyzed (including its thermal diffusion), and the leaky-dielectric approximation, where $\rho_e^{(j)} = 0$ and the net free charge is accounted for only through a charge surface concentration, as explained below.

Electrohydrodynamic phenomena such as electrospray and electrospinning are frequently analyzed by integrating over time this charge-conservative model with the VoF method. In these phenomena, electric charges migrate across the bulk towards the interface. The simulation converges to a solution in which $\rho_e^{(j)}$ vanishes except in a thin layer next to the interface. This situation corresponds to that explicitly imposed in the leaky-dielectric model. For this reason, the model is erroneously called leaky-dielectric. The conservative model described above provides accurate predictions when the charge is accumulated next to the interface. However, it may fail to correctly capture the physics of the transient regime to that state, where charge relaxation effects may be relevant.

[1] In Eqs. (2.24)–(2.27), ε is the fluid electrical permittivity. This symbol stands for the relative electrical permittivity in the rest of the text.

As mentioned above, the comparison between the system's spatial and temporal scales and the intrinsic ones λ_D and t_e indicates when an electrokinetic model must be used. Electrokinetic effects must be considered when, for instance, the size of the system is comparable to the Debye layer thickness λ_D [23]. In this case, the electrical conductivity exhibits a strong spatial dependence, and the Ohmic model fails to describe the transport of free charges across the fluid medium. The electrokinetic model may also be necessary to describe the disintegration of microdroplets and the pinching of fluid threads, where an interface can be created at a rate comparable to the inverse of the electric relaxation time t_e. Specific examples of phenomena demanding an electrokinetic model are the following: the initial phase of ejections from charged liquid surfaces [36, 38], the oscillation of liquid menisci with periodic emissions of charged liquid droplets [39, 40], some flows driven by AC electric fields [41–44], the free surface pinching in charged liquid jets [45], or the disintegration of tiny drops [46].

Electrokinetic effects are not the only cause that invalidates the Ohmic model. Equation (2.27) implicitly assumes that the conduction of electrical charges in the bulk is isotropic. However, macromolecules significantly stretched along the streamwise direction may limit the validity of that assumption in, for instance, the critical cone-jet transition region of electrospinning. In fact, Rubio et al. [47] have observed a significant decrease in the conductivity of a stretched viscoelastic filament due to the ion mobility reduction in the stretching direction.

2.2.3 The Leaky-Dielectric Model

Interfaces play a critical role in tip streaming due to the large surface-to-volume ratios in this phenomenon. Interfaces contain most of the physics of the problem and must be modeled accurately to obtain reliable predictions. In the presence of externally applied electric fields, interfaces act as barriers preventing the diffusion of free ions. The accumulation of charges and the jump of electrical permittivity at those surfaces substantially affect the interfacial forces and their equilibrium. The net interfacial force arising in low-conductivity liquids drives the electrified tip streaming.

As explained in the previous section, the free charge is built up within the Debye layer under an applied electric field. This layer is much thinner than the rest of the fluid domain in electrified tip streaming despite the smallness of the fluid shapes produced in this phenomenon. This disparity of spatial scales demands extremely high spatial resolutions in numerical simulations. For this reason, the Debye layer is not resolved but is topologically reduced to a surface and modeled through the boundary conditions. This constitutes the key element of the so-called Taylor-Melcher leaky-dielectric model [30, 48–50].

The fundamental approximation of the Taylor-Melcher leaky-dielectric model is to assume that all the net free charge accumulates at the interface within an *equilibrium* Debye layer much thinner than the system size. This approximation verifies only if $\lambda_D \ll L_c$ and $t_e \ll t_c$, where L_c and t_c are the characteristic length and time of the system. Two additional requisites are demanded: (i) the applied electric field is much larger than the electric field across the droplet caused by the thermal voltage, and (ii) the applied electric field is much smaller than that associated with the thermal voltage in the Debye layer [23].

If the above conditions hold, then the net free charge contained in the Debye layer can be described in terms of the surface charge density σ_e exclusively ($\rho_e = 0$), and (ii) the electrical conductivity can be regarded as a constant throughout the liquid domain (the Ohmic conduction model). Schnitzer and Yariv [50] have derived the Taylor-Melcher leaky-dielectric model from the electrokinetic equations considering the adsorption kinetics of charges at the interface. In this derivation, the conductivity is not an intrinsic fluid property but a derived macroscale quantity depending upon the adsorption coefficients.

In the leaky-dielectric model, the electric field is calculated by integrating the Maxwell equations (2.12) with $\rho_e^{(j)} = 0$ and the interface boundary conditions (2.13) but with the surface charge density σ_e:

$$||\varepsilon^{(j)}\mathbf{E}^{(j)}|| \cdot \mathbf{n} = \sigma_e, \quad ||\mathbf{E}^{(j)}|| \times \mathbf{n} = \mathbf{0}. \tag{2.28}$$

The momentum equation (2.3) is solved with $\mathbf{F}_e^{(j)} = \mathbf{0}$, and the Maxwell stress (2.14) is added only to the balance of stresses (2.6) at the interface. It must be noted that the surface charge density σ_e produces a jump in the normal electric displacement field $\varepsilon^{(j)}\mathbf{E}^{(j)}$ at the interface, which translates into an electrostatic pressure. In electrified tip streaming, the surface charge does not completely screen the external electric field, allowing an interfacial electric field to exist. The combination of surface charge density and tangential electric field gives rise to shear stress, which plays a fundamental role in electrified tip streaming.

The conservation equation for the surface charge reads

$$\frac{\partial \sigma_e}{\partial t} + \nabla_s \cdot (\sigma_e \mathbf{v}_s) + \sigma_e(\nabla_s \cdot \mathbf{n})(\mathbf{v} \cdot \mathbf{n}) = -||K^{(j)}\mathbf{E}^{(j)}|| \cdot \mathbf{n}, \tag{2.29}$$

where $\mathbf{v}_s = \mathsf{I}_s \mathbf{v}^{(j)}$ the surface velocity. In the above equation, both conduction and diffusion *along* the interface have been neglected. It is worth mentioning that the charge convection $\nabla_s \cdot (\sigma_e \mathbf{v}_s)$ and surface dilatation $\sigma_e(\nabla_s \cdot \mathbf{n})(\mathbf{v} \cdot \mathbf{n})$ terms were absent in the original formulation of the leaky-dielectric model [48, 49]. The electrostatic and flow problems are decoupled without those terms and can be solved independently. This approximation can be made if the characteristic hydrodynamic time is much larger than the electric relaxation time [30]. However, the charge convection and surface dilatation terms are essential in flows around singularities, where intense acceleration can occur. Electrified tip streaming is a good example of this.

Melcher and Taylor [49] probably defined through their pioneering work the field of electrohydrodynamics, where the interaction between low-conductivity liquids and strong electric fields continues to yield new and intriguing phenomena [51]. The leaky-dielectric model has proved to be a valuable tool for examining the dynamic behavior of poorly conducting droplets in poorly conducting baths. In particular, it

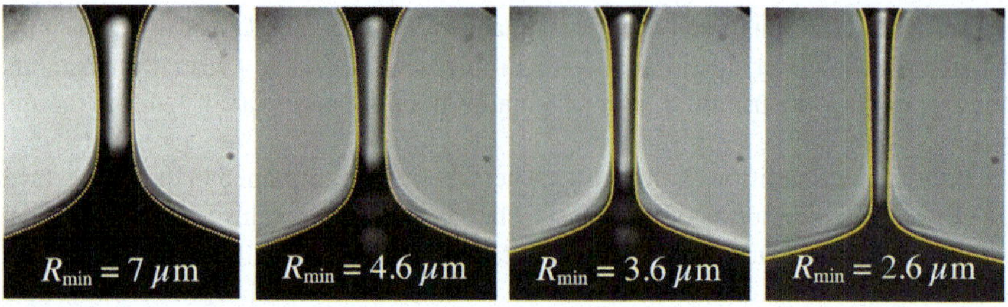

Fig. 2.3 Evolution of a low-conductivity liquid bridge between two electrodes. The solid line is the prediction of the leaky-dielectric model. $R_{\rm min}$ is the minimum free surface radius (reprinted from Ref. [64] with permission of AIP publishing)

provides accurate predictions for tip streaming flows such as the steady cone-jet mode of electrospray [52–57] and electrospinning [58, 59]. It has also been used to describe AC electrospray phenomena [60, 61], and has been extended to simulate ionic liquid menisci undergoing evaporation of ions [62, 63]. Under certain conditions, the leaky-dielectric model provides accurate results even when the characteristic hydrodynamic time is commensurate with the electric relaxation time (Fig. 2.3) [64].

The leaky-dielectric model imposes conditions on the infinitely thin interface separating two fluids. Strictly, this model can be solved only with sharp interface numerical methods, such as the boundary fitted [34] and interface tracking [65] procedures, in which the interface conditions can be prescribed as boundary conditions. Diffusive interface algorithms, such as the VoF method, integrate the charge-conservative model described in the previous section, whose solution may approximate that of the leaky-dielectric model.

2.3 Viscoelasticity

Tip streaming applications such as electrospinning [58, 59] or gaseous flow focusing spinning [2] involve dilute polymer solutions. These liquids exhibit a practically constant viscosity (shear thinning can be neglected) over a wide range of shear rates [66]. The major polymer effects are elasticity and the increase of the solution viscosity with respect to that of the solvent [66]. As explained in Chap. 1, the elasticity of quasi-monodisperse solutions is approximately quantified by a single characteristic time, which is related to the slowest relaxation process of the entire molecular chain [67].

In the FENE-P model [68, 69], $\boldsymbol{\tau}^{(j)} = \boldsymbol{\tau}_s^{(j)} + \boldsymbol{\tau}_p^{(j)}$, where $\tau_s^{(j)}$ is the solvent viscous stress calculated from the Navier-Poisson law (2.4) with the solvent viscosity $\mu_s^{(j)}$, and $\boldsymbol{\tau}_p^{(j)}$ is the polymer contribution. This contribution is calculated as

$$\boldsymbol{\tau}_p^{(j)} = \frac{\mu_p^{(j)}}{\lambda_s^{(j)}} \mathcal{F}(I_1)(\mathbf{A}^{(j)} - \mathbf{I}), \quad \mathcal{F}(I_1) = \frac{L^2 - 3}{L^2 - I_1}, \quad I_1 = \text{tr}(\mathbf{A}^{(j)}), \tag{2.30}$$

where $\mu_p^{(j)} = \mu^{(j)} - \mu_s^{(j)}$ is the polymer viscosity, $\lambda_s^{(j)}$ is the stress relaxation time, L^2 is the finite extensibility parameter, and $\mathbf{I}$ is the identity matrix. The conformation tensor $\mathbf{A}^{(j)}$ is a tensorial field calculated from the nonlinear relaxation law

$$\overset{\triangledown}{\mathbf{A}}^{(j)} = -\frac{1}{\lambda_s^{(j)}}(\mathcal{F}(I_1)\mathbf{A}^{(j)} - \mathbf{I}), \tag{2.31}$$

where $\overset{\triangledown}{\mathbf{A}}^{(j)}$ is the upper-convected time derivative of $\mathbf{A}^{(j)}$.

If finite extensibility effects are neglected ($L^2 \to \infty$, $\mathcal{F} = 1$), we obtain the popular Oldroyd-B model [70]

$$\boldsymbol{\tau}^{(j)} + \lambda_s^{(j)}\overset{\triangledown}{\boldsymbol{\tau}}^{(j)} = 2\mu^{(j)}(\mathsf{D}^{(j)} + \lambda_r^{(j)}\overset{\triangledown}{\mathsf{D}}^{(j)}), \tag{2.32}$$

where $\lambda_r^{(j)} = \lambda_s^{(j)}\mu_s^{(j)}/\mu^{(j)}$ is the stress retardation time. The Navier-Poisson law (2.4) for an incompresible Newtonian fluid is recovered for $\lambda_s^{(j)} = 0$.

The popularity of the Oldroyd-B model can be attributed to its relative simplicity and the fact that it can be derived from kinetic theory by assuming that the viscoelastic solution is an ideal system of Hookean dumbbells dissolved in a Newtonian liquid [68]. Following a pure continuum approach, it can also be obtained by assuming (i) a linear relationship between the polymer stress and the conformation tensor and (ii) a linear relaxation law for this tensor [71].

The Oldroyd-B and FENE-P models are believed to provide reasonable predictions in capillary extensional flows [72] when the stress relaxation time is adequately adjusted. For this reason, one expects to obtain reliable results from these approximations for tip streaming flows such as electrospinning [58, 73], gaseous flow focusing spinning [2], and viscoelastic selective withdrawal [74], in which the polymer is subject to strong extensional flows when crossing the tip of the tapering meniscus or film.

It must be pointed out that the solution viscosity used in the model is typically measured in a *shear* rheometer. This value may be inadequate to calculate the polymer stress in *extensional* flows. Therefore, inaccurate predictions may be obtained from these or similar models in the above-mentioned applications. Turkoz et al. [75] have found considerable discrepancies between the Oldroyd-B and experimental [76] self-similar dynamics for the final stages of the thinning of a viscoelastic filament. Ponce-Torres et al. [77] have provided experimental evidence of the dramatic failure of the Oldroyd-B model to describe the damping of a pendant viscoelastic droplet oscillation.

2.4 Surfactants

2.4.1 Surfactant Conservation Equations

Soluble surfactants play a fundamental and complex role in tip streaming [78]. Here, we assume that the surfactant is dissolved either in the inner ($j = i$) or the outer ($j = o$) phase. Surfactants are present as monomers in solution for bulk concentrations below the critical micelle concentration (CMC). Above that critical value, fluid-like aggregates called micelles form spontaneously. The volumetric concentration (mols per unit volume) of surfactant as monomers, $c^{(j)}(\mathbf{r}, t)$, and micelles, $m^{(j)}(\mathbf{r}, t)$, are calculated in a fluid dynamical problem from the advection-diffusion conservation equations [79]

$$\frac{\partial c^{(j)}}{\partial t} + \mathbf{v}^{(j)} \cdot \nabla c^{(j)} = \mathcal{D}_c^{(j)} \nabla^2 c^{(j)} - n \mathcal{J}_{cm}^{(j)}, \tag{2.33}$$

$$\frac{\partial m^{(j)}}{\partial t} + \mathbf{v}^{(j)} \cdot \nabla m^{(j)} = \mathcal{D}_m^{(j)} \nabla^2 m^{(j)} + \mathcal{J}_{cm}^{(j)}, \tag{2.34}$$

where $\mathcal{D}_c^{(j)}$ and $\mathcal{D}_m^{(j)}$ are the diffusion coefficients for the surfactant as monomers and micelles, respectively, n is the number of monomers that constitute a micelle, while $\mathcal{J}_{cm}^{(j)}$ stands for the net rate either of formation ($\mathcal{J}_{cm}^{(j)} > 0$) or breakup ($\mathcal{J}_{cm}^{(j)} < 0$) of mols of micelles per unit volume.

Surfactant molecules adsorb on the interface to form a monolayer. The surfactant surface concentration Γ (the surface coverage, measured in mols per unit area) verifies the advection-diffusion equation [79]

$$\frac{\partial \Gamma}{\partial t} + \nabla_s \cdot (\Gamma \mathbf{v}_s) + \Gamma(\nabla_s \cdot \mathbf{n})(\mathbf{v} \cdot \mathbf{n}) = \mathcal{D}_s \nabla_s^2 \Gamma + \mathcal{J}_{\Gamma c}^{(j)}, \tag{2.35}$$

where $\mathcal{D}_s$ is the *effective* surface diffusion coefficient and $\mathcal{J}_{\Gamma c}^{(j)}$ denotes the net flux of surfactant from the phase j to the interface due to the adsorption/desorption process. To derive this equation, one supposes the micelles do not adsorb directly onto the interface but completely dissociate into bulk monomers before the adsorption process [79].

The diffusive term $\mathcal{D}_s \nabla_s^2 \Gamma$ is obtained from the Fick law, which can be derived from first principles only under conditions that are rarely accurate in systems of practical interest [80]. However, it is assumed in almost all studies. Due to the small values taken by surface diffusion coefficient $\mathcal{D}_s$ for strong surfactants [81], the convection of these molecules over the interface is much more important than the diffusion mechanism in most microfluidic applications (the surface Péclet number is much greater than unity), and the latter is frequently neglected.

The sublayer next to the interface exchanges surfactant molecules with the interface. The adsorption/desorption net flux $\mathcal{J}_{\Gamma c}^{(j)}$ equals the diffusive flux evaluated on the side (sublayer) j of the interface, i.e.,

$$\mathcal{D}_c^{(j)} \nabla c^{(j)} \cdot \mathbf{n} = \pm \mathcal{J}_{\Gamma c}^{(j)}, \quad \mathcal{D}_m^{(j)} \nabla m^{(j)} \cdot \mathbf{n} = 0, \tag{2.36}$$

where the sign + and − applies to $j = o$ and i, respectively. This equation couples surfactant transport across the bulk and over the interface.

The conservation equations for the electric charge (2.29) and the surfactant molecules (2.35) only differ in the source term: the conduction of charge from the bulk and the adsorption/desorption of surfactant molecules at the interface. It is worth mentioning that Eq. (2.35) is frequently written in the form

$$\frac{\partial \Gamma}{\partial t} + \nabla_s \cdot (\Gamma \mathbf{v}) = \mathcal{D}_s \nabla_s^2 \Gamma + \mathcal{J}_{\Gamma c}^{(j)}, \tag{2.37}$$

where $\mathbf{v}$ is the 3D velocity on the interface. In this way, the convection $\nabla_s \cdot (\Gamma \mathbf{v}_s)$ and dilatational/compression $\Gamma(\nabla_s \cdot \mathbf{n})(\mathbf{v} \cdot \mathbf{n})$ terms are recast. The same occurs in the case of Eq. (2.29).

2.4.2 *Kinetic Model and Equation of State*

The sublayer next to the interface constitutes a source/sink of surfactant molecules for the interface during the system evolution. The net adsorption/desorption flux can be obtained from the simple interface kinetic model

$$\mathcal{J}_{\Gamma c}^{(j)} = \frac{k_a^{(j)} c_s^{(j)}}{\Gamma_\infty} (\Gamma_\infty - \Gamma) - k_d^{(j)} \Gamma, \tag{2.38}$$

where $k_a^{(j)}$ and $k_d^{(j)}$ are the adsorption and desorption constants, respectively, $c_s^{(j)}$ is the surfactant concentration evaluated at the sublayer j of the interface, and Γ_∞ is the maximum packing density. This *ideal* model assumes that the adsorption process is the result of the interaction between a surfactant molecule in the sublayer and a "vacancy" in the interface, and therefore the number of adsorbed molecules per unit time is proportional to the number of molecules in the sublayer given by c_s and the number of vacancies given by $\Gamma_\infty - \Gamma$. In addition, it is assumed that the probability per unit time for a molecule to desorb from the interface does not depend on Γ. Therefore, the number of desorbed molecules equals the surfactant surface concentration multiplied by the constant $k_d^{(j)}$.

Equation (2.38) yields the Langmuir isotherm at equilibrium ($\mathcal{J}_{\Gamma c}^{(j)} = 0$):

$$\frac{c_s^{(j)}}{k_d^{(j)} / k_a^{(j)}} = \frac{\Gamma_\infty \Gamma}{\Gamma_\infty - \Gamma}. \tag{2.39}$$

The Langmuir isotherm relates the sublayer and surface concentrations if $c_s^{(j)} \lesssim$ CMC. For $c_s^{(j)} \gtrsim$ CMC, the concentration Γ saturates to an approximately constant value.

At small enough concentrations, the Gibbs isotherm [82]

$$\Gamma = -\frac{1}{R_g T}\left(\frac{\partial \sigma}{\partial \ln c_s^{(j)}}\right)_{T,p} \tag{2.40}$$

relates the sublayer and surface concentrations to the interfacial tension at equilibrium (R_g is the universal gas constant).

The combination of the Langmuir and Gibbs isotherms leads to the Langmuir equation of state [81]

$$\sigma = \sigma_0 + R_g T \Gamma_\infty \ln\left(1 - \frac{\Gamma}{\Gamma_\infty}\right), \tag{2.41}$$

where σ_0 is the surface tension of the clean interface, and T is the temperature. Experiments show that $\sigma(\Gamma)$ reaches a plateau for sufficiently large values of Γ. This effect is not captured by Eq. (2.41), which must be replaced by an appropriate equation of state for values of Γ close to Γ_∞. This typically occurs if the volumetric concentration $c^{(j)}$ reaches values close to or larger than the CMC.

For a dilute monolayer ($\Gamma \ll \Gamma_\infty$), the Langmuir equation of state (2.40) reduces to the Henry isotherm,

$$\sigma = \sigma_0 - R_g T \Gamma. \tag{2.42}$$

corresponding to an ideal gas model in which the interaction between the adsorbed molecules is not considered.

More sophisticated kinetic models and equations of state can be used to improve the theoretical predictions. In the generalized Frumkin model, the net adsorption/desorption flux is calculated as [83]

$$\mathcal{J}_{\Gamma c}^{(j)} = \frac{k_a^{(j)} c_s^{(j)}}{\Gamma_\infty}(\Gamma_\infty - \Gamma) - k_d^{(j)} \Gamma \exp\left[\kappa (\Gamma/\Gamma_\infty)^n\right], \tag{2.43}$$

where κ and n are surfactant constants. The condition $\mathcal{J}_{\Gamma c}^{(j)} = 0$ at equilibrium yields the isotherm:

$$\frac{c_s^{(j)}}{k_d^{(j)}/k_a^{(j)}} = \exp\left[\kappa (\Gamma/\Gamma_\infty)^n\right] \frac{\Gamma_\infty \Gamma}{\Gamma_\infty - \Gamma}. \tag{2.44}$$

The combination of this equation and the Gibbs isotherm leads to the equation of state

$$\sigma = \sigma_0 + R_g T \Gamma_\infty \left[\ln\left(1 - \frac{\Gamma}{\Gamma_\infty}\right) - \frac{\kappa n}{n+1}\left(\frac{\Gamma}{\Gamma_\infty}\right)^{n+1}\right]. \tag{2.45}$$

The Frumkin model does not capture the behavior of $\sigma(\Gamma)$ for values of $\Gamma \lesssim \Gamma_\infty$ either. This deficiency can be overcome by empirical approximations of $\sigma(\Gamma)$ [84, 85]. When the Langmuir or Frumkin equation of state is used in tip streaming simulations, one must ensure that Γ does not exceed a cut-off value above which the equation state $\sigma(\Gamma)$ provides unrealistic values.

2.4.3 Transport of Surfactant

In many tip streaming realizations, the dispersed phase is injected from a reservoir with uniform surfactant concentration. In this case, Eq. (2.33) leads to $Dc^{(j)}/Dt = 0$. Therefore, the reservoir concentration c_∞ is convected by the fluid particles, and $c^{(j)}$ is uniform throughout most of the liquid domain. The diffusion coefficient $\mathcal{D}_c^{(j)}$ takes very small values in most experimental realizations. This implies that the spatial variation of the surfactant concentration is confined within a very thin diffusive boundary layer next to the interface. The thickness $\hat{\lambda}_D$ of this layer can be estimated as follows:

$$\frac{\mathcal{D}_c^{(j)} c_\infty}{\hat{\lambda}_D} \sim k_a^{(j)} c_\infty \Longrightarrow \hat{\lambda}_D = \frac{\mathcal{D}_c^{(j)}}{k_a^{(j)}}. \tag{2.46}$$

The thickness $\hat{\lambda}_D$ can be several orders of magnitude smaller than the system size, which poses a significant challenge for numerical simulations.

When the surfactant and hydrodynamic characteristic times are compared, several limits can be considered. The transfer of surfactant molecules from the bulk to the fresh interface created at the onset of tip streaming can be limited (controlled) either by the interface kinetics or by diffusion from the bulk or by both.

In the high-diffusivity (sorption-controlled) limit, the diffusion time is much smaller than those characterizing the interface kinetics. Then, one can assume that the volumetric concentration c_s at the interface is approximately the bulk value c_∞. Consequently, one does not need to solve the surfactant transport across the bulk, and the diffusion coefficient $\mathcal{D}_c^{(j)}$ does not enter the problem.

In the high-solubility (diffusion-controlled) limit, the times characterizing the interface kinetics are much smaller than the diffusion time. This implies that the volumetric concentration c_s at the interface is in local equilibrium with the surface concentration Γ [86], and Eq. (2.39) approximately verifies at any instant. This means that the surfactant kinetic properties enter the problem only through the ratio $k_d^{(j)}/k_a^{(j)}$. In this limit, the kinetic model (2.38) [or (2.43)] can be ignored, and $\mathcal{J}_{\Gamma c}^{(j)}$ in the conservation equation (2.35) is replaced by the diffusive flux $\mathcal{D}_c^{(j)}\nabla c^{(j)} \cdot \mathbf{n}$ [80].

When the diffusion and kinetic characteristic times are much smaller than any other characteristic time (the high-diffusivity and high-solubility limits combined), both Γ and σ take the equilibrium values $\Gamma(c_\infty)$ and $\sigma(c_\infty)$ corresponding to c_∞.

In many experiments, the times characterizing the interface kinetics are much longer than any hydrodynamic time (in particular, the inverse of the interface production rate). This means that surfactant can neither adsorb onto the interface nor desorb from it at this stage of the flow and, therefore, behaves as an insoluble surfactant. This considerably simplifies the problem because it eliminates Eqs. (2.33) and (2.38) from the analysis. Therefore, the (normally unknown) quantities $\mathcal{D}_c^{(j)}$, k_a, and k_d are not considered.

Following the onset of tip streaming, the microjetting regime is established, and the flow adopts a steady solution in most of the fluid domain. In this case, the adsorption/desorption time must be compared with the residence time of an interface element in the tip of the tapering meniscus. If the former is much larger and surface diffusion is ignored, the amount of surfactant transported by the interface element can be considered constant.

2.4.4 Solutocapillarity, Marangoni Stress, and Surfactant Viscosity

Interfaces between two immiscible fluids are rapidly formed and stretched in tip streaming. In this case, the high-solubility and high-diffusivity limits described above do not apply. Then, the surfactant surface concentration can significantly differ from that at equilibrium, and the interfacial tension can considerably vary over the interface. In this case, Eq. (2.6) involves the *local* value of the interfacial tension (the so-called solutocapillarity effect) and the Marangoni stress

$$\boldsymbol{\tau}_{Ma} = \boldsymbol{\nabla}_s \sigma \tag{2.47}$$

must be added to the left side of Eq. (2.6).

The gradient of the surface velocity can produce significant viscous surface stress. While Marangoni stress tends to eliminate inhomogeneities of surfactant concentration, viscous surface stress opposes surface velocity variations. The viscous surface stress obeys different constitutive relationships depending on the surfactant molecule nature. In fact, adsorbed surfactant monolayers at fluid surfaces usually exhibit rheological properties, and the surface viscosities frequently depend on the time scale and amplitude of the deformation owing to surface relaxations and nonlinear responses [87].

For an inertialess Newtonian interface, the viscous surface stress tensor can be calculated from the Boussinesq-Scriven constitutive equation [88] as

$$\mathbf{T}_s = \left(\kappa^s - \mu^s\right)\left(\boldsymbol{\nabla}_s \cdot \mathbf{v}\right) \mathsf{I}_s + 2\mu^s \mathsf{D}_s, \tag{2.48}$$

where κ^s and μ^s are the dilatational and shear surface viscosities, respectively, while $\mathsf{D}_s = 1/2\,[\boldsymbol{\nabla}_s \mathbf{v} \cdot \mathsf{I}_s + \mathsf{I}_s \cdot (\boldsymbol{\nabla}_s \mathbf{v})^T]$ is the surface strain rate tensor. Equation (2.48) is the surface counterpart of the Navier-Poisson law (2.4) for a compressible fluid. The viscous surface stress [89]

$$\boldsymbol{\tau}_s = \nabla_s \cdot \mathbf{T}_s = -\mathbf{n}\,(\nabla_s \cdot \mathbf{n})\,(\kappa^s - \mu^s)\,(\nabla_s \cdot \mathbf{v}) + \nabla_s\left[(\kappa^s - \mu^s)\,(\nabla_s \cdot \mathbf{v})\right] + 2\nabla_s \cdot (\mu^s \mathsf{D}_s) \tag{2.49}$$

is added to the left side of Eq. (2.6).

It must be pointed out that the velocity appearing in Eq. (2.49) is the 3D fluid velocity evaluated on the interface, including both the tangential and normal components to that surface [89–91]. Some studies have erroneously interpreted this velocity as the 2D velocity resulting from the projection of the fluid velocity onto the interface, which can lead to significant errors for high interface curvatures.

Intrinsic coordinates are convenient when calculating the viscous surface stress over an interface with an apex. The normal and tangential surface stresses, τ_{sn} and τ_{st}, in this coordinates read [92]

$$\tau_{sn} = \mu^s \overline{\kappa}\left[-F\left(\frac{v_t}{F}\right)' + \overline{\kappa} v_n\right] + \kappa^s \kappa\left[\frac{(F v_t)'}{F} + \kappa v_n\right], \tag{2.50}$$

$$\tau_{st} = \left[(\mu^s + \kappa^s)\left(\frac{F v_t' + F' v_t}{F} + \kappa v_n\right)\right]' + 2\left[\mu^s K - (\mu^s)'\frac{F'}{F}\right] v_t - 2\kappa_1(\mu^s v_n)', \tag{2.51}$$

where $\kappa = -(\nabla_s \cdot \mathbf{n})$ is twice the mean curvature of the interface, κ_1 and κ_2 are the curvatures along the meridians and parallels in the normal direction, respectively, $\overline{\kappa} = \kappa_1 - \kappa_2$, and $K = \kappa_1 \kappa_2$. In addition, v_n and v_t are the normal and tangential components of the velocity on the interface, and the prime denotes the derivative with respect to the intrinsic surface coordinate. Equations (2.50) and (2.51) reduce to those derived by Martínez-Calvo [93] in cylindrical coordinates.

The shear and dilatational surface viscosities depend on the surfactant surface concentration Γ [94]. Some studies [95, 96] have shown that the shear surface viscosity grows exponentially with the surface pressure $\Pi = \sigma_0 - \sigma$ for some typical surfactants; specifically,

$$\mu^s = \mu_0^s e^{(\Pi - \Pi_0)/\Pi_c}, \tag{2.52}$$

where μ_0^s and Π_0 are reference values, while Π_c is the surface pressure scale over which the viscosity changes. For many insoluble surfactants, Π_c is only a few mN/m [97]. Positive/negative values of this last quantity correspond to Π-thickening/thinning surfactants. It is frequently assumed that the ratio κ^s/μ^s takes a constant value independent from the surface pressure.

The simplest model for the dependence of the shear surface viscosity on Γ is the linear relationship

$$\mu^s = \mu_0^s \Gamma / \Gamma_0, \tag{2.53}$$

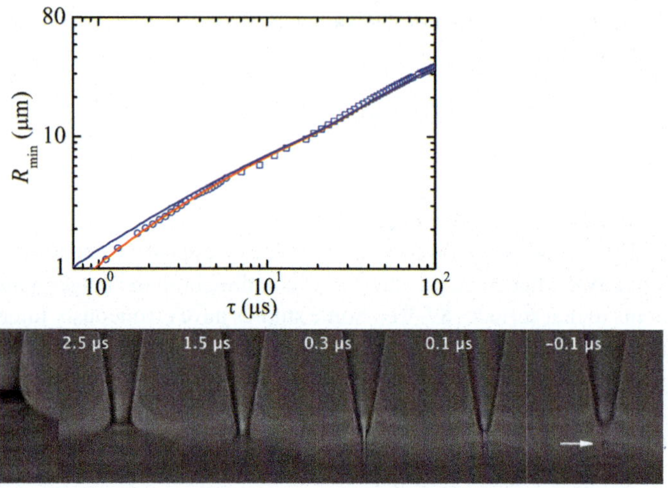

Fig. 2.4 Free surface minimum radius $R_{\rm min}$ as a function of the time to the pinching τ during the breakup of a pendant droplet of deionized water loaded with SDS at the concentration 0.8 cmc. The symbols are the experimental data, while the blue and red lines are the numerical result for zero surface viscosity and ($\mu^s = 1.2 \times 10^{-9}$ Pa s m, $\kappa^s = 0$), respectively. The lower images show the free surface shape before the pinching. A tiny subsatellite droplet forms due to the surfactant effects

where μ_0^s and Γ_0 are reference values. It has been shown that this approximation leads to realistic predictions [84].

The lack of precise information about the values taken by the surface viscosities, as well as the mathematical complexity of the calculation of the viscous stresses, has motivated that most of the experimental and theoretical works in microfluidics do not take into account those stresses. However, they may considerably affect the dynamics of interfaces even for surface viscosities much smaller than the bulk ones [84]. This may occur for two reasons: (i) viscous surface stresses may significantly alter the transport of surfactants over the interface, which may have significant consequences in the resulting solutocapillarity effect and Marangoni stress; and (ii) their relevance increases as the surface-to-volume ratio increases, as happens, for instance, during the interface breakup [84].

It is believed that foam and emulsion stability can be caused by the surface shear viscosity of the surfactant used to stabilize them. In fact, surface viscosity can significantly increase the drainage time during the coalescence of two bubbles/droplets [98]. However, there is no clear evidence that soluble and small-molecule surfactants have measurable surface shear viscosities [99]. As shown in Fig. 2.4, the shear surface viscosity may be of the order of 10^{-9} Pa s m. Without surface viscosity effects, the stabilization is attributed to the Marangoni stress [85]. Marangoni convection resists the outwards radial flow in the gap between two approaching interfaces, thus preventing coalescence [100].

2.5 The 1D (Slender-Jet) Approximation

The theoretical analysis can be greatly simplified when the inner fluid (typically a jet) adopts a slender shape along the streamwise direction z. In this case, the inner velocity profile is approximately parallel and uniform in a jet cross-section [101]. The effects of the outer medium are neglected in most problems. If one also considers the leaky-dielectric approximation, the 1D (slender-jet) model for viscoelastic flow becomes [58, 73, 102–105]

$$Q = \pi F^2 w, \tag{2.54}$$

$$I = \pi F^2 K^{(i)} E_z + 2\pi F w \sigma_e, \tag{2.55}$$

$$\begin{aligned} \rho^{(i)}(w_t + ww') &= \rho^{(i)} g - \left(\frac{\sigma}{F}\right)' + \frac{T'}{\pi F^2} + \frac{\sigma_e \sigma_e'}{\varepsilon^{(o)}} + (\varepsilon^{(i)} - \varepsilon^{(o)}) E_z E_z' + \frac{2\sigma_e E_z}{F}, \\ &+ \frac{2\sigma'}{F} + \frac{(9\mu^s F w')'}{2F^2} + \frac{(\kappa^s F w')'}{2F^2}, \end{aligned} \tag{2.56}$$

where Q and I are the flow rate and electric current transported by the jet, respectively, $F(z)$ and $w(z)$ are the interface contour and jet velocity, respectively, $\sigma_e(z)$ is the charge surface density, $E_z(z)$ is the axial component of the electric field, and $T = \pi F^2(\tau_{zz} - \tau_{rr})$ is the tensile force in the jet. The sub-index t and the prime stand for the derivative with respect to the time and the z coordinate, respectively.

The elements τ_{zz} and τ_{rr} of the deviatoric stress tensor are given by the corresponding 1D version of the constitutive relationship: Navier-Poisson law [106], FENE-P model [107], Oldroyd-B model [58], etc. Under the condition $\varepsilon^{(i)} E_n^{(i)} \ll \varepsilon^{(o)} E_n^{(o)}$, the charge surface density is calculated as $\sigma_e = \varepsilon^{(o)} E_n$. The normal outer electric field E_n can be obtained from the expression $E_n = \mathcal{A}/F$, where $\mathcal{A}(z)$ is a prescribed charge line distribution [108].

The last three terms in Eq. (2.56) correspond to the Marangoni and viscous surface stress caused by a surfactant monolayer [105]. Interestingly, the shear and dilatational viscosity terms have the same form, and, therefore, the relevant parameter to this order becomes $9\mu^s + \kappa^s$. Something similar occurs in the analysis of films, in which the two surface viscosities are indistinguishable from each other because they enter the problem through the single parameter $\mu^s + \kappa^s$ [109].

We use the 1D version of the surfactant conservation equation (2.35) to calculate $\Gamma(z)$. For an insoluble surfactant, this equation reads

$$\Gamma_t + w\Gamma' + \frac{\Gamma}{2} w' - \frac{\mathcal{D}_s}{F} \left(F\Gamma'\right)' = 0. \tag{2.57}$$

The dependence of the surface tension σ upon Γ can be obtained, for instance, from the Langmuir equation of state (2.45). The linear model (2.53) has been used to calculate the surface viscosities [105].

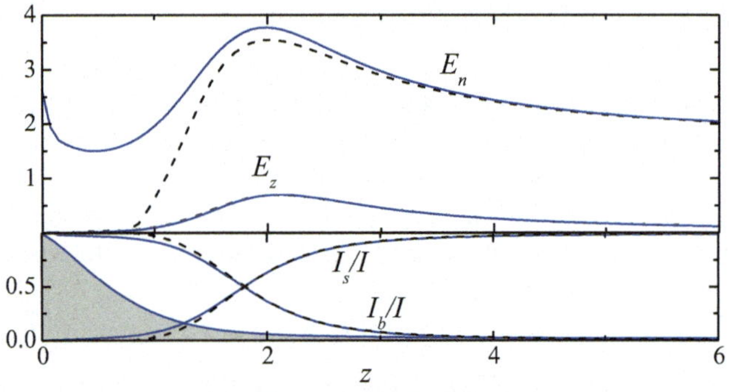

Fig. 2.5 The cone-jet mode of electrospray. Free surface contour (shaded area), axial and normal components of the electric field, E_z and $E_n = \sigma_e/\varepsilon^{(o)}$, and bulk and surface contributions to the electric current, $I_b = \pi F^2 K^{(i)} E_z$ and $I_s = 2\pi F w \sigma_e$, calculated from the 2D leaky-dielectric model (solid blue lines) and the 1D approximation (black dashed lines). The free surface contour and E_z calculated from the two approximations overlap. The quantities have been made dimensionless as indicated in Ref. [110]

Figure 2.5 compares the results for the cone-jet mode of electrospray obtained from the 2D and 1D leaky-dielectric models [110]. Significant discrepancies arise in the electric field normal to the cone, while the 1D approximation accurately predicts the rest of the variables. The electrostatic driving pressure $\varepsilon^{(o)} E_n^2 = \sigma_e^2/\varepsilon^{(o)}$ and shear electric field $\varepsilon^{(o)} E_n E_z$ take small values in the cone, which explains why the liquid acceleration (interface shape) is practically the same in the two approximations.

The 1D approximation provides useful predictions for jets emitted at large enough flow rates in the gravitational [111], electrospray [54], electrospinning [58], coflowing [112] and flow focusing [113] configurations (in the last two cases, the dynamical effects of the outer medium are to be taken into account). However, the use of this approximation in some realizations of tip streaming is questionable because significant 2D effects can be erroneously neglected in the inception of the emitted jet.

For steady flow ($w_t = 0$), Eq. (2.56) admits a twofold interpretation. Its spatial integration in an Eulerian frame of reference leads to the kinetic energy supplied to or withdrawn from the jet by the different factors over the integration interval. On the other hand, one can focus on a liquid slice moving throughout the fluid domain. Using this Lagrangian frame of reference, the right-hand side of Eq. (2.56) represents the driving and resistant forces acting on that slice.

In the absence of surface tension, electric field, surfactant, and viscoelasticity, one arrives at the following self-similar universal equation for steady flow:

$$v d_x v - v d_x (v^{-1} d_x v) = 1 \,, \tag{2.58}$$

where $v = w/v_0$ and $x \equiv z/L_0$ are the velocity and distance to the feeding capillary scaled with $v_0 \equiv (L_0 g)^{1/2}$ and $L_0 \equiv 3^{2/3}(\mu^{(i)2}/\rho^{(i)2} g)^{1/3}$, respectively [114]. The exact steady solution of (2.58) is [114, 115]

$$v = \frac{2^{-1/3}\text{Ai}^2(r)}{\text{Ai}'^2(r) - r\text{Ai}^2(r)} , \quad r = 2^{-1/3}(x - k) , \tag{2.59}$$

where Ai and Ai$'$ are the Airy function and its derivative, respectively, and $k = 2.9455\ldots$ is the first zero of $\text{Ai}(-k/2^{1/3})$. If the shape and velocity of the jet emitted in gaseous flow focusing are appropriately scaled, the results approximately match the solution (2.59) [113].

References

1. Gañán-Calvo AM (1998) Generation of steady liquid microthreads and micron-sized monodisperse sprays in gas streams. Phys Rev Lett 80:285–288
2. Ponce-Torres A, Ortega E, Rubio M, Rubio A, Vega EJ, Montanero JM (2019) Gaseous flow focusing for spinning micro and nanofibers. Polymer 178(121):623
3. Oliveira JE, Moraes EA, Costa RGF, Afonso AS, Mattoso LHC, Orts WJ, Medeiros ES (2011) Nano and submicrometric fibers of poly(d, l-lactide) obtained by solution blow spinning: process and solution variables. J Appl Polym Sci 122:3396–3405
4. Wang B, Yao Y, Peng J, Lin Y, Liu W, Luo Y, Xiang R, Li R, Wu D (2009) Preparation of poly(ester imide) ultrafine fibers by gas-jet/electrospinning. J Appl Polym Sci 114:883–891
5. Gañán-Calvo AM (2005) Enhanced liquid atomization: from flow-focusing to flow-blurring. Appl Phys Lett 86(214):101
6. Daristotle JL, Behrens AM, Sandler AD, Kofinas P (2016) A review of the fundamental principles and applications of solution blow spinning. ACS Appl Mater Interfaces 8:34,951–34,963
7. Chapman HN et al (2011) Femtosecond X-ray protein nanocrystallography. Nature 470:73–79
8. Gañán-Calvo AM (2007) Electro-flow focusing: the high-conductivity low-viscosity limit. Phys Rev Lett 98(134):503
9. Jiang J, Wang X, Li W, Liu J, Liu Y, Zheng G (2018) Electrohydrodynamic direct-writing micropatterns with assisted airflow. Micromachines 9:456
10. Cruz-Mazo F, Herrada MA, Gañán-Calvo AM, Montanero JM (2017) Global stability of axisymmetric flow focusing. J Fluid Mech 832:329–344
11. Gañán-Calvo AM, Gordillo JM (2001) Perfectly monodisperse microbubbling by capillary flow focusing. Phys Rev Lett 87(274):501
12. Gañán-Calvo AM (2004) Perfectly monodisperse microbubbling by capillary flow focusing: an alternate physical description and universal scaling. Phys Rev E 69(027):301
13. Rodríguez-Rodríguez J, Sevilla A, Martínez-Bazán C, Gordillo JM (2015) Generation of microbubbles with applications to industry and medicine. Annu Rev Fluid Mech 47:405–429
14. Gibbs JW (1928) The collected works of J. Willard Gibbs, Longmans, New York, USA
15. Rider WJ, Kothey DB (1998) Reconstructing volume tracking. J Comput Phys 141:112–152
16. Thompson JF, Warsi ZUA (1982) Boundary-fitted coordinate systems for numerical solution of partial differential equations-a review. J Comput Phys 47:1–108
17. Brackbill JU, Kothe DB, Zemach C (1992) A continuum method for modeling surface tension. J Comput Phys 100:335–354
18. Neto C, Evans DR, Bonaccurso E, Butt HJ, Craig VSJ (2005) Boundary slip in Newtonian liquids: a review of experimental studies. Rep Prog Phys 68:2859–2897
19. Denn MM (2001) Extrusion instabilities and wall slip. Annu Rev Fluid Mech 33:265–287
20. Dussan EB (1979) On the spreading of liquids on solid surfaces: static and dynamic contact lines. Annu Rev Fluid Mech 11:371–400
21. Zeleny J (1914) The electrical discharge from liquid points, and a hydrostatic method of measuring the electric intensity at their surfaces. Phys Rev 3:69–91

22. Taylor G (1964) Disintegration of water drops in electric field. Proc R Soc Lond A 280:383–397
23. Gañán-Calvo AM, López-Herrera JM, Herrada MA, Ramos A, Montanero JM (2018) Review on the physics of electrospray: from electrokinetics to the operating conditions of single and coaxial Taylor cone-jets, and AC electrospray. J Aerosol Sci 125:32–56
24. Doshi J, Reneker DR (1995) Electrospinning process and applications of electrospun fibers. J Electrost 35:151–60
25. Yarin AL, Koombhongse S, Reneker DH (2001) Bending instability in electrospinning of nanofibers. J Appl Phys 89:3018–3026
26. Theron SA, Zussman E, Yarin AL (2004) Experimental investigation of the governing parameters in the electrospinning of polymer solutions. Polymer 45:2017–2030
27. Bhardwaj N, Kundu SC (2010) Electrospinning: a fascinating fiber fabrication technique. Biotechnol Adv 28:325–347
28. Yarin AL (2011) Coaxial electrospinning and emulsion electrospinning of core-shell fibers. Polym Advan Technol 22:310–317
29. Gañán-Calvo AM, Lasheras JC, Dávila J, Barrero A (1994) The electrostatic spray emitted from an electrified conical meniscus. J Aerosol Sci 25:1121–1142
30. Saville DA (1997) Electrohydrodynamics: the Taylor-Melcher leaky dielectric model. Annu Rev Fluid Mech 29:27–64
31. Stratton JA (1941) Electromagnetic theory. McGraw-Hill, New York
32. Conroy DT, Matar K, Craster RV, Papageorgiou DT (2011) Breakup of an electrified viscous thread with charged surfactants. Phys Fluids 23(022):103
33. López-Herrera J, Herrada M, Gañán-Calvo A (2023) Electrokinetic modelling of cone-jet electrosprays. J Fluid Mech A 19:964
34. Herrada MA, Montanero JM (2016) A numerical method to study the dynamics of capillary fluid systems. J Comput Phys 306:137–147
35. Ferrera C, López-Herrera JM, Herrada MA, Montanero JM, Acero AJ (2013) Dynamical behavior of electrified pendant drops. Phys Fluids 25(012):104
36. Gañán-Calvo AM, López-Herrera JM, Rebollo-Muñoz N, Montanero JM (2016) The onset of electrospray: the universal scaling laws of the first ejection. Sci Rep 6(32):357
37. Dastourani H, Jahannama MR, Eslami-Majd A (2018) A physical insight into electrospray process in cone-jet mode: role of operating parameters. Int J Heat Fluid Flow 70:315–335
38. Gomez A, Tang K (1994) Charge and fission of droplets in electrostatic sprays. Phys Fluids 6:404–414
39. Jaworek A, Krupa A (1999) Classification of the modes of EHD spraying. J Aerosol Sci 30:873–893
40. Hijano AJ, Loscertales IG, Ibáñez SE, Higuera FJ (2015) Periodic emission of droplets from an oscillating electrified meniscus of a low-viscosity, highly conductive liquid. Phys Rev E 91(013):011
41. Yeo LY, Gagnon Z, Chang HC (2005) AC electrospray biomaterials synthesis. Biomaterials 26:6122–6128
42. Chetwani N, Maheshwari S, Chang HC (2008) Universal cone angle of AC electrosprays due to net charge entrainment. Phys Rev Lett 101(204):501
43. Maheshwari S, Chang HC (2009) Assembly of multi-stranded nanofiber threads through AC electrospinning. Adv Mater 21:349–354
44. Chetwani N, Cassou C, Go D, Chang HC (2010) High-frequency AC electrospray ionization source for mass spectrometry of biomolecules. J Am Soc Mass Spectrom 21:1852–1856
45. López-Herrera JM, Gañán-Calvo AM, Popinet S, Herrada MA (2015) Electrokinetic effects in the breakup of electrified jets: a volume-of-fluid numerical study. Int J Multiphase Flow 71:14–21
46. Pillai R, Berry JD, Harvie DJE, Davidson MR (2016) Electrokinetics of isolated electrified drops. Soft Matter 12:3310–3325
47. Rubio M, Sadek S, Vega EJ, Gañán-Calvo AM, Montanero JM (2021) Electrical conductivity of a stretching viscoelastic filament. Materials 14:1–12

48. Taylor GI (1966) Conical free surfaces and fluid interfaces. In: Applied mechanics. Springer, Berlin, pp 790–796
49. Melcher JR, Taylor GI (1969) Electrohydrodynamics: a review of the role of interfacial shear stresses. Annu Rev Fluid Mech 1:111–146
50. Schnitzer O, Yariv E (2015) The Taylor-Melcher leaky dielectric model as a macroscale electrokinetic description. J Fluid Mech 773:1–13
51. Vlahovska PM (2019) Electrohydrodynamics of drops and vesicles. Annu Rev Fluid Mech 51:305–330
52. Higuera FJ (2010) Numerical computation of the domain of operation of an electrospray of a very viscous liquid. J Fluid Mech 648:35–52
53. Herrada MA, López-Herrera JM, Gañán-Calvo AM, Vega EJ, Montanero JM, Popinet S, (2012) Numerical simulation of electrospray in the cone-jet mode. Phys Rev E 86(026):305
54. Ponce-Torres A, Rebollo-Muñoz N, Herrada MA, Gañán-Calvo AM, Montanero JM (2018) The steady cone-jet mode of electrospraying close to the minimum volume stability limit. J Fluid Mech 857:142–172
55. Borra JP (2018) Review on water electro-sprays and applications of charged dropswith focus on the corona-assisted cone-jet mode for High Efficiency Air Filtration by wet electro-scrubbing of aerosols. J Aerosol Sci 125:208–236
56. Gamero-Castaño M, Magnani M (2019) The minimum flow rate of electrosprays in the cone-jet mode. J Fluid Mech 876:553–572
57. Jiang Z, Gan Y, Shi Y (2020) An improved model for prediction of the cone-jet formation in electrospray with the effect of space charge. J Aerosol Sci 139(105):463
58. Carroll CP, Joo YL (2006) Electrospinning of viscoelastic Boger fluids: modeling and experiments. Phys Fluids 18(053):102
59. Reneker DH, Yarin AL (2008) Electrospinning jets and polymer nanofibers. Polymer 49:2387–2425
60. Yeo LY, Lastochkin D, Wang SC, Chang HC (2004) A new AC electrospray mechanism by Maxwell-Wagner polarization and capillary resonance. Phys Rev Lett 92(133):902
61. Demekhin E, Polyanskikh S, Ramos A (2011) Taylor cones in a leaky dielectric liquid under an AC electric field. Phys Rev E 84(035):301(R)
62. Higuera FJ (2008) Model of the meniscus of an ionic-liquid ion source. Phys Rev E 77(026):308
63. Coffman CS, Martínez-Sánchez M, Lozano PC (2019) Electrohydrodynamics of an ionic liquid meniscus during evaporation of ions in a regime of high electric field. Phys Rev E 99(063):108
64. Rubio M, Ponce-Torres A, Herrada MA, Gañán-Calvo AM, Montanero JM (2021) Effect of an axial electric field on the breakup of a leaky-dielectric liquid filament. Phys Fluids 33(092):114
65. Anthony CR et al (2023) Sharp interface methods for simulation and analysis of free surface flows with singularities: Breakup and coalescence. Annu Rev Fluid Mech 55:707–747
66. James DF (2009) Boger fluids. Annu Rev Fluid Mech 41:129–142
67. Clasen C, Plog JP, Kulicke WM, Owens M, Macosko C, Scriven LE, Verani M, McKinley GH (2006) How dilute are dilute solutions in extensional flows? J Rheol 50:849–881
68. Bird RB, C R, Armstrong, Hassager O (1987) Dynamics of polymeric liquids volume I: fluid mechanics; volume II: kinetic theory. Wiley, New York
69. Entov VM, Hinch EJ (1997) Effect of a spectrum of relaxation times on the capillary thinning of a filament of elastic liquid. J Non-Newtonian Fluid Mech 72:31–53
70. Oldroyd JG (1950) On the formulation of rheological equations of state. Proc Roy Soc Lond 200:523–541
71. Snoeijer JH, Pandey A, Herrada MA, Eggers J (2020) The relationship between viscoelasticity and elasticity. Proc R Soc A 476(20200):419
72. Bazilevskii AV, Entov VM, Rozhkov AN (2001) Breakup of an Oldroyd liquid bridge as a method for testing the rheological properties of polymer solutions. Polym Sci Ser A 42:716–726

73. Feng JJ (2002) The stretching of an electrified non-Newtonian jet: a model for electrospinning. Phys Fluids 14:3912–3926
74. Zhoua D, Feng JJ (2010) Selective withdrawal of polymer solutions: computations. J Non-Newtonian Fluid Mech 165:839–851
75. Turkoz E, López-Herrera JM, Eggers J, Arnold CB, Deike L (2018) Axisymmetric simulation of viscoelastic filament thinning with the Oldroyd-B model. J Fluid Mech 851:R2
76. Clasen C, Eggers J, Fontelos MA, Li J, McKinley GH (2006) The beads-on-string structure of viscoelastic threads. J Fluid Mech 556:283–308
77. Ponce-Torres A, Acero AJ, Herrada MA, Montanero JM (2018) On the validity of the Jeffreys (Oldroyd-B) model to describe the oscillations of a viscoelastic pendant drop. J Non-Newtonian Fluid Mech 260:69–75
78. Anna SL (2016) Droplets and bubbles in microfluidic devices. Annu Rev Fluid Mech 48:285–309
79. Craster RV, Matar OK, Papageorgiou DT (2009) Breakup of surfactant-laden jets above the critical micelle concentration. J Fluid Mech 629:195–219
80. Manikantan H, Squires TM (2020) Surfactant dynamics: hidden variables controlling fluid flows. J Fluid Mech 892:P1
81. Tricot YM (1997) Surfactants: static and dynamic surface tension, liquid film coating, vol 1. Chapman and Hall, pp 100–136
82. Chang CH, Franses EI (1995) Adsorption dynamics of surfactants at the air/water interface: a critical review of mathematical models, data, and mechanisms. Colloids Surf A 100:1–45
83. Moyle TM, Walker LM, Anna SL (2012) Predicting conditions for microscale surfactant mediated tipstreaming. Phys Fluids 24(082):110
84. Ponce-Torres A, Montanero JM, Herrada MA, Vega EJ, Vega JM (2017) Influence of the surface viscosity on the breakup of a surfactant-laden drop. Phys Rev Lett 118(024):501
85. Ponce-Torres A, Rubio M, Herrada MA, Eggers J, Montanero JM (2020) Influence of the surface viscous stress on the pinch-off of free surfaces loaded with nearly-inviscid surfactants. Sci Rep 10(16):065
86. Jin F, Stebe KJ (2007) The effects of a diffusion controlled surfactant on a viscous drop injected into a viscous medium. Phys Fluids 19(112):103
87. Langevin D (2014) Rheology of adsorbed surfactant monolayers at fluid surfaces. Annu Rev Fluid Mech 46:4765
88. Scriven LE (1960) Dynamics of a fluid interface equation of motion for Newtonian surface fluids. Chem Eng Sci 12:98–108
89. Martínez-Calvo A, Sevilla A (2018) Temporal stability of free liquid threads with surface viscoelasticity. J Fluid Mech 846:877–901
90. Wee H, Wagoner BW, Garg V, Kamat PM, Basaran OA (2021) Pinch-off of a surfactant-covered jet. J Fluid Mech 908:A38
91. Jaensson NO, Anderson PD, Vermant J (2021) Computational interfacial rheology. J Non-Newtonian Fluid Mech 290(104):507
92. Herrada MA, Ponce-Torres A, Rubio M, Eggers J, Montanero JM (2022) Stability and tip streaming of a surfactant-loaded drop in an extensional flow influence of surface viscosity. J Fluid Mech 934:A26
93. Martínez-Calvo A (2020) Dynamics of complex capillary flows: stability, rupture, and influence of surfactants. Doctoral dissertation
94. Luo ZY, Shang XL, Bai BF (2019) Influence of pressure-dependent surface viscosity on dynamics of surfactant-laden drops in shear flow. J Fluid Mech 858:91–121
95. Kim K, Choi SQ, Zell ZA, Squires TM, Zasadzinski JA (2013) Effect of cholesterol nanodomains on monolayer morphology and dynamics. Proc Natl Acad Sci 110:E3054–E3060
96. Samaniuk JR, Mermant J (2014) Micro and macrorheology at fluid-fluid interfaces. Soft Matt 10:7023–7033
97. Manikantan H, Squires TM (2017) Pressure-dependent surface viscosity and its surprising consequences in interfacial lubrication flows. Phys Rev Fluids 2(023):301

98. Ozan SC, Jakobsen HA (2019) On the role of the surface rheology in film drainage between fluid particles. Int J Multiphase Flow 120(103):103
99. Zell ZA, Nowbahar A, Mansard V, Leal LG, Deshmukh SS, Mecca JM, Tucker CJ, Squires TM (2014) Surface shear inviscidity of soluble surfactants. Proc Natl Acad Sci 111:3677–3682
100. Edwards DA, Brenner H, Wasan DT (1991) Interfacial transport processes and rheology. Butterworth-Heinemann
101. Eggers J (1993) Universal pinching of 3D axisymmetric free-surface flow. Phys Rev Lett 71:3458–3460
102. Eggers J (1997) Nonlinear dynamics and breakup of free-surface flows. Rev Mod Phys 69:865–929
103. Gañán-Calvo AM (1999) The surface charge in electrospraying: its nature and its universal scaling laws. J Aerosol Sci 30:863–872
104. Feng JJ (2003) Stretching of a straight electrically charged viscoelastic jet. J Non-Newtonian Fluid Mech 116:55–70
105. Wee H, Wagoner BW, Kamat PM, Basaran OA (2020) Effects of surface viscosity on breakup of viscous threads. Phys Rev Lett 124(204):501
106. Gañán-Calvo AM (1999) The electrohydrodynamic atomization of liquids today. J Aerosol Sci 30:S547–S548
107. Gadkari S (2017) Influence of polymer relaxation time on the electrospinning process: numerical investigation. Polymers 9:501
108. Gañán-Calvo AM (2004) On the general scaling theory for electrospraying. J Fluid Mech 507:203–212
109. Hu T, Fu Q, Yang L (2020) Falling film with insoluble surfactants: effects of surface elasticity and surface viscosities. J Fluid Mech 889:A16
110. López-Herrera JM, Herrada MA, Montanero JM, Rebollo-Muñoz N, Gañán-Calvo AM (2013) On the validity and applicability of the one-dimensional approximation in cone-jet electrospray. J Aerosol Sci 61:60–69
111. Javadi A, Eggers J, Bonn D, Habibi M, Ribe NM (2013) Delayed capillary breakup of falling viscous jets. Phys Rev Lett 110(144):501
112. Gordillo JM, Sevilla A, Campo-Cortés F (2014) Global stability of stretched jets: conditions for the generation of monodisperse micro-emulsions using coflows. J Fluid Mech 738:335–357
113. Gañán-Calvo AM, Ferrera C, Montanero JM (2011) Universal size and shape of viscous capillary jets: application to gas-focused microjets. J Fluid Mech 670:427–438
114. Clarke NS (1966) A differential equation in fluid mechanics. Mathematika 12:51–53
115. Montanero JM, Herrada MA, Ferrera C, Vega EJ, Gañán-Calvo AM (2011) On the validity of a universal solution for viscous capillary jets. Phys Fluids 23(122):103

Chapter 3
Theoretical Methods

Abstract This chapter briefly describes the theoretical approaches commonly used to gain insight into the tip streaming phenomenon. These approaches allow one to solve or at least obtain information from the governing equations presented in the previous chapter.

We discuss the main characteristics of direct numerical simulations and global and local stability analyses. The differences between these two last approaches are explained to emphasize the importance of the global stability analysis for tip streaming. Concepts such as global modes, asymptotic stability, and short-term response are also discussed.

The temporal and spatiotemporal local stability analyses provide information on the behavior of the jets emitted in the microjetting mode of the tip streaming configurations. This chapter closes with some results obtained from those analyses and directly related to those configurations, including the effects of electric fields, surfactants, and viscoelasticity.

Keywords Direct numerical simulation · Global stability analysis · Local stability analysis

3.1 Direct Numerical Simulations

3.1.1 Discretization Schemes

The methods for the spatial discretization of the hydrodynamic equations are traditionally classified into three main categories: finite difference, finite volume, and finite element [1]. The finite volume method has become very popular in studying tip streaming and other similar problems. Open-source software such as OPEN FOAM, GERRIS [2], and its successor BASILISK use this method and have been widely used in this field.

J. M. Montanero, *Tip Streaming of Simple and Complex Fluids*, Fluid Mechanics and Its Applications 137, https://doi.org/10.1007/978-3-031-52768-5_3

In the finite element scheme, one writes the flow variables as a sum of certain "basis functions" multiplied by coefficients whose values are calculated to satisfy the hydrodynamic equations. Spectral methods [3, 4] differ from the finite element approach in the basis functions. In the finite element method, these functions vanish except on small subdomains, while they are different from zero over the whole domain in the spectral techniques. Chebyshev polynomials [5] are frequently used in the direction normal to the interface as the basis functions of the spectral method.

The Chebyshev spectral collocation technique [6] provides very accurate results when the gradients of the hydrodynamic quantities increase significantly in the vicinity of the interface because it naturally accumulates the grid points in that region. It can also be applied to the other two axes of the problem or combined with other discretization methods, such as the (forward/central/backward) finite differences.

The boundary element method has emerged over the last few decades as a useful alternative technique. It requires only the interface discretization to calculate the variables along the interface. Once the surface values of these variables are obtained, integral equations are used to calculate the solution throughout the fluid domain. The boundary element method is more computationally efficient than other numerical techniques where there is a small surface-to-volume ratio. This method has been successfully used, for instance, to solve the Laplace and Stokes equations for the electric and velocity fields of the cone-jet mode of electrospray [7]. Boundary element methods have been combined with finite centered differences when the liquid inertia is considered [8].

3.1.2 Interface Treatment

A direct numerical simulation integrates the hydrodynamic equation over time, starting from a given initial condition. Interfacial forces play a critical role in tip streaming. For this reason, these forces and the resulting interface evolution must be calculated accurately. The numerical procedures devised to deal with the interface evolution are typically classified into interface-capturing [9, 10] and interface-tracking [11] methods.

An interface-capturing method, such as the VoF method, uses a fixed spatial domain and introduces an auxiliary scalar function to determine the interface location (Fig. 3.1). This location corresponds to the topological region where that function takes a specific value. The temporal evolution of the auxiliary function is calculated from an advection equation on the Eulerian grid, which makes the interface location suffer from numerical diffusion. To improve the simulation accuracy, one uses the adaptive mesh refinement [12], which accumulates the grid points in the vicinity of the interface. The interfacial tension force is incorporated into the momentum equation through the continuum surface force approximation [13] (see Chap. 2). False leakage of liquid mass and charge across the interface must be avoided through appropriate treatment of that surface. Different methods to reconstruct and sharpen the interface at each time step can increase the magnitude of spurious currents inherent to the

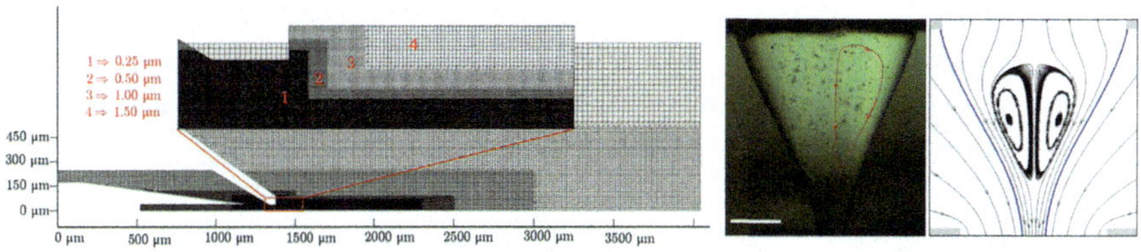

Fig. 3.1 (Left) VoF simulation of transonic flow focusing (reprinted with permission from Ref. [14]). (Right) Comparison between an experiment and a VoF simulation for liquid-liquid flow focusing (reprinted from Ref. [15] with permission of Elsevier)

VoF method. Gerris [2] can adaptively mesh over nearly five orders of magnitude in spatial scale and has proved to be very accurate in surface tension-driven problems.

In an interface-tracking technique, the mesh follows the interface by advecting with the flow a discrete set of points distributed along the interface. The Arbitrary Lagrangian-Eulerian method [16] allows the fluid domain mesh to move arbitrarily, although the mesh velocity must still satisfy the boundary conditions. A salient example of this technique is the fully implicit method of lines algorithm developed by Basaran and coworkers [17]. This technique is based on the Galerkin finite-element method for spatial discretization, working with various meshing techniques and adaptive finite difference methods for time integration. This numerical procedure has been refined over the years to analyze different capillary systems, including tip streaming (Fig. 3.2). The historical development and applications of this *sharp interface method* have been reviewed by Anthony et al. [17]. The later versions have incorporated the elliptic mesh generation technique, facilitating the simulation of singularity problems. In the most efficient implementation, the meshes can also be densified in regions near singularities by subdividing elements into smaller ones.

The interface evolution can also be considered by introducing boundary-fitted coordinates [19, 20]. In this case, one defines a time-dependent coordinate transformation such that the physical space occupied by the fluids is mapped onto a fixed computational domain, obviating the need to track the interface. Typically, the boundaries are aligned with a constant coordinate line in the mapped domain, simplifying the treatment of boundary conditions. The interface position is explicitly characterized by a contour function whose temporal evolution is calculated during the simulation. This approach avoids the numerical diffusion of the interface and takes no approximation to compute the surface tension force, which makes the method very appropriate for capillarity-dominated flows. However, the simulation cannot surpass topological changes such as drop formation or coalescence, where the coordinate transformation cannot be applied.

Herrada and Montanero [21] have developed a numerical method combining boundary-fitted coordinates and the Chebyshev spectral collocation technique. With this discretization, grid points are concentrated near solid surfaces and interfaces to resolve boundary layers. The resulting algebraic equations are solved with the iterative Newton-Raphson technique. The elements of the Jacobian of the discretized

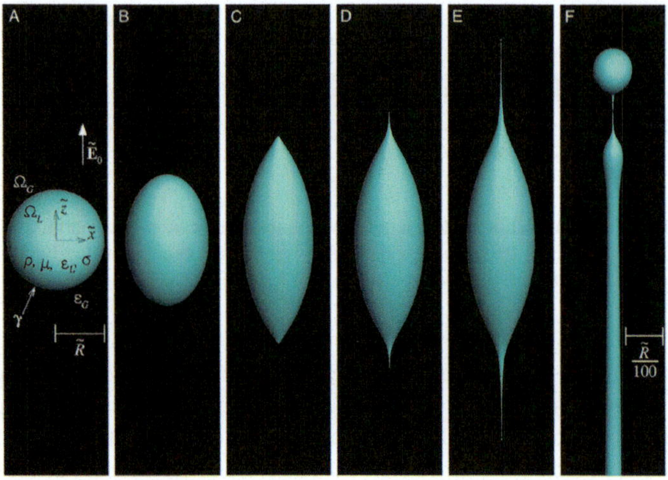

Fig. 3.2 Electrohydrodynamic tip streaming from a low-conductivity liquid droplet subject to a uniform external electric field (reprinted with permission from Ref. [18])

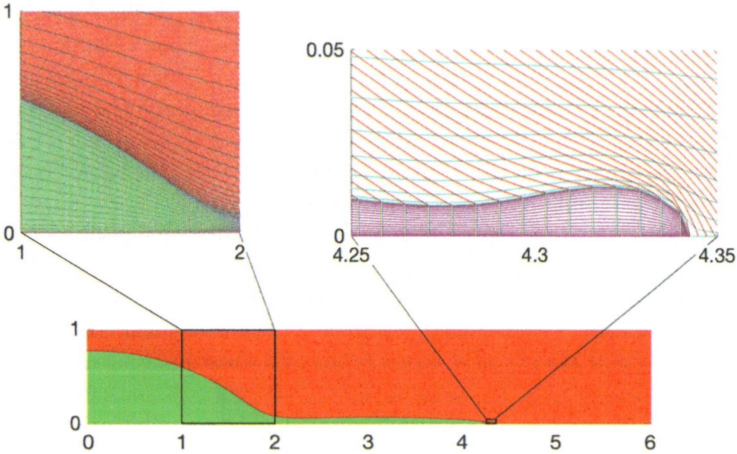

Fig. 3.3 Details of the grid used to simulate the tip streaming in a surfactant-loaded droplet submerged in an extensional flow [23]. The inner and outer fluid domains are mapped onto two quadrangular domains through non-singular mapping. A quasi-elliptic transformation is applied in the outer bath

system of equations are symbolic functions calculated before running the simulation. These functions are evaluated numerically in the Newton-Raphson iterations to find the solution at each time step, considerably reducing the computing time. This method has proved to provide very accurate results in analyzing tip streaming flows. The later versions have incorporated the elliptic mesh generation technique [22] to improve the accuracy of the simulations (Fig. 3.3).

3.2 Global Linear Stability Analysis

Tip streaming is characterized by disparate temporal and spatial scales. For this reason, the direct numerical simulation of the 3D model constitutes a difficult task, even if simplifications such as the leaky-dielectric approximation or the surfactant insolubility condition are considered. Nevertheless, relevant information can be extracted by conducting the linear stability analysis of the base flow. In this analysis, one avoids the time integration of the model by splitting the calculation into two parts: the steady base flow and its linear modes. These modes describe the base flow response to small-amplitude perturbations, which determines the system stability in most cases.

Here, the term "global" alludes to the fact that the entire base flow is considered. In the *local* stability analysis, one studies the response of a portion of this base flow; typically, the region where the velocity profile becomes quasi-uniform and quasi-parallel, and the perturbation can be described as the superposition of the normal modes. As explained in Sect. 3.3, this analysis allows one to determine necessary (but not sufficient) conditions for microjetting.

Tip streaming is an evasive phenomenon that occurs for particular parameter conditions. Determining the regions in the parameter space where tip streaming arises and keeps running is highly relevant at both the fundamental and practical levels. One is interested in predicting and understanding the conditions (i) for the onset of tip streaming and (ii) for the steady microjetting regime to become unstable (Fig. 3.4). The global linear stability analysis is the appropriate tool for this purpose.

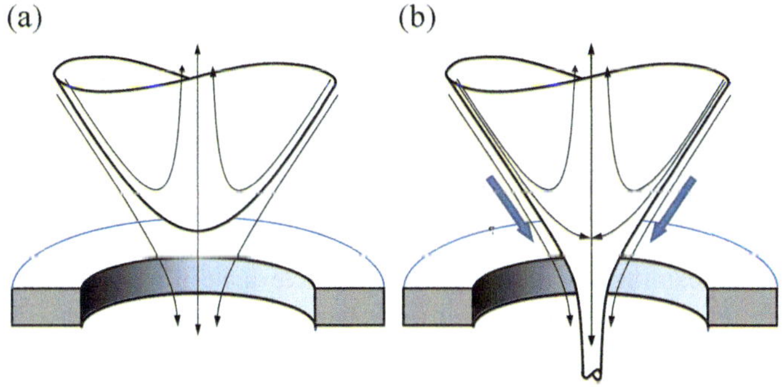

Fig. 3.4 Sketch of the two problems in the global stability analysis of flow focusing: **a** stability of the base flow transiting to tip streaming and **b** stability of microjetting

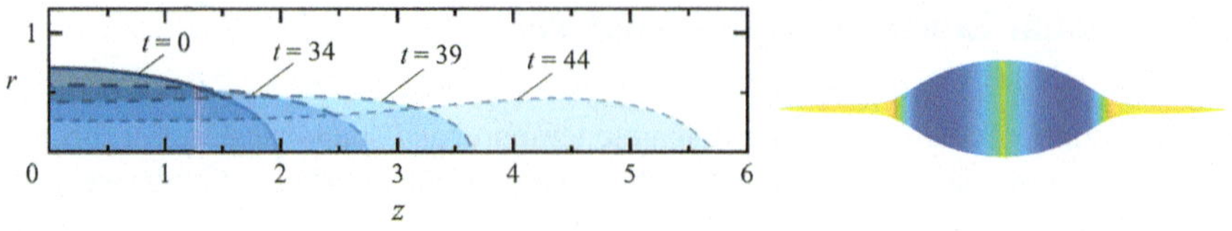

Fig. 3.5 Breakup modes of a droplet submerged in an extensional flow when the critical outer velocity gradient is exceeded [23]. (Left) the mother drop splits into two daughter droplets with sizes that are commensurate with that of the mother. (Right) Tip streaming when an insoluble surfactant monolayer is deposited on the interface

3.2.1 The Onset of Tip Streaming

Tip streaming arises when a precursor steady flow destabilizes. One can explore the response of that flow to small amplitude (infinitesimal) perturbations to predict the conditions for the onset of tip streaming. In this global linear stability analysis, we typically assume that the part of the parameter hyper-surface on which the growth rate of those perturbations vanishes (marginal stability) corresponds to the critical conditions for tip streaming to arise. Whether this phenomenon finally occurs or the system evolves towards a different state depends on additional factors, not well-known in many cases. For instance, Fig. 3.5 shows two breakup modes of a droplet submerged in an extensional flow when the critical outer velocity gradient is exceeded. In the absence of surfactant, the mother drop splits into two daughter droplets with sizes that are commensurate with that of the mother. However, the steady extensional flow instability gives rise to tip streaming when an insoluble surfactant monolayer is deposited on the interface [23].

3.2.2 The Stability of Microjetting

Jetting via tip streaming (microjetting) is of great relevance in droplet-based microfluidics due to its numerous technological applications. The existence of this regime demands stability under small-amplitude perturbations of the base flow sustaining the liquid ejection. In this case, one assumes that steady jetting has been established and interrogates the flow about its response to infinitesimal perturbations. The steady microjetting mode is stable if all the perturbations are damped by viscosity. One of the central problems in the theoretical analysis of tip streaming is to determine the parameter conditions for which this condition is verified.

In many microfluidic applications, such as gravitational or coflowing jetting, the fluid source is a slender meniscus hanging on a feeding capillary. This configuration can be regarded as a simple upstream extension of the emitted jet where the velocity field is quasi-parallel as well. Then, the system's stability can be approximately explained in terms of that of the jet (local stability analysis), as explained below.

This consideration does not apply to tip streaming, where the liquid source is a fluidic structure (in most cases, an acute cone-like meniscus) fundamentally different from the emitted jet. The tapering meniscus, film, or droplet can exhibit complex flow patterns, including boundary layers, stagnation points, and recirculation cells. For this reason, the jet's stability becomes a necessary but insufficient condition for jetting via tip streaming. Determining the necessary and sufficient condition requires the stability analysis of the entire base flow (global stability analysis). This analysis can be conducted using the 2D (axisymmetric) model or the 1D approximation.

Capillary jets eventually break into droplets or suffer other convective instabilities. This implies that the global stability analysis of a jetting base flow unlimited in the downstream direction must show the existence of unstable convective modes independently from the operating conditions. In practice, we set a boundary (cutoff) in the downstream direction to define the finite fluid domain considered in the analysis. The analyzed fluid domain must contain an ejected fluid thread much longer than its diameter. The flow is assumed to operate in the jetting regime if all the global modes of that finite system decay on time.

Of course, there is a certain degree of arbitrariness when selecting the fluid domain cutoff. In some cases, the cutoff does not significantly affect the eigenvalues for a sufficiently large interval of jet lengths. However, there are examples where the stability analysis output depends on the numerical domain length. This situation typically corresponds to experiments where ambiguity exists because there is no clear distinction between the jetting and dripping modes. In hydrodynamic focusing, this problem can be avoided by making the inner and outer stream cross a cylindrical discharge tube. In this way, the jet does not slow down beyond the focusing orifice, which removes the unstable modes associated with the jet and allows one to examine the meniscus stability. In addition, the linear eigenmodes become practically insensitive to the discharge tube length for relatively small values of this parameter.

3.2.3 Global Modes

In the global stability analysis of an axisymmetric base flow, the general 3D time-dependent flow is written as

$$\Phi^{(j)}(r, z, \theta; t) = \Phi_0^{(j)}(r, z) + \hat{\Phi}^{(j)}(r, z, \theta; t), \tag{3.1}$$

$$\Xi(z, \theta; t) = \Xi_0(z) + \hat{\Xi}(z, \theta; t), \tag{3.2}$$

where z is the symmetry axis, $\Phi^{(j)}$ is any bulk quantity, and Ξ is the value of any interfacial quantity at the surface element corresponding to the coordinates (z, θ). In addition, the subscript "0" and the "hat" denote the base flow solution and the infinitesimal perturbation, respectively. The decomposition (3.1) and (3.2) is introduced into the governing equations to obtain a linearized system of equations for the perturbation $\{\hat{\Phi}^{(j)}, \hat{\Xi}\}$. Due to the homogeneity of these equations in θ and t, their general solution can be written as

$$\hat{\Phi}^{(j)}(r, z, \theta; t) = \sum_{n,m} \phi_{n,m}^{(j)}(r, z)\, e^{i(m\theta - \omega_{n,m} t)}, \tag{3.3}$$

$$\hat{\Xi}(z, \theta; t) = \sum_{n,m} \xi_{n,m}^{(j)}(z)\, e^{i(m\theta - \omega_{n,m} t)}, \tag{3.4}$$

where $\omega_{n,m}$ is a complex eigenfrequency and m the azimuthal mode number. The solution for given values of n and m is called a global mode and corresponds to a motion in which the entire system oscillates harmonically with the same complex frequency $\omega_{n,m}$ and a fixed phase relation [24, 25].

3.2.4 Stability Criterion

The global modes are calculated as the eigenfunctions of the linearized Navier-Stokes operator as applied to a given configuration (base flow). The base flow is linearly and *asymptotically* stable if the spectrum of eigenfrequencies, $\{\omega_{n,m}\}$, is in the stable complex half-plane ($\{\text{Im}[\omega_{n,m}]\} < 0$). In this case, any initial small-amplitude perturbation will decay exponentially on time for $t \to \infty$ (as long as the linear approximation applies). On the contrary, the system becomes unstable when the growth rate of the dominant mode (i.e., that with the largest growth rate) becomes positive. In this case, the projection of an arbitrary spontaneous perturbation onto that mode grows exponentially on time.

The critical modes can be categorized as non-oscillatory ($\text{Re}[\omega_{n,m}] = 0$) and oscillatory ($\text{Re}[\omega_{n,m}] \neq 0$). The latter case corresponds to an oscillation with an exponentially growing amplitude (Hopf bifurcation). The microjetting mode of tip streaming becomes unstable due to an oscillatory perturbation in most realizations. When the flow rate decreases below a certain critical value, the growth rate of the dominant oscillatory mode becomes positive, and the steady jetting regime destabilizes under infinitesimal perturbations [26–30]. The oscillatory character of the dominant mode is linked to the presence of the jet. Capillary modes are convected by the jet, which translates into an oscillatory behavior in the Eulerian frame of reference.

Non-oscillatory critical modes are expected when the instability is confined in a region of the fluid domain. The jet is absent in the steady flow whose instability leads to tip streaming. Therefore, one expects the critical mode not to exhibit an oscillatory nature in this case. Non-oscillatory critical modes have also been found in coflowing [31], flow focusing [29], and coaxial electrospray [32] under specific parameter conditions.

3.2.5 Examples

It may be instructive to examine the dimensionless equations solved to calculate an eigenmode. For the sake of simplicity, we consider an axisymmetric problem without surfactants, viscoelasticity, electric field, and gravitational effects.

Consider the steady base flow given by the expressions ${\bf v}_{\rm BF}{}^{(j)}(r,z)$, $p_{\rm BF}^{(j)}(r,z)$ and $f_{\rm BF}(r,z)$ corresponding to the velocity field, pressure field, and interface position, respectively. The global eigenmode $(m=0,n)$ is given by the perturbations $\hat{\bf v}^{(j)}(r,z)e^{-i\omega_{0,n}t}$, $\hat{p}^{(j)}(r,z)e^{-i\omega_{0,n}t}$ and $\hat{f}(r,z)e^{-i\omega_{0,n}t}$ of those quantities. These perturbations obey the linearized equations:

$$\nabla\cdot\hat{\bf v}^{(j)}=0, \tag{3.5}$$

$$\rho^{\delta_{jo}}\,({\rm Oh}_i)^{-2}\left(-i\omega_{0,n}\hat{\bf v}^{(j)}+\hat{\bf v}^{(j)}\cdot\nabla{\bf v}_{\rm BF}{}^{(j)}+{\bf v}_{\rm BF}{}^{(j)}\cdot\nabla\hat{\bf v}^{(j)}\right)+\nabla\hat{p}^{(j)}-\mu^{\delta_{jo}}\nabla^2\hat{\bf v}^{(j)}=0, \tag{3.6}$$

$$||-\hat{p}^{(j)}+2\mu^{\delta_{jo}}\partial\hat{v}_n^{(j)}/\partial n||+\hat{\kappa}=0,\qquad ||\mu^{\delta_{jo}}(\partial\hat{v}_n^{(j)}/\partial\tau+\partial\hat{v}_t^{(j)}/\partial n)||=0, \tag{3.7}$$

$$||\hat{\bf v}^{(j)}||=0,\quad -i\omega_{0,n}\hat{f}+\hat{\bf v}^{(j)}\cdot\nabla f_{\rm BF}+{\bf v}_{\rm BF}^{(j)}\cdot\nabla\hat{f}=0, \tag{3.8}$$

where δ_{jk} is the Kronecker delta, and $\hat{\kappa}$ is the perturbation of the local mean curvature. Due to the nonlinear character of the curvature, $\hat{\kappa}$ involves the base flow interface position $f_{\rm BF}(r,z)$.

In the above equations, the velocity and pressure fields have been made dimensionless using the visco-capillary velocity $U_{\mu\sigma}^{(i)}=\sigma/\mu^{(i)}$ and the capillary pressure $\mu^{(i)}U_{\mu\sigma}^{(i)}/L_c=\sigma/L_c$ as the corresponding characteristic scales. The lengths are expressed in terms of the characteristic length L_c of the tip streaming configuration (e.g., the feeding capillary radius in electrospray or flow focusing). The eigenfrequency $\omega_{0,n}$ is expressed in terms of the characteristic frequency $U_{\mu\sigma}^{(i)}/L_c$. In addition, ${\rm Oh}_i=\mu^{(i)}/(\rho^{(i)}U_{\mu\sigma}^{(i)}L_c)^{1/2}$ is the Ohnesorge number (the inverse of the visco-capillary Reynolds number), and $\rho=\rho^{(o)}/\rho^{(i)}$ and $\mu=\mu^{(o)}/\mu^{(i)}$ are the density and viscosity ratio, respectively. Finally, n and τ denote the axis normal and tangential to the interface, respectively.

The problem is completed with the rest of the boundary conditions characterized by the dimensionless numbers $\{\mathscr{P}_i\}$. For given values of $\{\rho,\mu,{\rm Oh}_i;\,\mathscr{P}_i\}$, there is an infinite discrete set of eigenfrequencies $\{\omega_{0,n}\}$ $(n=1,2,\ldots)$ for which the above homogeneous system of equations admits a non-trivial solution. If ω_0^* is the eigenfrequency with the largest imaginary part (growth rate), the critical conditions for the onset of the instability are the values of $\{\rho,\mu,{\rm Oh}_i;\,\mathscr{P}_i\}$ for which ${\rm Im}[\omega_0^*]=0$.

Interesting conclusions can be drawn by examining the linearized equations in simple but relevant cases. Suppose (i) the outer phase produces negligible dynamic effects, (ii) the base flow of the inner phase $j=i$ vanishes (${\bf v}_{\rm BF}{}^{(i)}={\bf 0}$), and (iii) the instability is non-oscillatory ($\omega_0^*=0$ at the marginal stability). In this simple case, $f_{\rm BF}(r,z)$ is not affected by ρ, μ, and ${\rm Oh}_i$. We conclude that the solution to the linearized equations for $j = i$ becomes independent of those parameters. Therefore, the critical conditions can be expressed only in terms of $\{\mathscr{P}_i\}$, which do not involve the inner density and viscosity in most cases. This means that these quantities do not influence the onset of the instability.

The above conclusion applies to the stability of pendant droplets and liquid bridges [33]. It also applies to an electrified droplet giving rise to tip streaming [34, 35]. In this case, the Maxwell stress in Eq. (3.7), the surface charge conservation equation, and the electric field equations do not affect this conclusion.

If the velocity fields are made dimensionless using the corresponding visco-capillary velocity $U_{\mu\sigma}^{(j)} = \sigma/\mu^{(j)}$, the linearized equations for the Stokes (viscosity-dominated) flow at the non-oscillatory instability become:

$$\nabla \cdot \hat{\mathbf{v}}^{(j)} = 0, \qquad -\nabla \hat{p}^{(j)} + \nabla^2 \hat{\mathbf{v}}^{(j)} = 0, \tag{3.9}$$

$$|| - \hat{p}^{(j)} + 2\partial \hat{v}_n^{(j)}/\partial n|| + \hat{\kappa} = 0, \qquad ||(\partial \hat{v}_n^{(j)}/\partial \tau + \partial \hat{v}_t^{(j)}/\partial n)|| = 0, \tag{3.10}$$

$$||\mu^{\delta_{ji}} \hat{\mathbf{v}}^{(j)}|| = 0, \quad \hat{\mathbf{v}}^{(j)} \cdot \nabla f_{\mathrm{BF}} + \mathbf{v}_{\mathrm{BF}}^{(j)} \cdot \nabla \hat{f} = 0. \tag{3.11}$$

These equations depend on μ only through the curvature $\hat{\kappa}$, which involves the base flow interface position $f_{\mathrm{BF}}(r, z)$, and the kinematic boundary conditions (3.11).

As explained in Chap. 6, the velocity field in a droplet loaded with an insoluble surfactant and submerged in an extensional flow practically vanishes ($\mathbf{v}_{\mathrm{BF}}{}^{(i)} \simeq \mathbf{0}$). In this case, $f_{\mathrm{BF}}(r, z)$ and $\hat{\kappa}$ are not significantly affected by μ. This means that μ influences the droplet stability only through the velocity continuity condition $||\mu^{\delta_{ji}} \hat{\mathbf{v}}^{(j)}|| = 0$. For this reason, the critical condition for the onset of tip streaming becomes practically independent of the droplet viscosity [23].

3.2.6 Sources of Errors

In the stability analysis of the microjetting regime, "soft" boundary conditions, such as the so-called outflow or traction-free boundary condition, can be applied at the downstream numerical domain cutoff mentioned above. This choice may lead to significant errors in the calculation of convective modes. Nevertheless, global modes of convective nature are subdominant in viscous systems, such as the liquid-liquid coflowing configuration [31, 36]. For this reason, the outlet boundary conditions are practically irrelevant in these systems, provided that the cutoff length is much larger than the injector diameter [31, 36, 37].

In most tip streaming realizations, the calculation of the eigenmodes requires more refined grids than those used to compute the steady base flow. Spurious eigenmodes may arise due to the lack of spatial resolution of grids which can accurately resolve the base flow. To eliminate those spurious results, one conducts a grid sensitivity analysis when solving the eigenvalue problem.

The global stability analysis of surfactant-loaded and leaky-dielectric droplets involves the conservation equation for the surfactant adsorbed onto the interface and the electric surface charge, respectively (see Chap. 2). Thermal surface diffusion can be neglected in both cases due to the smallness of the corresponding diffusion coefficients. However, the absence of a diffusive term in those equations may lead to spurious eigenmodes, which do not disappear when the grid is refined. These eigenmodes typically exhibit spatial fluctuations with characteristic lengths of the order of the distance between the grid points. Adding a small diffusive term to the conservation equation eliminates those spurious eigenmodes.

3.2.7 Asymptotic Stability

Global modes are calculated as the eigenfunctions of the linearized Navier-Stokes operator as applied to a given base flow. If the spectrum of eigenvalues is in the stable complex half-plane, the base flow is linearly and *asymptotically* stable. The term "asymptotically" emphasizes that we are exploring the system evolution for sufficiently large times (within the range of validity of the linear approximation), when that evolution is governed only by the dominant eigenmode. The following example illustrates this idea.

Consider a cylindrical capillary in a linear uniaxial extensional flow. The micro-jetting mode is achieved when the inner phase is injected through the capillary at a constant flow rate. In this case, the inner phase forms a cone-like tapering meniscus that ejects a very thin jet from its tip [38]. We analyze the temporal evolution of a small initial perturbation of the interface location next to the feeding capillary [38].

Figure 3.6 shows the evolution of the interface displacement $F(z,t) - F_0(z)$ at the meniscus tip, as indicated in the inset of the figure. Here, $F(z,t)$ and $F_0(z)$ are the instantaneous and unperturbed interface locations, respectively, while z is the flow symmetry axis. The parameter conditions correspond to an asymptotically stable flow close to the stability limit. The perturbation triggers a train of capillary waves that propagate downstream, producing a significant oscillation of the interface position in the meniscus tip. Viscosity dampens the perturbation. For sufficiently large t, the oscillation amplitude is sufficiently small for the nonlinear terms to become negligible. In addition, the contribution of subdominant eigenmodes becomes negligible, and the dominant mode essentially governs the system's linear dynamics.

The comparison between the direct numerical simulation and the prediction

$$F - F_0 = a_0\, e^{\omega_i (t-t_0)} \cos[\omega_r (t - t_0)] \tag{3.12}$$

of the global stability analysis shows excellent agreement (Fig. 3.6). Here, a_0 and t_0 are fitting parameters, and ω_i and ω_r are the damping rate and frequency of the dominant mode. Figure 3.7 shows the whole interface deformation calculated from the direct numerical simulation and the deformation corresponding to the dominant mode.

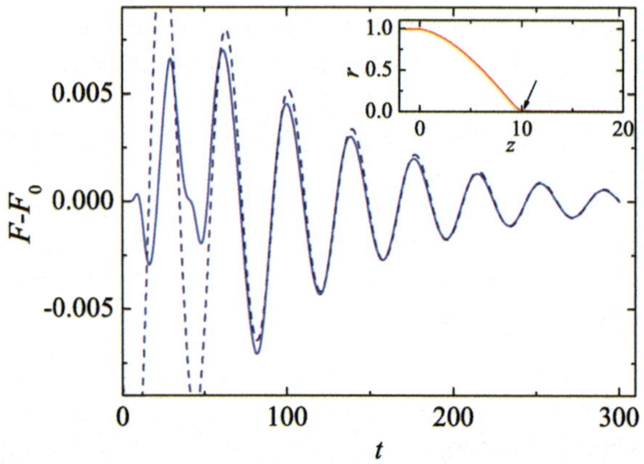

Fig. 3.6 Evolution of the free surface displacement, $F(z, t) - F_0(z)$, at the meniscus tip ($z = 10.19$, $F_0(z) = 0.0174$) calculated from the direct numerical simulation (solid line). The dashed line is the global stability analysis's prediction (3.12). The arrow in the inset shows the analyzed interface point. The values of the governing parameters can be found in Ref. [38]. The time and spatial coordinates have been made dimensionless with the visco-capillary time and the feeding capillary radius, respectively

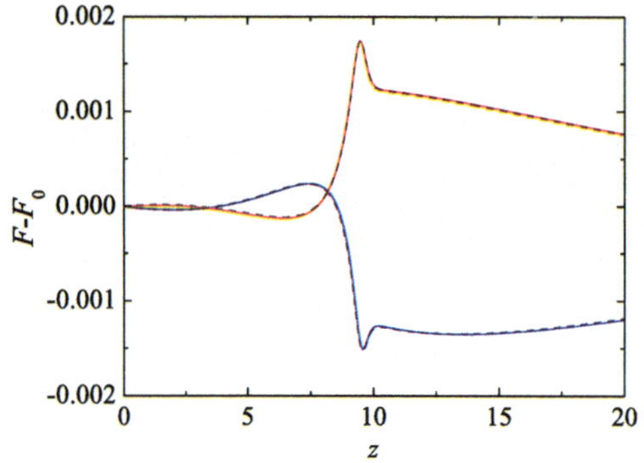

Fig. 3.7 Free surface displacement, $F(z, t) - F_0(z)$, at $t = 200.1$ (blue solid line) and 216.1 (red solid line) calculated from the direct numerical simulation. The dashed lines are the deformation corresponding to the dominant mode. The values of the governing parameters can be found in Ref. [38]. The time and spatial coordinates have been made dimensionless with the visco-capillary time and the feeding capillary radius, respectively

3.2.8 Short Term Response

It is frequently believed that asymptotic linear stability is a sufficient condition for the linear stability of the base flow. However, this is not necessarily true. If the linearized Navier-Stokes operator is non-normal, the system's short-term dynamics can result from a "constructive interference" of stable global modes, which can lead to a bifurcation before those modes are damped out [24, 39, 40]. In other words, the super-

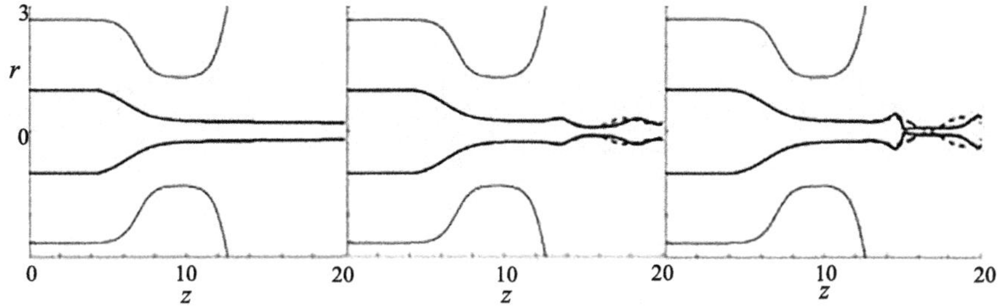

Fig. 3.8 Free surface deformation calculated with the linearized (dash line) and nonlinear (solid line) hydrodynamic equations at three instants for an asymptotically stable base flow of gaseous flow focusing (reproduced with permission, Ref. [26])

position of decaying small-amplitude perturbations introduced into a microfluidic configuration can destabilize the flow before those perturbations disappear, which prevents the system from reaching the microjetting regime. In this case, asymptotic stability becomes a necessary but insufficient condition for microjetting.

The global stability analysis must be completed with direct numerical simulations of the system to examine its response to initial perturbations (the initial value problem). The nonlinear terms can be dropped when integrating the hydrodynamic equations to speed up the calculation [26]. This allows one to see whether the superposition of linear global modes makes the resulting perturbation grow within the small-amplitude response regime. In any case, this is a complex problem because the outcome can significantly depend on the type and location of the initial perturbation, something difficult to determine in an experiment.

The gaseous flow focusing configuration is an excellent example of the situation mentioned above. Cruz-Mazo et al. [26] have shown a transient growth of linear perturbations before the asymptotic (exponential) regime is reached. This growth leads to dripping for small applied pressure drops. Figure 3.8 shows the free surface deformation at three instants for an asymptotically stable base flow. The superposition of decaying linear global modes triggered by the perturbation gives rise to the free surface pinch-off within the numerical domain.

As mentioned above, modal instability is not the only possible path for amplifying perturbations. Disturbances can grow through non-modal (non-exponential) mechanisms even when the flow is exponentially stable. Figure 3.9 shows another example: the amplification of perturbations on a capillary jet base flow with an eigenvalue spectrum corresponding to an asymptotically stable flow.

3.2.9 *Mechanisms of Instability*

Global stability analysis can be used not only to determine the conditions for instability but also to unveil the physical mechanisms responsible for that instability.

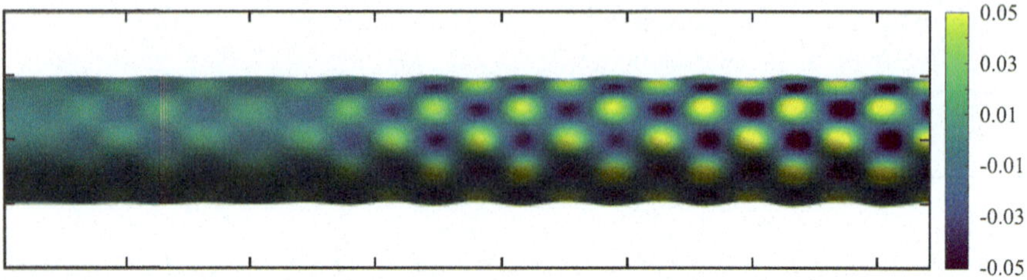

Fig. 3.9 Isosurfaces of the local displacement of a capillary jet interface (reproduced with permission, Ref. [41])

This can be done by analyzing the eigenfunction $\{\phi_{n,m}^{(j)}(r,z), \xi_{n,m}^{(j)}(z)\}$ corresponding to the critical mode. This analysis may indicate, for instance, where the instability of microjetting originates. In this case, one can distinguish two regions of the fluid domain: the tapering meniscus and the emitted jet. The inspection of the interface shape perturbation shows whether the tapering meniscus becomes unstable or it remains unperturbed while the perturbation grows in the jet.

The analysis of the energy transfer from the steady base flow to the critical eigenmode (the so-called energy budget analysis) also provides insights into the instability mechanism and helps to understand the underlying physics. For this purpose, one multiplies the linearized momentum equation by the perturbation velocity vector. The resulting equation is integrated over the volumes occupied by each phase, which allows one to calculate the rate of change of the perturbation kinetic energy. The resulting terms represent the production rates of the kinetic energy transferred from the base flow to the perturbation.

Consider a problem characterized by the set of parameters $\{\mathscr{P}_i\}$. The linearized equations governing the evolution of the perturbations depend on $\{\mathscr{P}_i\}$ through two routes: (i) the dependence of the base flow Φ_{BF} on those parameters, and (ii) the explicit dependence of the linearized equations on $\{\mathscr{P}_i\}$; i.e., the critical eigenfrequency obeys the formal relationship $\omega^* = \omega^*(\Phi_{\text{BF}}(\{\mathscr{P}_i\}), \{\mathscr{P}_i\})$. Increasing the value of some parameter $\mathscr{P}_1$ can destabilize the flow through its influence on the base flow Φ_{BF} even if it has a stabilizing effect on the linearized solution for a given (fixed) base flow Φ_{BF}. Most studies on tip streaming stability have focused on the first route mentioned above, analyzing how different factors modify the base flow and, consequently, the system stability.

3.2.10 *Dripping and Self-sustained Oscillations*

After a linear transient regime, the nonlinear terms of the governing equations come into play, and the unstable perturbation no longer follows the exponential growth described above. This occurs even for Stokes (viscosity-dominated) flows because the capillary pressure always involves nonlinear terms.

Two scenarios can be expected when the nonlinear terms arise in an oscillatory instability. In the first case, the perturbation grows until the interface breaks up. This process leads to either axisymmetric dripping or lateral jet fragmentation for $m = 0$ and $m \neq 0$, respectively. In the other case, the nonlinear terms limit the growth of the critical mode, and a periodic oscillatory behavior is obtained. This behavior is usually called oscillatory jetting or whipping, depending on whether the precursor critical mode is axisymmetric ($m = 0$) or not ($m \neq 0$).

In oscillatory jetting and whipping, there is always a finite interval of the control parameter for which there is no interference among unstable modes, and instability is controlled only by the critical one. In this case, the amplitude and frequency of the nonlinear oscillation are expected to coincide approximately with those of the linear mode responsible for the instability.

The self-sustained oscillations arising in the oscillatory jetting regime must be clearly distinguished from those externally excited by uncontrolled perturbations occurring in the experiments. For instance, when perturbations originated by the liquid pump were carefully eliminated in experiments with confined selective withdrawal, the apparently self-sustained meniscus oscillations disappeared [30] (see Chap. 9).

In the terminology of the bifurcation theory, the oscillatory jetting corresponds to a stable limit cycle originated by a supercritical Hopf bifurcation, in which the steady jetting solution transforms into an axisymmetric periodic oscillation beyond the critical point. This point corresponds to the marginal stability condition obtained in the linear stability analysis. These self-excited oscillations have been observed in homogeneous [42], unconfined [43, 44] and confined gravitational [45] jets, as well as in wake flows in the presence of the surface tension [46].

The oscillation amplitude increases as the distance $\varepsilon = (\alpha - \alpha_c)/\alpha_c$ increases (α is the control parameter and α_c its critical value obtained from the global stability analysis). For sufficiently large values of ε, the oscillatory jetting regime destabilizes and yields the dripping mode. It must be noted that the oscillatory regime occurs for a tiny interval of α ($\varepsilon \ll 1$), which explains why the critical values obtained from the global stability analysis agree with the experimental ones at the jetting-to-dripping transition.

The instability of some configurations producing tip streaming, such as a droplet submerged in an extensional flow, has a non-oscillatory character ($\omega^*_{0r} = 0$); i.e., the growth of the critical mode is given by an imaginary eigenfrequency. In this case, the system undergoes a saddle-node transition, which leads to transient tip streaming in the nonlinear phase. The critical point corresponds to a turning point when one represents a certain output parameter (e.g., the droplet deformation) versus the control parameter (the velocity gradient) [23]. We will come back to this point in Chap. 6.

3.3 Local Linear Stability Analysis

3.3.1 *Dispersion Relationship*

The global stability analysis requires the calculation of the entire base flow producing tip streaming and the linear eigenmodes that determine its stability. Due to the disparity of scales arising in tip streaming, this is a relatively complex numerical calculation. When the stability analysis is conducted *locally*, the problem becomes analytically or semi-analytically tractable. In this case, one considers only a portion of the base flow and analyzes the growth of infinitesimal perturbations with wavelengths much smaller than the length of that portion in the streamwise direction (the symmetry axis). This is commonly referred to as the WKBJ approximation and allows one to examine the stability of slowly spatially varying base flows.

In most cases, the selected base flow portion corresponds to the jet sufficiently far from the emission point (the meniscus tip in tip streaming). Then, the base flow is assumed to be quasi-parallel and quasi-uniform in the streamwise direction, which greatly simplifies the analysis. In this approximation, the evolution of perturbations in the linear regime at the considered flow station can be described as the superposition of the normal modes

$$\Phi^{(j)}(r, \theta, z; t) = \phi^{(j)}(r)\, e^{i(m\theta + kz - \omega t)}, \tag{3.13}$$

where $\Phi^{(j)}$ represents any variable of the problem, (r, θ, z) is a cylindrical coordinate system whose z-axis is the base flow symmetry axis (the streamwise direction), $\phi^{(j)}$ is the spatial structure of the linear mode, m is the azimuthal mode number, and $k = k_r + ik_i$ and $\omega = \omega_r + i\omega_i$ are the perturbation wavenumber and frequency, respectively.

The fulfillment of the hydrodynamic equations and boundary conditions determines the spatial structure of the linear mode, $\phi^{(j)}(r)$, and, more importantly, leads to the (dimensionless) dispersion relationship

$$\mathscr{D}_m(k, \omega; \{\mathscr{P}_i\}) = 0, \tag{3.14}$$

which gives the eigenfrequency ω of a mode with azimuthal and axial wavenumber m and k, respectively, as a function of the parameters $\{\mathscr{P}_i\}$ $(i = 1, 2, \ldots, N)$ characterizing the problem.

The dispersion relationship can be applied throughout the flow considering the local values of the parameters $\{\mathscr{P}_i\}$ of the problem. The total growth of the perturbation results from the integration of the Eulerian growth rate $\omega_i(z)$ along Lagrangian paths, taking into account the variation of k and $\{\mathscr{P}_i\}$ along those paths [47].

The relative simplicity of the calculation of the dispersion relationship has favored the application of the local stability analysis to a plethora of problems, many of them with little connection with experiments or applications.

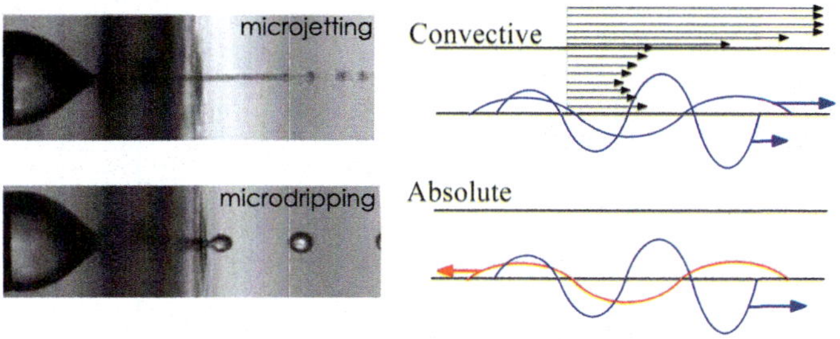

Fig. 3.10 Relationship between the jetting-to-dripping and convective-to-absolute instability transitions. The images on the left show the transition from microjetting to microdripping in gaseous flow focusing for small applied pressure drops (reproduced with permission, Ref. [54]). The sketches on the right side represent the evolution of growing capillary waves in a convectively and absolutely unstable jet

3.3.2 The Convective-to-Absolute Instability Transition

The breakup mode adopted by a fluid thread can be predicted in terms of the so-called convective-to-absolute instability transition, a concept widely used in instabilities of shear flows and wakes [48]. In convectively unstable jets, capillary waves are swept away downstream by the current, which keeps a considerable portion of the jet free of perturbations (jetting regime) (Fig. 3.10). Conversely, growing perturbations travel downstream and upstream along absolutely unstable jets, precluding the jet formation (dripping mode). The jetting-to-dripping transition has been successfully linked to the convective-to-absolute instability transition in both the liquid [49, 50] and gaseous [51] cases. However, significant discrepancies between the conditions leading to absolute instability and dripping have been observed in liquid sheets [52] and round jets [53].

Suppose a thin fluid thread is emitted from the tip of a steady meniscus in a tip streaming realization. In the microdripping mode, the thread breaks up into droplets next to the orifice. In contrast, a steady thread, long compared to its diameter, forms in the microjetting regime. This thread eventually breaks up into droplets due to the capillary instability. The microjetting-to-microdripping transition has been predicted from the convective-to-absolute instability transition on the relatively scarce occasions this transition has been calculated (Fig. 3.10) [54, 55].

The critical conditions leading to the convective-to-absolute instability transition are determined by the spatiotemporal analysis of the dispersion relationship (3.14). This analysis explores the system's response to perturbations characterized by a *complex* axial wavenumber k observed by a fixed observer anchored at the nozzle. The dispersion relationship is typically derived using the frame of reference traveling with the jet. To change the frame of reference from a traveling observer to a fixed one, one just needs to replace the wave frequency ω by $\omega' - v_j k$ in the dispersion relation (3.14), where v_j is the jet velocity.

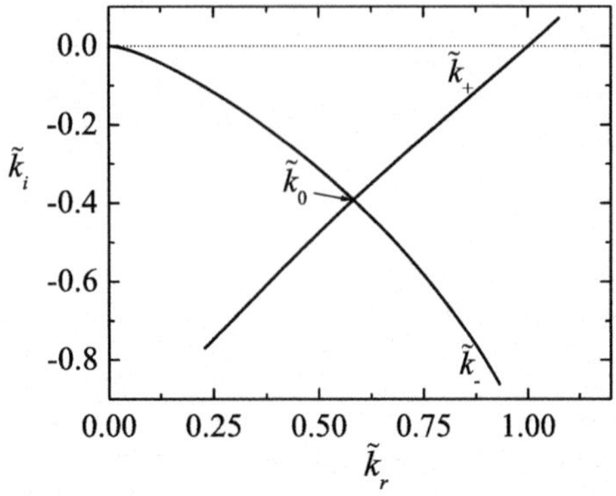

Fig. 3.11 Spatial branches $\widetilde{k}_{\pm}(\omega'^*)$ for a jet of radius R_j moving at the velocity v_j in a vacuum. These branches originated from the solution $\widetilde{k}_0 = 0.5828 - 0.3921i$ and $\omega_0'^* = 0.5757$ of the dispersion relation. As can be observed, the branch $\widetilde{k}_+$ crosses the $\widetilde{k}_r$ axis at $\widetilde{k}_r = 1$ (Rayleigh stability limit). Here, $\widetilde{k} = kR_j$ and $\omega'^* = \omega' R_j/v_j$ (reproduced from Ref. [57]. Copyright (2008) by the American Physical Society)

For fixed values of the control parameters $\{\mathscr{P}_i\}$ $(i = 2, \ldots, N)$, one calculates the critical value $\mathscr{P}_{1c}$ for which Brigg's pinch condition [48, 56] is satisfied. This condition establishes that there must be at least one pinching of a k^+ and a k^- spatial branch with $\omega_i' = 0$, where the k^+ is the path of $\mathscr{D}_m = 0$ in the complex k plane which moves into the $k_i > 0$ half-plane as ω_i' increases, while the k^- branch always remains in the $k_i < 0$ half-plane as ω_i' increases (Fig. 3.11).

van Saarloos [58] proposed an alternative criterion for determining the convective-to-absolute instability transition based on the analysis of the propagation front velocity. Specifically, the system becomes absolutely unstable when the rear front velocity of a localized initial distortion becomes zero. Many researchers in capillary flows have used this criterion because it is intuitive and immediately understandable in physical terms. Montanero and Gañán-Calvo [59] showed the equivalence between this and the classical Briggs's saddle-point criterion [48, 56].

3.3.3 Temporal and Spatial Stability Analyses

Practically all of the jets produced in microfluidic applications eventually break up due to the capillary instability. This means that they are unstable in a strict sense. In this context, the adjective "stable" means *convectively unstable*, which indicates that perturbations are convected downstream, leaving an intact jet portion long as compared with its diameter. Here, the adjective "intact" means that the jet shape is nearly indistinguishable from that of the unperturbed base flow. We will come back to this point in the next section.

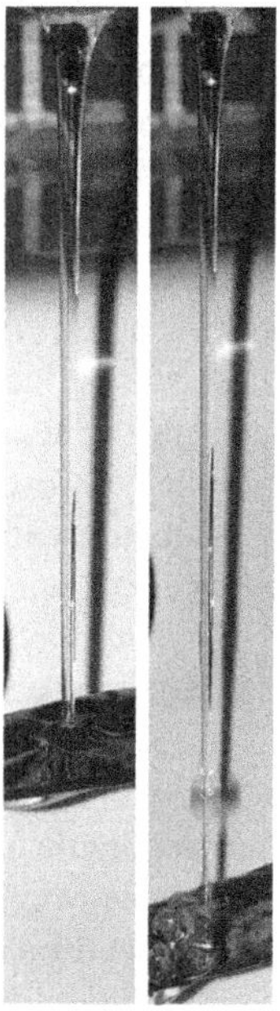

Fig. 3.12 (Left) Gravitational-capillary water jet intercepted at a distance from the tap smaller than the breakup length. The jet is strictly stable. One may observe the classical small wavelength ripples formed right upstream of the interception point. (Right) The jet is intercepted beyond the breakup region, which does not alter its convectively unstable nature (reprinted from Ref. [61] with permission of IOP Publishing)

It is worth mentioning that there are very specific situations in which strictly stable infinite jets can be formed; for instance, when the jet is confined in a rectangular channel under certain conditions [60]. Stability in a strict sense can also be observed when a jet hits a downstream steady boundary condition that precludes the growth of perturbations (Fig. 3.12).

If a jet is convectively unstable, the temporal stability analysis allows one to predict the critical aspects of the breakup process. In this analysis, the growth rate ω_i is calculated as a function of the real wavenumber k and the parameters $\{\mathscr{P}_i\}$ characterizing the problem. One is typically interested in whether a specific factor (electric field, viscoelasticity, surfactant, ...) stabilizes or destabilizes the jet. The growth rates, the range of unstable wavenumbers, and the most unstable wavenumber generally decrease if the considered factor stabilizes the jet. The opposite occurs under the action of a destabilizing factor. We will show some examples in Sect. 3.4.

The most interesting quantities of the temporal stability analysis of a capillary jet are probably the growth rate $\omega_i^{\max}$ and wavelength $\lambda^{\max}$ corresponding to the most unstable mode. These quantities allow one to estimate the jet breakup length l_b and droplet diameter d_d as

$$l_b \sim v_j/\omega_i^{\max}, \quad \pi R_j^2 \lambda^{\max} \simeq \pi d_d^3/6, \tag{3.15}$$

where v_j and R_j are the jet's mean velocity and radius, respectively.

Equation (3.15)-left has been used to calculate the breakup length of, for instance, gravitational [47] and electrified [62] jets. This expression implicitly assumes that the perturbation responsible for the breakup is born next to the jet inception region and convected by the jet. The details of the ejection procedure and geometry are expected to affect the initial amplitude of that perturbation, which suggest that the breakup length should somewhat depend on the ejection characteristics. However, the breakup length for different "smooth" ejectors in well-controlled experiments seem to depend only on the liquid properties and operating parameters [63], which questions the idea that the critical perturbation originates in the ejector. Umemura [64] has hypothesized that the excess surface energy at the breakup feeds the perturbations causing the next breakup event, which would explain the rather deterministic manner in which unforced capillary jets spontaneously break up.

Equation (3.15)-right takes into account that the fluid volume distribution after the jet breakup is essentially decided in the linear regime, i.e., before nonlinear effects come into play. These effects explain the appearance of satellite droplets during the jet breakup, which is undesirable in most tip streaming realizations. The volume contained in the satellite droplets hardly affects the validity of Eq. (3.15)-right.

Equation (3.15)-right is the expression most commonly used to estimate the droplet diameter in the jetting regime. Castro-Hernández et al. [65] have proposed an alternative way to derive that expression. They have shown how to correct it to calculate the droplet diameter following the breakup of widening jets in the coflowing configuration.

The perturbations in the temporal analysis are characterized by a real wavenumber k. This implies that they are spatially periodic and grow at the same rate both next to the nozzle and downstream. This unrealistic assumption is eliminated in the spatial stability analysis, where the complex waver number is calculated as a function of the real eigenfrequency. The spatial and temporal stability analyses are equivalent if the speed of the jet is much larger than that of the small-amplitude capillary waves [66].

3.3.4 Summary

Figure 3.13 summarizes the ideas introduced in this section. Direct numerical simulations calculate the nonlinear transient dynamics of the tip streaming realization. It allows one to obtain the value of any hydrodynamic quantity, $\Phi(r, \theta, z; t)$, at any location as a function of time starting from a given initial condition.

In the global stability analysis, one calculates the *steady* and *axisymmetric* base flow, $\Phi_0(r, z)$, and determines the evolution of the small-amplitude perturbations, $\phi(r, z)$, around that flow. This analysis considers the entire system. The axial length $\hat{\lambda}$ characterizing the perturbation does not need to be much smaller than the system size L. The base flow corresponds to either (i) the steady solution leading to the onset of tip streaming or (ii) the microjetting realization whose stability is examined. In the first case, one obtains the critical conditions for the onset of tip streaming (either microdripping or microjetting). In the last case, the global stability analysis allows us to determine the parameter conditions for which the microjetting realization becomes

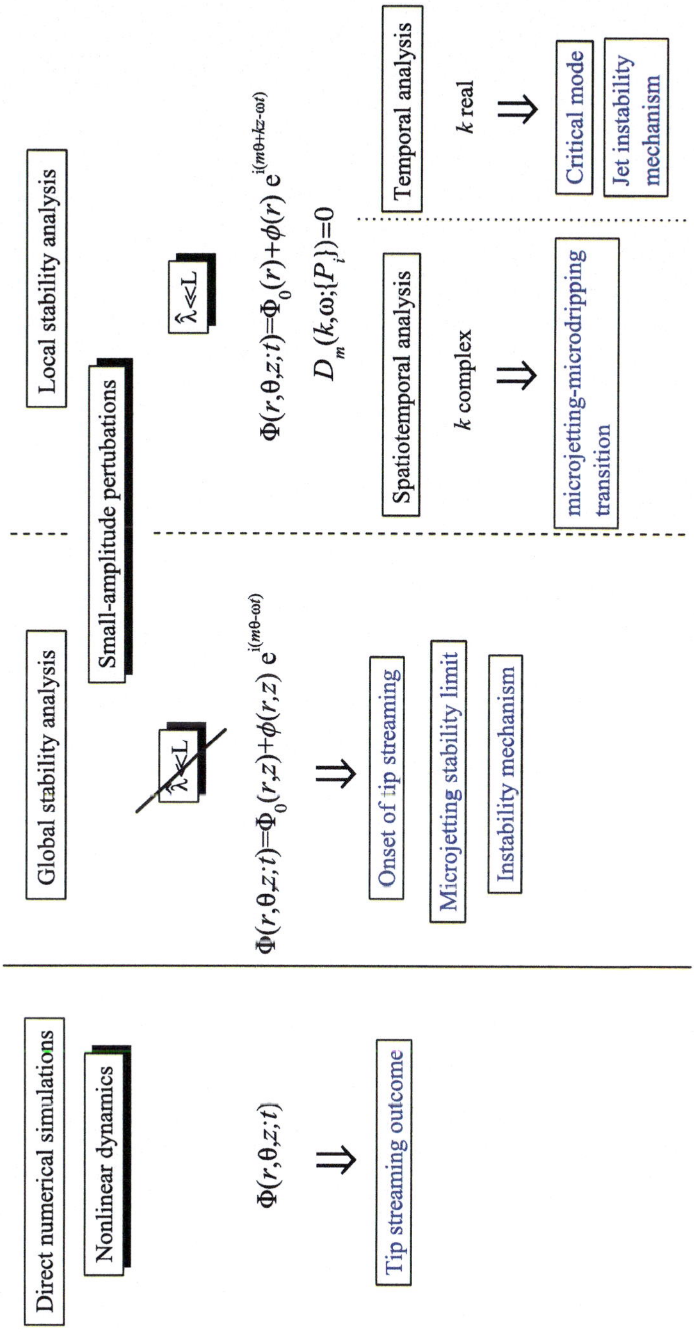

Fig. 3.13 Summary of the theoretical approaches for studying tip streaming

unstable. We can investigate the physical mechanisms that enhance or suppress the transition to tip streaming or the stability of microjetting.

The local stability analysis assumes that the wavelength $\hat{\lambda}$ of the small-amplitude perturbation is much smaller than any hydrodynamic axial length L of the flow, which prevents the stability analysis of any tip streaming flow. This approach allows us to examine only the behavior of the jets emitted in the microjetting regime. In most cases, one derives semi-analytically the dispersion relationship (3.14) and calculates the eigenfrequency ω as a function of the wavenumber k for an azimuthal number m and given values of the set of governing parameters $\{\mathscr{P}_i\}$. If k is a complex number (spatiotemporal analysis), one can determine the critical conditions for the jet convective-to-absolute instability transition, which corresponds to the microjetting-to-microdripping transition in many tip streaming configurations. If k is a real number (temporal analysis), the solution of the dispersion analysis allows one to identify the critical mode responsible for the jet breakup and the instability mechanisms.

3.4 Results of Local Linear Stability Analysis

Montanero and Gañán-Calvo [61] reviewed the main results of the local linear stability analysis for both simple and complex fluids. Starting from Rayleigh's linear stability analysis [67], several effects such as viscous damping [68], a non-negligible fluid environment [69], confinement [50, 60, 70–73], and an outer shell [74–77] were described. In this section, we review only the results directly related to the microjetting mode of tip streaming, including the effects of a high-speed outer gaseous stream, electric field, surfactant, and viscoelasticity.

3.4.1 Aerodynamic Effects

Jets moving with a quasi-uniform velocity profile can be described from the Lagrangian frame of reference moving with the jet. In this case, the jet velocity no longer enters the problem, and viscous effects are quantified by the Ohnesorge number $\text{Oh}_j = \mu_j(\rho_j\sigma R_j)^{-1/2}$, where μ_j, ρ_j, R_j are the jet's viscosity density, and radius, respectively.

In tip streaming realizations such as gaseous flow focusing, the big difference between the velocities of a liquid jet and the outer gas stream can significantly alter their velocity profiles. The consideration of realistic velocity distributions unveils unstable non-axisymmetric modes that do not appear in the stability analysis of inviscid parallel and uniform streams. Those distributions are essential to explain atomization experiments at large enough Weber numbers [78]. Gañán-Calvo et al. [79] have analytically shown that a shear boundary layer on the inner side of the interface can be regarded as a destabilizing factor for distances from the boundary

Fig. 3.14 Whipping instability in a viscous microjet produced by gaseous flow focusing (left) (reprinted from Ref. [82] with permission of IOP Publishing) and electrospray (right) (reprinted from Ref. [83] with permission of AIP Publishing)

layer origin of the order of the jet's radius. The boundary layer does not affect the jet's stability for larger distances, and Rayleigh's dispersion relation is recovered.

The existence of an outer fluid medium triggers the transition from the varicose to the whipping (bending, kink) mode (Fig. 3.14) for large enough values of the Weber number (defined in terms of the jet velocity relative to that of the outer medium) [80]. In this case, surface tension has always a stabilizing effect, and the destabilizing factor is purely aerodynamic: a perturbation at the interface causes the outer fluid to accelerate as it passes a crest, lowering the pressure at that point and encouraging the crest to increase in size (as in wind-generated ripples on a liquid free surface). It must be pointed out that an accurate calculation of the perturbation growth rate requires the consideration of the gas viscosity even for small values of this parameter [81]. The existence of an outer boundary layer significantly affects the whipping of a jet discharging into a still gaseous atmosphere.

In absolute whipping [84], the lateral jet oscillations propagate upstream, affecting the tapering meniscus of tip streaming. The conditions leading to absolute whipping have been barely studied [84]. The oscillations frequently observed in the microjetting mode of the plate-orifice configuration of gaseous flow focusing [85] do not generally propagate upstream and, therefore, do not alter the tapering meniscus stability. The gaseous radial flow in front of the discharge orifice constitutes a barrier for whipping perturbations. In the context of tip streaming, absolute aerodynamic whipping has been observed only in liquid jets focused by high-speed gaseous currents inside converging nozzles [82].

The violent lashes characterizing the whipping regime are responsible for broader droplet size distributions with smaller average values. Aerodynamic whipping precludes the gentle deposition of the emitted jet or the droplets resulting from its breakup. For this reason, it is an undesirable effect in most tip streaming realizations working with Newtonian liquids.

3.4.2 Electric Field

The linear stability analysis of electrified jets involves two intrinsic dimensionless parameters: the relative electrical permittivity and dimensionless charge relaxation time,

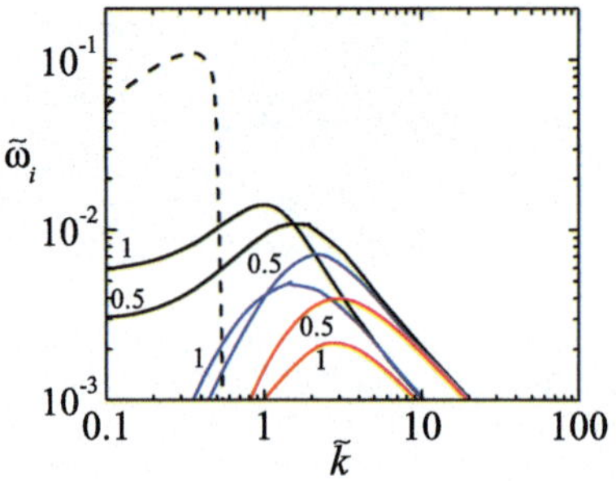

Fig. 3.15 Temporal growth rate $\widetilde{\omega}_i$ as a function of the wavenumber $\widetilde{k}$ for an inviscid jet with $\varepsilon = 78$ in an axial electric field of intensity $\mathscr{E}_j = 5 \times 10^{-3}$ [86]. The labels indicate the values of the electric relaxation time τ_{ej}. The dashed line corresponds to the dielectric case ($\tau_{ej} = \infty$). The black, blue, and red lines correspond to the modes $m = 0$, 1 and 2, respectively. Here, $\widetilde{\omega}_i = \omega_i \hat{t}_{\rho\sigma}$ and $\widetilde{k} = kR_j$ (reprinted from Ref. [86] with permission of IOP Publishing)

$$\varepsilon = \frac{\varepsilon_j}{\varepsilon_o}, \quad \tau_{ej} = \frac{\varepsilon_j}{K_j \hat{t}_{\rho\sigma}}, \tag{3.16}$$

where ε_j and ε_o are the electrical pemittivity of the liquid jet and outer medium, respectively, K_j is the jet electrical conductivity, and $\hat{t}_{\rho\sigma} = (\rho_j R_j^3/\sigma)^{1/2}$ is the inertio-capillary time defined in terms of the jet density ρ_j and radius R_j. The charge relaxation time τ_{ej} is also the inverse of the dimensionless electrical conductivity considered in the study of the cone-jet mode of electrospray [27]. The perfect conductor and dielectric limits correspond to $\tau_{ej} \to 0$ and $\tau_{ej} \to \infty$, respectively.

If the jet is subjected to the action of an external axial electric field (like in electrospray), this last effect is quantified through the dimensionless electric strength

$$\mathscr{E}_j = \frac{R_j \varepsilon_o E^2}{4\pi\sigma}, \tag{3.17}$$

where E is the axial electric field intensity. This parameter is also called the electric Bond number and the Taylor number.

A low-conductivity jet in an axial electric field is the most relevant configuration to the electrospray cone-jet mode. In this case, the surface charge evolves during the growth of the perturbation trying to screen the external electric field. The dispersion relationship calculated from the leaky-dielectric model showed that this surface charge relaxation makes disturbances grow in an oscillatory manner ($\omega_r \neq 0$), contrary to what happens in the perfect dielectric and perfect conductor case [86]. The interaction between the electrical and viscous stresses in the electrohydrodynamic interface boundary layer of nearly-inviscid threads can destabilize both asymmetric and axisymmetric perturbations (Fig. 3.15). Mestel [87, 88] derived the dispersion relationships corresponding to the low- and high-viscosity limits. The electric field [89] and the axial polymeric stress of the base state [90] inhibit the axisymmetric mode of a viscoelastic jet.

Jets moving in a hydrodynamically passive dielectric medium may also develop the whipping instability for sufficiently large values of the electric intensity (3.17) (Fig. 3.14) [69, 91, 92]. This instability results from the electrostatic repulsion between bent and straight parts of the jet [93]. Both aerodynamic and electrical whipping (Fig. 3.14) is enhanced by the jet's viscosity, which damps out the varicose mode in favor of bending perturbations. As mentioned above, whipping almost always exhibits a convective character and is observed far beyond the tapering cone of electrospray. In fact, the convective-to-absolute instability transition for the lateral mode $m = 1$ has been found under very specific conditions [77, 94].

Whipping usually has a positive effect in the electrospinning of polymeric fibers because the slashes considerably reduce the diameter of the precursor jet before its solidification. Nevertheless, whipping can be suppressed by combining low and uniform electrical fields with the mechanical pulling exerted by the collector [95]. In electro-coflowing [93], electro-flow focusing [96] for ambient pressure mass spectrometry [97], and electrohydrodynamic direct-writing [98] assisted with airflow [99], whipping is controlled by making the surrounding fluid coflow with the jet.

3.4.3 Surfactant

In many tip streaming realizations, the characteristic time of surfactant adsorption/desorption is much greater than any characteristic hydrodynamic time, and surfactant solubility can be neglected [100, 101]. Figure 3.16 shows the temporal growth rate as a function of the wavenumber for a liquid thread loaded with an insoluble surfactant characterized by its elasticity.

The perturbation growth rate of a liquid thread loaded with an insoluble surfactant decreases as the surfactant elasticity (strength)

$$\beta = -\frac{\Gamma_0}{\sigma_0}\frac{d\sigma}{d\Gamma}\bigg|_{\Gamma=\Gamma_0} \tag{3.18}$$

increases [101] (Fig. 3.16). Here, Γ_0 and $\sigma_0 = \sigma(\Gamma_0)$ are the initial surfactant density and surface tension, respectively. The wavenumber of the most unstable mode, $\widetilde{k}^{\max}$, does not depend monotonically on β. In the diffusion-limited case, the stabilizing effects of a soluble surfactant are similar to those described for an insoluble one [102].

The above results can be explained in terms of the role played by the Marangoni stress. When the jet interface is loaded with a surfactant monolayer, both the outwards surfactant convection and the dilatation of the interface element make the surfactant surface density decrease in the necking region. This surfactant depletion increases the surface tension, producing Marangoni convection in the opposite direction.

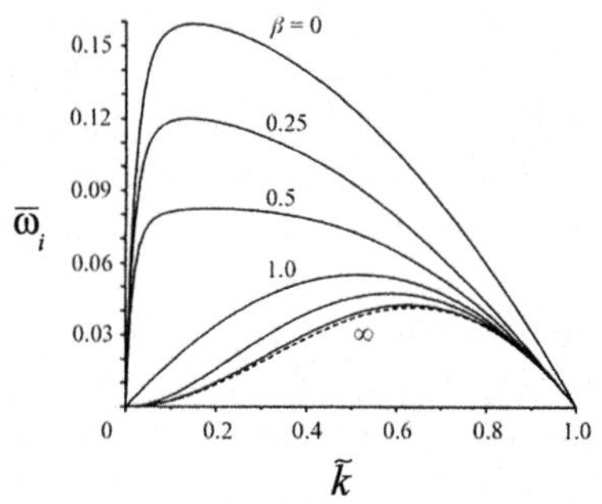

Fig. 3.16 Temporal growth rate $\overline{\omega}_i$ as a function of the wavenumber $\widetilde{k}$ for $\mathrm{Oh}_j = 10$ and different values of the surfactant elasticity β. Here, $\overline{\omega}_i = \omega_i \hat{t}_{vc}$ and $\widetilde{k} = kR_j$ (reprinted from Ref. [101] with permission of IOP Publishing)

3.4.4 Viscoelasticity

In most experiments, the viscoelastic stress built up during the jet emission relaxes over a short distance. In the linear regime, the viscoelasticity of a relaxed fluid is frequently modeled using the Jeffreys constitutive equation, which can be seen as the natural extension of the Maxwell model to include the strain rate relaxation time. It can also be regarded as a particularization of the generalized linear viscoelastic model when the relaxation modulus function takes a given form. More interestingly, the Jeffreys model results from the linearization of the popular Oldroyd-B constitutive equation [103].

The Jefferys (linearized Oldroyd-B) model reads:

$$\tau_{ij} + \lambda_s \frac{\partial \tau_{ij}}{\partial t} = 2\mu_0 \left(D_{ij} + \lambda_r \frac{\partial D_{ij}}{\partial t} \right) , \tag{3.19}$$

where τ_{ij} is the total extra stress, λ_s the stress relaxation time, μ_0 the solution viscosity, and λ_r the strain relaxation (retardation) time. Consider harmonic oscillations of (real) frequency ω. In this case, Eq. (3.19) reduces to

$$\tau_{ij} = 2\mu_{\text{eff}} D_{ij}, \quad \text{where} \quad \mu_{\text{eff}} = \mu_0 \frac{1 - i\omega\lambda_r}{1 - i\omega\lambda_s} \tag{3.20}$$

is the effective viscosity. The real and imaginary parts of the effective viscosity are

$$\mu'_{\text{eff}} = \mu_0 \frac{1 + \omega^2 \lambda_s \lambda_r}{1 + \omega^2 \lambda_s^2} \quad \text{and} \quad \mu''_{\text{eff}} = \mu_0 \frac{\omega(\lambda_s - \lambda_r)}{1 + \omega^2 \lambda_s^2}, \tag{3.21}$$

respectively, which correspond to the loss G'' and storage G' modulii divided by the oscillation frequency ω, respectively. In most cases, $\lambda_r \ll \lambda_s$ and $(\omega\lambda_s)^2$ is at most of order 10^{-1}. Thus,

$$\mu'_{\text{eff}} \simeq \mu_0 \left[1 - (\omega\lambda_s)^2\right]. \tag{3.22}$$

According to the approximation (3.22), viscoelasticity has a destabilizing effect on relaxed jets: the axisymmetric capillary mode grows in a viscoelastic jet faster than

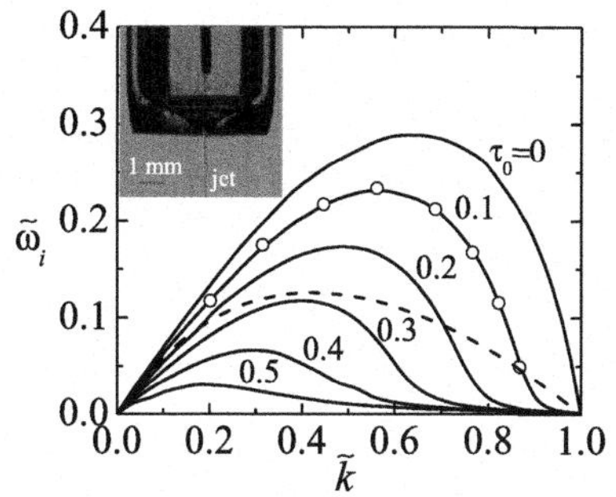

Fig. 3.17 Temporal growth rate $\widetilde{\omega}_i$ as a function of the wavenumber $\widetilde{k}$ for the stress relaxation time $\widetilde{\lambda}_s = 100$, the retardation time $\widetilde{\lambda}_r = 10$, the Ohnesorge number (defined in terms of the Newtonian viscosity) $\text{Oh}_j = 0.849$, and different values of the unrelaxed axial stress τ_0. The circles correspond to the solution obtained by Goren and Gottlieb [109] for $\tau_0 = 0.1$. The dashed line is the solution for a Newtonian jet ($\widetilde{\lambda}_s = \widetilde{\lambda}_r = \tau_0 = 0$). All the quantities have been made dimensionless with R_j, the inertio-capillary time $\hat{t}_{\rho\sigma}$ and the capillary pressure σ/R_j, i.e. $\widetilde{k} = kR_j$, $\widetilde{\omega}_i = \omega_i \hat{t}_{\rho\sigma}$, $\widetilde{\lambda}_s = \lambda_s/\hat{t}_{\rho\sigma}$ and $\widetilde{\lambda}_r = \lambda_r/\hat{t}_{\rho\sigma}$. The inset shows a microjet of a polyacrylamide aqueous solution at a concentration of 1000 ppm produced with gaseous flow focusing [112, 115] (reprinted from Ref. [110] with permission of IOP Publishing)

in a Newtonian one with the same Ohnesorge number owing to the apparent shear thinning ($\mu'_{\text{eff}} < \mu_0$) associated with elasticity [104–108]. This destabilizing effect translates into an increase of the critical Weber number for the convective-to-absolute instability transition [59].

The above results have relatively little consequences in the microjetting mode of tip streaming because the viscoelastic stress relaxes far away from the jet emission point in most applications. Suppose the unperturbed state is subject to unrelaxed axial elastic tension. In that case, the nonlinear terms of the rheological model contribute to the linearized equations, and the results depend on the specific nonlinear model even in the linear regime. The linear stability analysis of the Oldroyd-B model shows that the unrelaxed stress significantly reduces the growth rate of the varicose mode [109, 110] (Fig. 3.17), which explains the stability of long viscoelastic threads extruded in, for instance, electrospinning [111] or flow focusing [112, 113], even for quasi-Newtonian liquids [114]. Micrometer filaments with lengths up to 1 cm and Weber numbers of the order of 10^{-4} were formed in front of the discharge orifice of a gaseous flow focusing device [112, 115].

The dissolved polymers stretch in the tapering meniscus (droplet) tip, producing axial stress in the emitted jet. In some cases, macromolecules assemble along the jet axis, which prevents their relaxation to the coiling state even when the external elongational stresses have disappeared [114]. Super-stable jets of aqueous polymer solutions have been formed in transonic flow focusing, even for stress relaxation times of tens of microns [114]. In this case, the jet breakup occurs at distances from the ejector of the order of tens of thousands of the jet diameter.

Viscoelasticity produces a complex effect on the convective-to-absolute instability transition in the presence of unrelaxed stress. The unrelaxed axial stress increases the speed of the capillary waves traveling over the jet's surface [116], which favors absolute instability despite the reduction of the growth rates [110].

The combined effects of viscoelasticity and a surfactant monolayer have been studied on very few occasions. Alhushaybari and Uddin [117] studied the convective-absolute instability transition of a viscoelastic liquid jet falling under gravity in the presence of an insoluble surfactant. He and Wylie [118] analyzed the temporal linear instability of a surfactant-covered viscoelastic thread surrounded by another immiscible viscoelastic fluid. They concluded that the surfactant decreases the growth rate and increases the wavenumber of the dominant mode. Li and He [119] considered the effect of the surfactant viscosity as well.

References

1. Ferziger JH, Peric M (2013) Computational methods for fluid dynamics. Springer
2. Popinet S (2003) Gerris: a tree-based adaptive solver for the incompressible Euler equations in complex geometries. J Comput Phys 190:572–600
3. Karniadakis G, Sherwin SJ (1999) Spectral/hp element methods for CFD. Oxford University Press
4. Canuto C, Hussaini MY, Quarteroni A (2007) Spectral methods: fundamentals in single domains. Springer
5. Boyd JP (1989) Chebyshev and Fourier spectral methods. Springer
6. Khorrami MR (1989) Application of spectral collocation techniques to the stability of swirling flows. J Comput Phys 81:206–229
7. Higuera FJ (2006) Stationary viscosity-dominated electrified capillary jets. J Fluid Mech 558:143–152
8. Gamero-Castaño M, Magnani M (2019) Numerical simulation of electrospraying in the cone-jet mode. J Fluid Mech 859:247–267
9. Hirt CW, Nichols BD (1981) Volume of Fluid (VOF) method for the dynamics of free boundaries. J Comput Phys 39:201–225
10. Chang YC, Hou TY, Merriman B, Osher S (1996) A level set formulation of Eulerian interface capturing methods for incompressible fluid flows. J Comput Phys 124:449–464
11. Rider WJ, Kothey DB (1998) Reconstructing volume tracking. J Comput Phys 141:112–152
12. Verfurth R (1994) A posteriori error estimation and adaptive mesh-refinement techniques. J Comput Appl Math 50:67–83
13. Brackbill JU, Kothe DB, Zemach C (1992) A continuum method for modeling surface tension. J Comput Phys 100:335–354
14. Zahoor R, Belsak G, Bajt S, Sarler B (2018) Simulation of liquid micro-jet in free expanding high-speed co-flowing gas streams. Microfluidics and Nanofluidics 22:87
15. Mu K, Qiao R, Guo J, Yang C, Wu Y, Si T (2021) Parametric study on stability and morphology of liquid cone in flow focusing. Int J Multiphase Flow 135(103):507
16. Donea J, Huerta A, Ponthot JP, Rodríguez-Ferrán A (2004) Arbitrary Lagrangian–Eulerian methods. In: Encyclopedia of computational mechanics. Wiley, pp 413–437
17. Anthony CR et al (2023) Sharp interface methods for simulation and analysis of free surface flows with singularities: breakup and coalescence. Annu Rev Fluid Mech 55:707–747
18. Collins RT, Sambath K, Harris MT, Basaran OA (2013) Universal scaling laws for the disintegration of electrified drops. Proc Nat Acad Sci 110:4905–4910
19. Thompson JF, Warsi ZUA (1982) Boundary-fitted coordinate systems for numerical solution of partial differential equations-a review. J Comput Phys 47:1–108

20. Thompson JF, Warsi ZUA, Mastin CW (1985) Numerical grid generation: foundations and applications. Elsevier
21. Herrada MA, Montanero JM (2016) A numerical method to study the dynamics of capillary fluid systems. J Comput Phys 306:137–147
22. Dimakopoulos Y, Tsamopoulos J (2003) A quasi-elliptic transformation for moving boundary problems with large anisotropic deformations. J Comput Phys 192:494–522
23. Herrada MA, Ponce-Torres A, Rubio M, Eggers J, Montanero JM (2022) Stability and tip streaming of a surfactant-loaded drop in an extensional flow: influence of surface viscosity. J Fluid Mech 934:A26
24. Chomaz J (2005) Global instabilities in spatially developing flows. Annu Rev Fluid Mech 37:357–392
25. Theofilis V (2011) Global linear instability. Annu Rev Fluid Mech 43:319–352
26. Cruz-Mazo F, Herrada MA, Gañán-Calvo AM, Montanero JM (2017) Global stability of axisymmetric flow focusing. J Fluid Mech 832:329–344
27. Ponce-Torres A, Rebollo-Muñoz N, Herrada MA, Gañán-Calvo AM, Montanero JM (2018) The steady cone-jet mode of electrospraying close to the minimum volume stability limit. J Fluid Mech 857:142–172
28. Blanco-Trejo S, Herrada MA, Gañán-Calvo AM, Montanero JM (2019) Electrospray cone-jet mode for weakly viscoelastic liquids. Phys Rev E 100(043):114
29. Cabezas MG, Rubio M, Rebollo-Muñoz N, Herrada MA, Montanero JM (2021) Global stability analysis of axisymmetric liquid-liquid flow focusing. J Fluid Mech 909:A10
30. López M, Cabezas MG, Montanero JM, Herrada MA (2022) On the hydrodynamic focusing for producing microemulsions via tip streaming. J Fluid Mech 934:A47
31. Gordillo JM, Sevilla A, Campo-Cortés F (2014) Global stability of stretched jets: conditions for the generation of monodisperse micro-emulsions using coflows. J Fluid Mech 738:335–357
32. López-Herrera J, Herrada M, Gamero-Castaño M, Gañán-Calvo AM (2020) A numerical simulation of coaxial electrosprays. J Fluid Mech 885:A15
33. Montanero JM, Ponce-Torres A (2020) Review on the dynamics of isothermal liquid bridges. Appl Mech Rev 72(010):803
34. Beroz J, Hart AJ, Bush JM (2019) Stability limit of electrified droplets. Phys Rev Lett 122(244):501
35. Rubio M, Rodríguez-Díaz P, López-Herrera JM, Herrada MA, Gañán-Calvo AM, Montanero JM (2023) The role of charge relaxation in electrified tip streaming. Phys Fluids 35(017):131
36. Augello L, Fani A, Gallaire F (2018) The influence of the entry region on the instability of a coflowing injector device. J Phys Condens Matter 30(284):003
37. Tammisola O, Lundell F, Soderberg LD (2012) Surface tension-induced global instability of planar jets and wakes. J Fluid Mech 713:632–658
38. Rubio M, Montanero JM, Eggers J, Herrada MA (2024) Stable production of fluid jets with vanishing diameters via tip streaming. J Flui Mech 893: A4
39. Schmid PJ (2007) Nonmodal stability theory. Annu Rev Fluid Mech 39:129–162
40. de Luca L, Costa M, Caramiello C (2002) Energy growth of initial perturbations in two-dimensional gravitational jets. Phys Fluids 14:289–299
41. Hwang H, Moin P, Hack MJP (2021) A mechanism for the amplification of interface distortions on liquid. J Fluid Mech 911:A51
42. Danaila I, Dusek J, Anselmet F (1998) Nonlinear dynamics at a Hopf bifurcation with axisymmetry breaking in a jet. Phys Rev E 57:R3695–R3698
43. Sauter US, Buggisch HW (2005) Stability of initially slow viscous jets driven by gravity. J Fluid Mech 533:237–257
44. Rubio-Rubio M, Sevilla A, Gordillo JM (2013) On the thinnest steady threads obtained by gravitational stretching of capillary jets. J Fluid Mech 729:471–483
45. Martínez-Calvo A, Rubio-Rubio M, Sevilla A (2018) The nonlinear states of viscous capillary jets confined in the axial direction. J Fluid Mech 834:335–358

46. Schmidt S, Tammisola O, Lesshafft L, Oberleithner K (2021) Global stability and nonlinear dynamics of wake flows with a two-fluid interface. J Fluid Mech 919:A96
47. Javadi A, Eggers J, Bonn D, Habibi M, Ribe NM (2013) Delayed capillary breakup of falling viscous jets. Phys Rev Lett 110(144):501
48. Huerre P, Monkewitz PA (1990) Local and global instabilites in spatially developing flows. Annu Rev Fluid Mech 22:473–537
49. Lin SP (2003) Breakup of liquid sheets and jets. Cambridge University Press, New York
50. Guillot P, Colin A, Utada AS, Ajdari A (2007) Stability of a jet in confined pressure-driven biphasic flows at low Reynolds numbers. Phys Rev Lett 99(104):502
51. Gañán-Calvo AM, Herrada MA, Garstecki P (2006) Bubbling in unbounded coflowing liquids. Phys Rev Lett 96(124):504
52. de Luca L (1999) Experimental investigation of the global instability of plane sheet flows. J Fluid Mech 399:355–376
53. Dizes SL (1997) Global modes in falling capillary jets. Eur J Mech B/Fluids 16:761–778
54. Si T, Li F, Yin XY, Yin XZ (2009) Modes in flow focusing and instability of coaxial liquid-gas jets. J Fluid Mech 629:1–23
55. Vega EJ, Montanero JM, Herrada MA, Gañán-Calvo AM (2010) Global and local instability of flow focusing: the influence of the geometry. Phys Fluids 22(064):105
56. Briggs RJ (1964) Electron-stream interaction with plasmas. MIT Press, Cambridge
57. Montanero JM, Gañán-Calvo AM (2008) Stability of coflowing capillary jets under non-axisymmetric perturbations. Phys Rev E 77(046):301
58. van Saarloos W (1987) Dynamical velocity selection: marginal stability. Phys Rev Lett 58:2571–2574
59. Montanero JM, Gañán-Calvo AM (2008) Viscoelastic effects on the jetting-dripping transition in co-flowing capillary jets. J Fluid Mech 610:249–260
60. Cabezas MG, Herrada MA, Montanero JM (2019) Stability of a jet moving in a rectangular microchannel. Phys Rev E 100(053):104
61. Montanero JM, Gañán-Calvo AM (2020) Dripping, jetting and tip streaming. Rep Prog Phys 83(097):001
62. Ismail AS, Yao J, Xia HH, Stark JPW (2018) Breakup length of electrified liquid jets: scaling laws and applications. Phys Rev Appl 10(064):010
63. Gañán-Calvo AM, Chapman HN, Heymann M, Wiedorn MO, Knoska J, Gañán-Riesco B, López-Herrera JM, Cruz-Mazo F, Herrada MA, Montanero JM, Bajt S (2021) The natural breakup length of a steady capillary jet: application to serial femtosecond crystallography. Crystals 11:990
64. Umemura A (2016) Self-destabilising loop of a low-speed water jet emanating from an orifice in microgravity. J Fluid Mech 25:146–180
65. Castro-Hernández E, Gundabala V, Fernández-Nieves A, Gordillo JM (2009) Scaling the drop size in coflow experiments. New J Phys 11(075):021
66. Keller JB, Rubinov SI, Tu YO (1973) Spatial instability of a jet. Phys Fluids 16:2052–2055
67. Rayleigh L (1878) On the instability of jets. Proc London Math Soc s1-10:4–13
68. Tomotika S (1935) On the instability of a cylindrical thread of a viscous liquid surrounded by another viscous fluid. Proc R Soc Lond 150:322–337
69. Eggers J, Villermaux E (2008) Physics of liquid jets. Rep Prog Phys 71(036):601
70. Hickox CE (1971) Instability due to density and viscosity stratification in an axisymmetric pipe flow. Phys Fluids 14:251–262
71. Guillot P, Colin A, Ajdari A (2008) Stability of a jet in confined pressure-driven biphasic flows at low Reynolds number in various geometries. Phys Rev E 78(016):307
72. Janssen PJA, Meijer HEH, Anderson PD (2012) Stability and breakup of confined threads. Phys Fluids 24(012):102
73. Kashid MN, Kowalinski W, Renken A, Baldyga J, Kiwi-Minsker L (2012) Analytical method to predict two-phase flow pattern in horizontal micro-capillaries. Chem Eng Sci 74:219–232
74. Sanz A, Meseguer J (1985) One-dimensional linear analysis of the compound jet. J Fluid Mech 159:55–68

75. Chauhan A, Maldarelli C, Papageorgiou DT, Rumschitzk DS (2000) Temporal instability of compound threads and jets. J Fluid Mech 420:1–25
76. Lee SY, Snider C, Park K, Robinson JP (2007) Compound jet instability in a microchannel for mononuclear compound drop formation. J Micromech Microeng 17:1558
77. Herrada MA, Montanero JM, Ferrera C, Gañán-Calvo AM (2010) Analysis of the dripping-jetting transition in compound capillary jets. J Fluid Mech 649:523–536
78. Gordillo JM, Pérez-Saborid M, Gañán-Calvo AM (2001) Linear stability of co-flowing liquid-gas jets. J Fluid Mech 448:23–51
79. Gañán-Calvo AM, Herrada MA, Montanero JM (2014) How does a shear boundary layer affect the stability of a capillary jet? Phys Fluids 26(061):701
80. Lasheras JC, Hopfinger EJ (2000) Liquid jet instability and atomization in a coaxial gas stream. Annu Rev Fluid Mech 32:275–308
81. Gordillo JM, Pérez-Saborid M (2005) Aerodynamic effects in the break-up of liquid jets: on the first wind-induced break-up regime. J Fluid Mech 541:1–20
82. Acero AJ, Ferrera C, Montanero JM, Gañán-Calvo AM, (2012) Focusing liquid microjets with nozzles. J Micromech Microeng 22(065):011
83. Rubio M, Sadek SH, Gañán-Calvo AM, Montanero JM (2021) Diameter and charge of the first droplet emitted in electrospray. Phys Fluids 33(032):002
84. Herrada MA, Ferrera C, Montanero JM, Gañán-Calvo AM (2010) Absolute lateral instability in capillary coflowing jets. Phys Fluids 22(064):104
85. Gañán-Calvo AM (1998) Generation of steady liquid microthreads and micron-sized monodisperse sprays in gas streams. Phys Rev Lett 80:285–288
86. Saville DA (1971) Electrohydrodynamic stability: effects of charge relaxation at the interface of a liquid jet. J Fluid Mech 48:815–827
87. Mestel AJ (1994) Electrohydrodynamic stability of a slightly viscous jet. J Fluid Mech 274:93–113
88. Mestel AJ (1994) Electrohydrodynamic stability of a highly viscous jet. J Fluid Mech 312:311–326
89. Xie L, Yang L, Qin L, Fu Q (2017) Temporal instability of charged viscoelastic liquid jets under an axial electric field. Eur J Mech/Fluids 66:60–70
90. Carroll CP, Joo YL (2008) Axisymmetric instabilities of electrically driven viscoelastic jets. J Non-Newtonian Fluid Mech 153:130–148
91. Hohman MM, Shin M, Rutledge G, Brenner MP (2001) Electrospinning and electrically forced jets. I. Stability theory. Phys Fluids 13:2201–2220
92. Yang W, Duan H, Li C, Deng W (2014) Crossover of varicose and whipping instabilities in electrified microjets. Phys Rev Lett 112(054):501
93. Guerrero J, Rivero J, Gundabala VR, Perez-Saborid M, Fernandez-Nieves A (2014) Whipping of electrified liquid jets. Proc Natl Acad Sci 111:13,763–13,767
94. Li F, Gañán-Calvo AM, López-Herrera JM, Yin XY, Yin XZ (2013) Absolute and convective instability of a charged viscoelastic liquid jet. J Non-Newtonian Fluid Mech 196:58–69
95. Kiselev P, Rosell-Llompart J (2012) Highly aligned electrospun nanofibers by elimination of the whipping motion. J Appl Polym Sci 125:2433–2441
96. Gañán-Calvo AM (2007) Electro-flow focusing: the high-conductivity low-viscosity limit. Phys Rev Lett 98(134):503
97. Forbes TP, Sisco E (2014) Chemical imaging of artificial fingerprints by desorption electro-flow focusing ionization mass spectrometry. Analyst 139:2982
98. Huang Y, Bu N, Duan Y, Pan Y, Liu H, Yin Z, Xiong Y (2013) Electrohydrodynamic direct-writing. Nanoscale 5:12,007–12,017
99. Jiang J, Wang X, Li W, Liu J, Liu Y, Zheng G (2018) Electrohydrodynamic direct-writing micropatterns with assisted airflow. Micromachines 9:456
100. Kwak S, Pozrikidis C (2001) Effect of surfactants on the instability of a liquid thread or annular layer. Part I: quiescent fluids. Int J Multiphase Flow 27:1–37
101. Timmermans MLE, Lister JR (2002) The effect of surfactant on the stability of a liquid thread. J Fluid Mech 459:289–306

102. Hansen S, Peters GWM, Meijer HEH (1999) The effect of surfactant on the stability of a fluid filament embedded in a viscous fluid. J Fluid Mech 382:331–349
103. Bird RB, C R, Armstrong, Hassager O (1987) Dynamics of polymeric liquids volume I: fluid mechanics; volume II: kinetic theory. Wiley, New York
104. Middleman S (1965) Stability of a viscoelastic jet. Chem Eng Sci 20:1037–1040
105. Goldin M, Yerushalmi J, Pfeffer R, Shinnar R (1969) Breakup of a laminar capillary jet of a viscoelastic fluid. J Fluid Mech 38:689–711
106. Brenn G, Liu Z, Durst F (2000) Linear analysis of the temporal instability of axisymmetrical non-Newtonian liquid jets. Int J Multiphase Flow 26:1621–1644
107. Funada T, Joseph DD (2003) Viscoelastic potential flow analysis of capillary instability. J Non-Newtonian Fluid Mech 111:87–105
108. Ye HY, Yang LJ, Fu QF (2016) Instability of viscoelastic compound jets. Phys Fluids 28(043):101
109. Goren S, Gottlieb M (1982) Surface-tension-driven breakup of viscoelastic liquid threads. J Fluid Mech 120:245–266
110. Mohamed AS, Herrada MA, Gañán-Calvo AM, Montanero JM (2015) Convective-to-absolute instability transition in a viscoelastic capillary jet subject to unrelaxed axial elastic tension. Phys Rev E 92(023):006
111. Reneker DH, Yarin AL (2008) Electrospinning jets and polymer nanofibers. Polymer 49:2387–2425
112. Ponce-Torres A, Vega EJ, Castrejón-Pita AA, Montanero JM (2017) Smooth printing of viscoelastic microfilms with a flow focusing ejector. J Non-Newtonian Fluid Mech 249:1–7
113. Rubio A, Galindo F, Vega EJ, Montanero JM, Cabezas MG (2022) Viscoelastic transition in transonic flow focusing. Phys Rev Fluids 7(074):201
114. Rubio A, Vega EJ, Gañán-Calvo AM, Montanero JM (2022) Unexpected stability of micrometer weakly viscoelastic jets. Phys Fluids 34(062):014
115. Ponce-Torres A, Montanero JM, Vega EJ, Gañán-Calvo AM (2016) The production of viscoelastic capillary jets with gaseous flow focusing. J Non-Newtonian Fluid Mech 229:8–15
116. Gill SJ, Gavis J (1956) Tensile stress in jets of viscoelastic fluids. I. J Polym Sci 20:287–298
117. Alhushaybari A, Uddin J (2020) Absolute instability of free-falling viscoelastic liquid jets with surfactants. Phys Fluids 32(013):102
118. He D, Wylie JJ (2021) Temporal instability of a viscoelastic liquid thread surrounded by another viscoelastic fluid in presence of insoluble surfactant and inertia. J Non-Newtonian Fluid Mech 288(104):468
119. Li F, He D (2023) Dynamics of a surfactant-laden viscoelastic thread in the presence of surface viscosity. J Fluid Mech 966:A35

Chapter 4
Experimental Methods

Abstract The tip streaming of simple fluids is affected by the densities and viscosities of the phases involved in the phenomenon, as well as by the surface tensions of the interfaces formed between them. In addition, electrical conductivities and permittivities are essential when electrohydrodynamic forces drive this phenomenon. As explained in Chap. 1, complex fluids are referred to as those including surfactant molecules and polymer chains. The tip streaming of these fluids involves additional physical properties that must be known to analyze the phenomenon. The first section of this chapter describes the experimental techniques commonly used to characterize complex fluids for studying tip streaming.

The analysis of the tip streaming flow relies on visualization techniques in practically all the experiments. The method most frequently used for this purpose is the observation of the interface contour, in many cases using high-speed imaging. The last section of this chapter reviews the most relevant aspects of this methodology. We also briefly describe some specific characteristics of particle image velocimetry.

Keywords Surface tension · Extensional relaxation time · High-speed imaging · Flow visualization

4.1 Fluids Characterization

4.1.1 Surface Tension

The surface tension can be readily measured with a force tensiometer equipped with a Du Nouy ring [1]. This procedure requires a considerable volume of liquid and may not be appropriate for liquid-liquid interfaces. For this reason, the surface tension of an interface between immiscible liquids is frequently measured with the well-established drop shape method [2, 3]. The idea is to analyze the experimental shape of an axisymmetric liquid-fluid interface and compare it with the solution of the

J. M. Montanero, *Tip Streaming of Simple and Complex Fluids*, Fluid Mechanics and Its Applications 137, https://doi.org/10.1007/978-3-031-52768-5_4

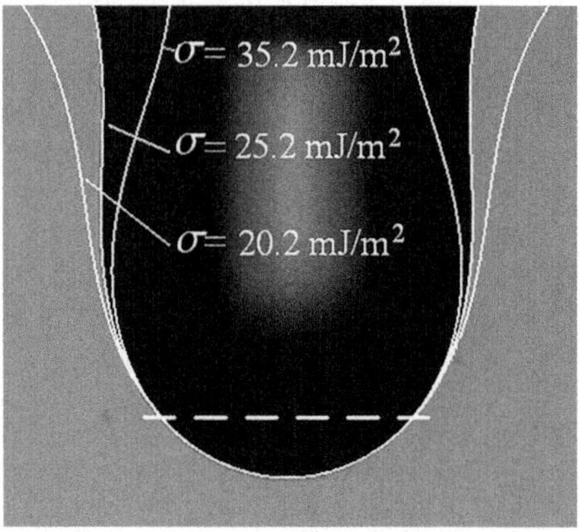

Fig. 4.1 Profiles of a cyclohexane droplet calculated for a given apex curvature and different surface tension values (reprinted from Ref. [7] with permission of Elsevier)

Young-Laplace equation. The surface tension is determined by fitting the theoretical contour to the experimental one using the surface tension as an adjustable variable (Fig. 4.1). Several versions of this method with different degrees of sophistication have been developed over the last decades [4–6]. The error of the measurement can be reduced to less than one mN/m.

Despite its simplicity and reliability, the drop shape method may lead to significant errors under certain adverse circumstances. The problem becomes more complicated for density-matched liquids because the drop shape is practically independent of the interfacial tension value. There have been some attempts to find a solution to this long-standing problem by using alternative functional principles [8–12].

Impurities affect the interfacial tension, especially if this quantity takes high values. The surface tension of the water-air interface rarely takes the nominal value $\sigma = 72$ mN/m, contrary to what is assumed in many works. The contaminants may come from the liquid (those not entirely eliminated by bulk purification) [13–15] or from the surrounding gaseous medium (adsorbed surface layers enriched in surface-active contaminants) [16].

Care must also be taken when measuring the surface tension involving high-viscosity liquids with the drop-shape method. In this case, tiny residual velocities associated with the pendant droplet formation can produce viscous stresses comparable with the capillary pressure and distort the surface tension value inferred from the fitting of the hydrostatic solution.

4.1.2 Surfactant

4.1.2.1 Surface Tension Versus Surfactant Concentration

In tip streaming, the surfactant surface density Γ (also called the surface excess concentration) significantly varies along the interface due to the surface convection of

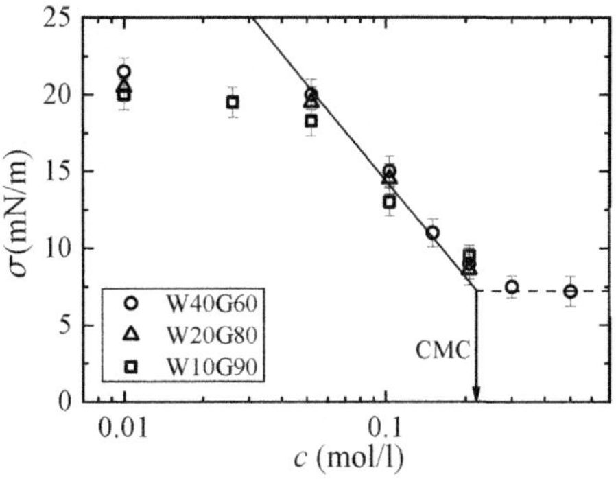

Fig. 4.2 Interfacial tension σ of the interface between n-Hexane and water/glycerol mixtures as a function of the sodium dodecyl sulfate (SDS) concentration c. The arrow indicates the estimated critical micelle concentration. The notation WXXGYY refers to a water (W)/glycerine (G) mixture with the proportion XX/YY% in weight (reprinted from Ref. [17] with permission of Elsevier)

surfactant molecules and the expansion/compression of the surface elements. Consequently, significant gradients of the interfacial tension σ may be found, especially for strong surfactants. The resulting local soluto-capillarity and interfacial elasticity (Marangoni stress) may play an essential role in the phenomenon. Therefore, the analysis of this phenomenon requires an accurate knowledge of the equation of state $\sigma(\Gamma)$.

The drop shape method described above provides the interfacial tension at equilibrium when a surfactant is dissolved in one of the liquid phases at the volume concentration c (Fig. 4.2). The curve $\sigma(c)$ allows one to estimate the critical micelle concentration (CMC), i.e., the concentration threshold for the formation of micelles. The standard procedure to determine the CMC value is to find the intersection between the straight line tangent to the inflection point of $\sigma(c)$ and the plateau for large c (Fig. 4.2). This method often provides a rough estimation due to the scattering in the experimental data.

Surfactant properties such as the surfactant surface density, the adsorption/desorption characteristic times, and the diffusion coefficient can be measured through their effects on the equilibrium and *dynamic surface tension* (i.e., the surface tension measured during the adsorption or desorption of surfactant molecules). As will be explained below, this approach makes use of the theoretical framework presented in Chap. 2.

The most commonly used procedures to measure the surfactant surface density Γ are the radiotracer method [18] and the neutron reflectivity technique [19]. However, as mentioned above, $\Gamma(c)$ can also be determined from an equilibrium surface tension experiment. In this case, the Gibbs isotherm (2.40) for ideal systems,

$$\Gamma = -\frac{1}{nR_gT}\frac{d\sigma}{d\ln c}, \tag{4.1}$$

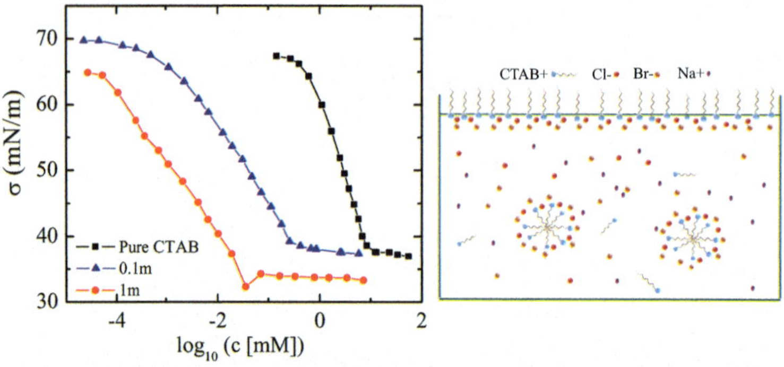

Fig. 4.3 (Left) Interfacial tension σ of the water-air interface as a function of the cetrimonium bromide (CTAB) concentration in the presence of NaCl [21]. (Right) Sketch of the molecules dissolved in water. CTAB is dissociated (the Br–ion is split off)

allows one to determine $\Gamma(\sigma)$ from the equation of state $\sigma(c)$. Here, n is the degree of dissociation of the surfactant, R_g is the universal gas constant, and T is the temperature. This equation is valid only for concentrations smaller than the CMC. According to Eq. (4.1), $\Gamma(\sigma)$ is calculated from the slope of the curve $\sigma(c)$ in a linear-log graph (Fig. 4.3-left). There is a certain debate about whether this procedure leads to accurate results, especially for ionic surfactants, due to the complexity associated with the electrostatic interaction between surfactant molecules (Fig. 4.3-right) [20].

A similar strategy, commonly used by experimentalists, is to assume the validity of a kinetic model to fit their parameters from the experimental equilibrium isotherm $\sigma(c)$. For instance, one considers the Frumkin kinetic model [Eqs. (2.44) and (2.45)] for a given value of n to calculate $\sigma(c)$; i.e.,

$$\left.\begin{aligned} \frac{c}{k_d/k_a} &= \exp\left[\kappa(\Gamma/\Gamma_\infty)^n\right]\frac{\Gamma_\infty \Gamma}{\Gamma_\infty - \Gamma} \\ \sigma &= \sigma_0 + R_g T \Gamma_\infty\left[\ln\left(1-\frac{\Gamma}{\Gamma_\infty}\right) - \frac{\kappa n}{n+1}\left(\frac{\Gamma}{\Gamma_\infty}\right)^{n+1}\right] \end{aligned}\right\} \Longrightarrow \sigma(k_d/k_a, \Gamma_\infty, \kappa; c). \tag{4.2}$$

One finds the value of $\{k_d/k_a, \Gamma_\infty, \kappa\}$ leading to the best fitting to experimental data.

4.1.2.2 Adsorption and Desorption Times and Diffusion Coefficient

The characteristic adsorption time can be obtained by measuring the dynamic surface tension $\sigma(t)$. This quantity typically exhibits an exponential dependence versus time when a fresh interface is exposed to the liquid bath. The inverse of the exponent is identified as the adsorption time. The dynamic surface tension is commonly measured using the Maximum Bubble Pressure method [22].

Consider a bubble formed quasi-instantaneously in a liquid bath where a surfactant is dissolved at the concentration c. Suppose that the adsorption kinetics is sufficiently fast for the transfer of surfactant from the bulk to the interface to be controlled by diffusion. Then, Fick's law describes the mass transfer to the clean bubble surface. If the diffusion is assumed to be spherically symmetric and convection effects are neglected, the surfactant surface concentration for small t (t is the surface age) is [23]

$$\Gamma(t) = 2c\left(\frac{\mathscr{D}t}{\pi}\right)^{1/2}, \tag{4.3}$$

where c is the initial surfactant concentration, and $\mathscr{D}$ is the diffusion coefficient. The combination of Eq. (4.3) and the Frumkin equation of state (2.41) for $\Gamma \to 0$,

$$\sigma_0 - \sigma = \Gamma R_g T, \tag{4.4}$$

leads to the expression [24]

$$\sigma = \sigma_0 - 2R_g T c\left(\frac{\mathscr{D}t}{\pi}\right)^{1/2}, \tag{4.5}$$

where σ_0 is the surface tension of the clean interface. The dynamic surface tension $\sigma(t)$ can be measured with the Maximum Bubble Pressure method [22] to infer the diffusion coefficient value from the data obtained for small t. This simplified procedure provides reasonably accurate results [23].

The volumetric and surface diffusion coefficients of most surfactants in tip streaming experiments take very small values [25]. The transfer of surfactant molecules from the bulk to the interface is limited only by diffusion, which means that this mechanism may affect tip streaming. However, the time scale of transient tip streaming is typically much smaller than the surfactant adsorption and desorption times, and the surfactant can be considered insoluble. Under this approximation, the problem is not influenced by the bulk transport of surfactant, and the volumetric diffusion coefficient does not enter the transitory problem. In a steady flow, the volumetric diffusion affects the thickness of the surfactant mass boundary layer at the interface. However, this aspect of the problem may hardly alter the steady flow and its stability, as observed in the two-phase flow in a finite-length microchannel [26].

4.1.2.3 Surface Viscosity

The onset of tip streaming involves large surface velocity gradients, which may produce significant surface viscous stresses in the presence of a surfactant monolayer. These stresses not only directly alter the balance of surface forces but also affect the surfactant distribution over the interface, which changes the capillary and Marangoni forces [27].

Small-molecule surfactants commonly used in microfluidic applications exhibit very low surface viscosities. For instance, Zell et al. [28] reported that the surface shear viscosity of sodium dodecyl sulfate (SDS) was below the sensitivity limit of their experimental technique ($\sim 10^{-8}$ Pa s m). For this reason, this property can only be measured by generating large surface velocity gradients in a controlled manner.

In the pinching of a Newtonian liquid free surface, the system spontaneously approaches a finite-time singularity, which offers a unique opportunity to observe the behaviour of fluids with arbitrarily small spatial and time scales. This property and its universal character (insensitivity to both initial and boundary conditions) make this problem a good candidate for measuring surface viscosities. Ponce-Torres et al. [29] estimated the surface viscosity of SDS by comparing the free surface minimum radius measured experimentally with that calculated from the simulation with the shear and dilatational viscosities as fitting variables (Fig. 2.4). The adsorption/desorption times of SDS are several orders of magnitude larger than the characteristic time of the breakup, which allows one to regard SDS as an insoluble surfactant.

4.1.3 Viscoelasticity

Many of the tip streaming experiments conducted with viscoelastic liquids are restricted to Boger fluids, i.e., dilute polymeric solutions with constant shear viscosity (the shear-thinning effect can be neglected). These fluids can be analyzed with relatively simple constitutive equations, such as the popular Oldroyd-B model (see Chap. 2). This model involves three parameters: the solvent viscosity μ_s, the (constant) polymer viscosity μ_p, and the stress relaxation time λ_s.

The polymer viscosity $\mu_p = \mu - \mu_s$ is calculated from the solution viscosity μ measured in a shear rheometer. This measurement must be conducted for a sufficiently large range of shear rates $\dot{\gamma}$ to ensure that the shear-thinning effect is negligible for the high values of $\dot{\gamma}$ typically arising in the tip streaming experiment. A stress-controlled rotational rheometer is commonly used for this purpose. The minimum and maximum shear rates are imposed by the rheometer sensitivity and the appearance of flow instabilities, respectively (Fig. 4.4).

In the filament thinning during the breakup of viscoelastic drops, liquid bridges, and jets, elastic stresses grow exponentially to balance the driving capillary pressure, while inertial, viscous, and gravitational effects are negligible [31]. In this elasto-capillary regime, the radius of a quasi-cylindrical thread decreases exponentially in time with the relaxation exponent $1/(3\lambda_e)$ (Fig. 4.5) [32, 33]. The motion in the liquid thread is essentially a uniaxial extensional flow characterized by the constant extensional rate $\dot{\varepsilon} = 2/(3\lambda_e)$. If the Oldroyd-B model is adopted to describe this flow, then the identity $\lambda_s = \lambda_e$ is obtained.

Different extensional rheometers produce the filament thinning described above to determine the value of λ_e (λ_s). The filament stretching extensional rheometry (FiSER) [35] and capillary breakup extensional rheometry (CaBER) [36] are probably the most popular ones. In the CaBER apparatus [32, 37], the fluid is introduced between

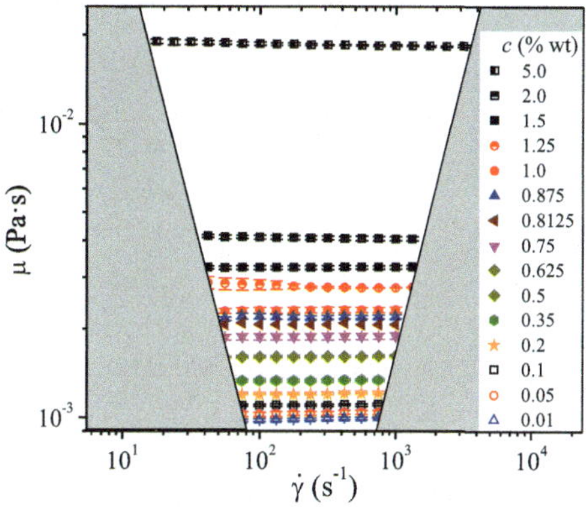

Fig. 4.4 Dependence of the solution shear viscosity μ upon the shear rate $\dot{\gamma}$ at 20 °C for PEO100K aqueous solutions at different concentrations c (reprinted with permission from Ref. [30]. Copyright (2022) by the American Physical Society)

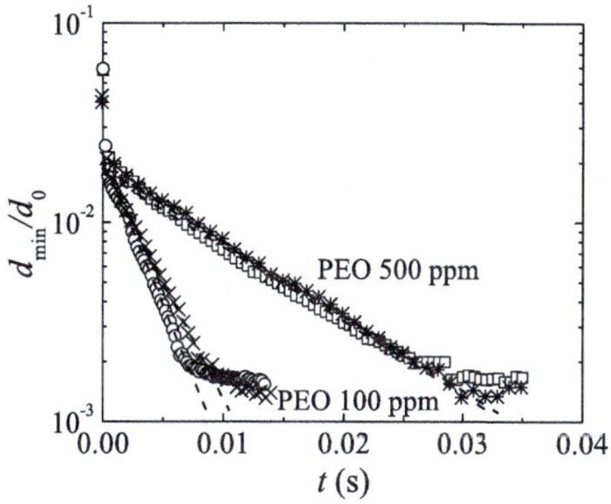

Fig. 4.5 Filament minimum diameter d_{min}/d_0 as a function of time for PEO aqueous solutions surrounded by air (○, □) and by 0.65. cSt silicone oil (×, ∗) [34]. The dashed lines indicate the fits to obtain the relaxation times. The filament is formed during the breakup of a liquid bridge. d_0 is the diameter of the liquid bridge supporting rods

two cylindrical plates to form a liquid bridge. Then, a rapid displacement of the upper plate is imposed to make the liquid bridge unstable. The liquid bridge breaks up under the balance of capillary, viscous, elastic, inertial, and gravitational forces [38]. This breakup process gives rise to the elasto-capillary regime that allows the measurement of the extensional relaxation time λ_e.

Most extensional rheometers operate appropriately only for sufficiently high viscosities ($\mu \gtrsim 20$ mPa s) and large enough values of λ_e ($\lambda_e \gtrsim 1$ ms) [39, 40]. This constitutes an important obstacle for our purposes because interesting phenomena arise in the tip streaming of low-viscosity ($\mu \sim 1$ mPa s) quasi-Newtonian ($\lambda_e \sim 10$ μs) liquids [30]. There have been notable efforts to reduce the lower operational limits of the solution viscosity and the extensional relaxation time. Some salient examples

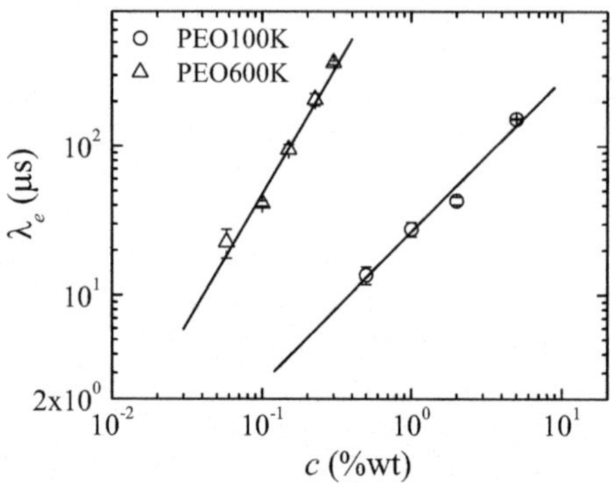

Fig. 4.6 Extensional relaxation time λ_e of PEO100K and PEO600K dissolved in water at different concentrations. The solid lines are the fittings $\lambda_e = 26.18\,c^{1.020}$ and $\lambda_e = 2707\,c^{1.733}$ to the experimental results for PEO100K and PEO600K, respectively (with λ_e and c measured in µs and %wt, respectively) (reprinted with permission from Ref. [30]. Copyright (2022) by the American Physical Society)

are the Rayleigh-Ohnesorge jet elongational rheometer (ROJER) [41], the optically-detected elastocapillary self-thinning dripping-onto-substrate (ODES-DoS) method [42], the dripping-onto-substrate version of CaBER rheometry (CaBER-DoS) [43], the slow retraction method (SRM) [34, 44], and the fast capillary thinning extensional rheometer "Cambridge Trimaster" (CTM) [45]. Rubio et al. [46] used a modified and miniaturized version of the slow retraction method to measure relaxation times of the order of tens of microseconds (Fig. 4.6).

4.1.4 Electric Fields

Electrohydrodynamic tip streaming is practically always analyzed in terms of either the charge-conservative model or the leaky-dielectric approximation (see Chap. 2). In both cases, the relevant electrical properties are the conductivity K and the relative permittivity ε. The conductivity of the leaky-dielectric liquids frequently used in tip streaming experiments originates from impurities present in a dielectric fluid. This implies that this quantity may be very sensitive to the purity of the manufactured liquid and the manipulation before conducting the experiment. For this reason, it is highly recommendable to measure the conductivity even though its value has been reported in the literature.

The dependency of the electrical conductivity upon the liquid temperature can be measured with the following procedure [47]. A cylindrical borosilicate capillary is submerged in a dielectric liquid bath kept at a fixed temperature controlled with a thermocouple. The capillary is filled with the tested liquid. One waits until the thermal equilibrium between that liquid and the surrounding bath is established. Then, a voltage difference is applied between the ends of the capillary, and the

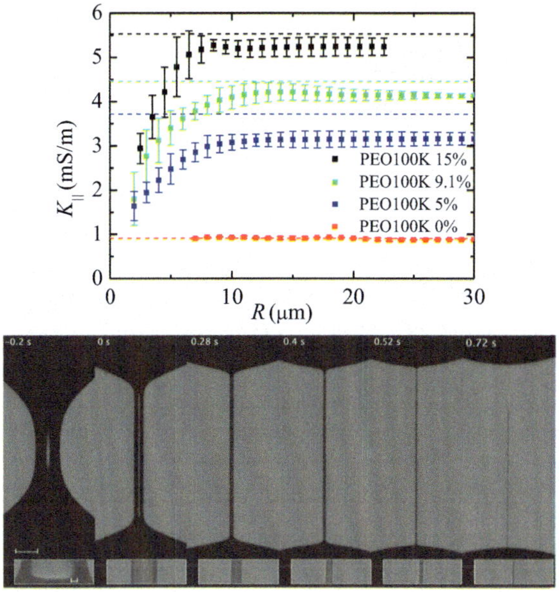

Fig. 4.7 (Upper graph) Electrical conductivity in the polymer stretching direction, $K_{\parallel}$, as a function of the filament radius R. The symbols correspond to results obtained for different liquids, while the dashed lines represent hydrostatic values. The liquids were solutions of PEO in a glycerin-water mixture 50/50% (w/w). (Lower images) Images of the filament during the breakup of a liquid bridge of PEO100K 9.1%. The insets show the images of the center of the filament used to measure the filament radius. The instant $t = 0.52$ s corresponds to the last point of $K_{\parallel}(R)$ [48]

resulting electric current is measured. The conductivity is determined from the slope of the linear relationship between the applied voltage and measured current.

The conductivity measured with the procedure described above may differ from that affecting the electrohydrodynamic tip streaming phenomenon. This phenomenon occurs for electric fields much more intense than those used to measure the conductivity, which may significantly change as the electric field increases. Rubio et al. [48] found a significant decrease in the longitudinal electrical conductivity $K_{\parallel}$ (i.e., that parallel to the flow) when a viscoelastic filament was stretched (Fig. 4.7). This effect was attributed to reduced ion mobility in the polymer stretching direction.

The liquid complex dielectric constant can be measured using a parallel plate capacitor, which consists of two conducting plates of area A separated by a distance d ($d \ll A^{1/2}$). The tested liquid fills the gap between the plates. A voltage difference is established between them. The complex dielectric constant ε^* is obtained from the relation [49, 50]:

$$\varepsilon^*(\omega) = \varepsilon' - i\varepsilon'' = \alpha \left(\frac{C_P}{C_o} - i \frac{1}{\omega C_o R_p} \right), \tag{4.6}$$

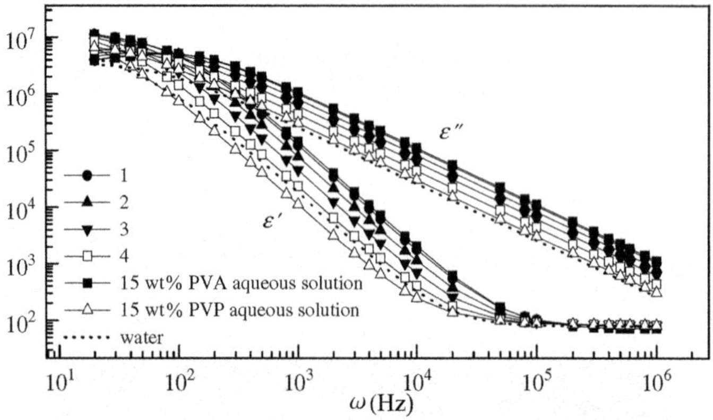

Fig. 4.8 Frequency dependence of the relative permittivity ε' and dielectric loss ε'' of water, aqueous solutions of PVA14000 and PVP40000, and coexisting aqueous solutions of PVA14000+PVP40000 at the volume percentage ratios (1) 80:20, (2) 60:40, (3) 40:60, and (4) 20:80 (reprinted from Ref. [50] with permission of Elsevier)

where ω is the angular frequency of the applied voltage, C_o and C_P are the capacitances of the free space and the tested liquid, respectively, R_p is the equivalent parallel resistance of the cell with the liquid, and α is the cell correction coefficient. The real part ε' is frequently referred to as the relative permittivity (the apostrophe is usually dropped), while ε'' is the dielectric loss.

The low-frequency process in polar solutions is caused by the electrode polarization, which results in large values of the permittivity ε' and dielectric loss ε'' (Fig. 4.8). We are normally interested in the value taken by ε' for sufficiently large frequency. The manufacturer normally provides this value for the liquids commonly used in tip streaming experiments. This quantity is hardly sensitive to impurities; therefore, that value is reliable.

4.2 Flow Visualization

4.2.1 Backlight Optical Imaging

Most of the experiments of tip streaming are analyzed using the backlight optical imaging technique. Due to the smallness of the observed fluid shape (large interface curvature) and the big difference between the indexes of refraction of the liquid and gaseous phases, the fluid shape scatters the light beam, causing the so-called silhouette effect. This makes the fluid shape appear in the image as a nearly dark object on a grey background (Fig. 4.9-left). In liquid-liquid tip streaming realizations, the indexes of refraction take similar values in the inner and outer phases. In this case, a dark contour delimits the fluid shape on a lighter background (Fig. 4.9-right).

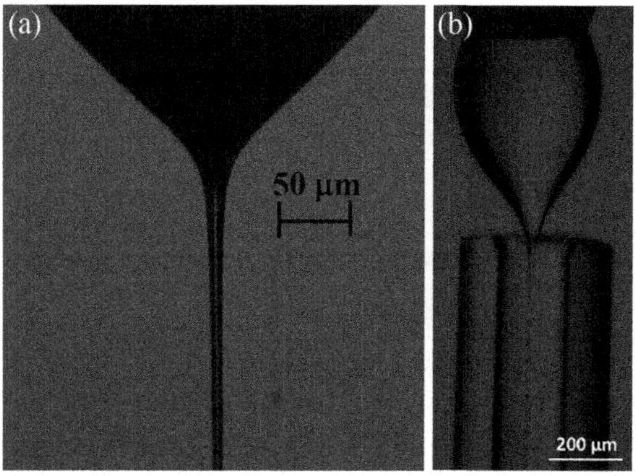

Fig. 4.9 **a** Silhouette effect in electrospray (reprinted with permission from Ref. [51]. Copyright (2012) by Americal Physical Society). **b** Image of a tapering meniscus of distilled water submerged in a bath of 10-cSt silicone oil in the confined selective withdrawal configuration [52]

The interface curvature at the droplet tip is frequently considered in analyzing the onset of tip streaming. The calculation of this quantity requires computing second-order derivatives of the interface location versus the axial or intrinsic coordinates. That computation enormously amplifies the noise in the detected interface contour; therefore, that contour must be very smooth and accurate.

The simplicity of the images described above allows one to apply numerical methods capable of accurately determining the contour position. The image is defined mathematically by the grey intensity matrix $I(i, j)$, which takes an integer value between 0 (black) and 255 (white) for each pixel (i, j). Figure 4.10 shows the grey intensity profiles along the horizontal white line indicated in the image. The liquid free surface must correspond to a step in the grey intensity in the direction perpendicular to those edges. However, the intensity changes over several pixels due to diffraction. The thickness of this transition depends on the quality of both the optical lenses and the focus of the image.

Most of the experimental works in microfluidics use simple contour detection algorithms to determine the interface position at the pixel level. These procedures are insufficient to carry out an accurate quantitative analysis, particularly to calculate the interface curvature. The following two-stage sub-pixel resolution method can be employed to improve the result.

First, one determines the pixels (i_p, j_p) corresponding to the interface contour using any standard method, such as the Canny algorithm [54]. At this stage, the intensity gradient determines the direction ξ normal to the contour element. In the second step, one interpolates the values of the grey intensity matrix around (i_p, j_p) along the ξ axis. The point corresponding to the interface location is calculated at the subpixel level using any local thresholding criterion [55–57]. For instance, the

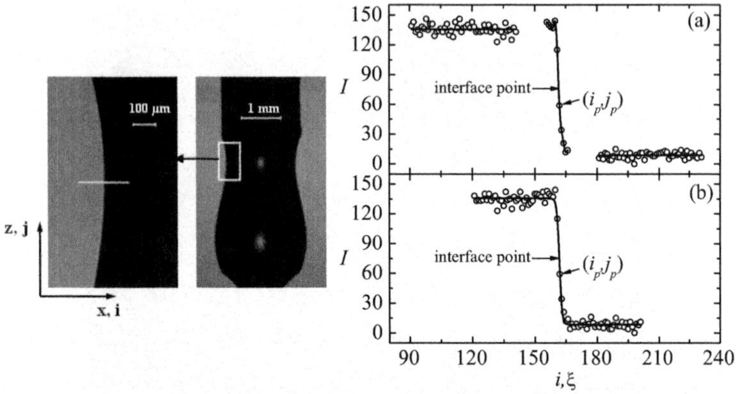

Fig. 4.10 Grey intensity profile along the horizontal white line indicated in the image. Elements in the standard **a** and modified **b** sub-pixel algorithms. Graph (**a**): the horizontal lines are the plateaus at the two sides of the grey profile, while the curve in the transient region corresponds to the spline interpolation of the grey intensity values. Graph (**b**): the solid line is the sigmoid function (4.7) fitted to the grey intensity values (reprinted with permission from Ref. [53])

interface point can be chosen as that for which the grey intensity equals the average value of the two plateaus (Fig. 4.10a).

Song and Springer [58] proposed an improved image processing technique to detect the free surface position at the sub-pixel level, which yields more precise results than the standard sub-pixel resolution method described above. Vega et al. [53] applied the same idea to study the free surface dynamics at high frame rates. In this method, the grey profile is seen as the sum of a sigmoidal function

$$g(\xi) = \frac{g_1 - g_2}{1 + \exp[(\xi - \xi_0)/W]} + g_2 \tag{4.7}$$

plus the noise due to the electronic components of the camera and, to a lesser extent, optical effects. Here, g_1 and g_2 correspond to the plateau grey values at the two sides of the edge (interface), W is a measure of the edge width, and ξ_0 is the mid-point of the profile. At the sub-pixel stage of the method, the function (4.7) is fitted to the grey intensity profile along the ξ axis (Fig. 4.10b) instead of using an interpolation technique (Fig. 4.10a). The edge point is given by ξ_0, for which (4.7) takes the value $(g_1 + g_2)/2$ (Fig. 4.10b).

Standard sub-pixel resolution methods [55–57] only filter the noise at the two sides of the edge to calculate the grey intensity threshold. In contrast, fitting (4.7) to the whole grey intensity profile also filters the noise in the transient region.

Simple smoothing algorithm can be used to reduce the fluctuations of the detected free surface position, such as the Savitzky-Golay filter [59] employed by Anastasiadis et al. [3] and Zuo et al. [5] to measure the surface tension from an image of a pendant drop.

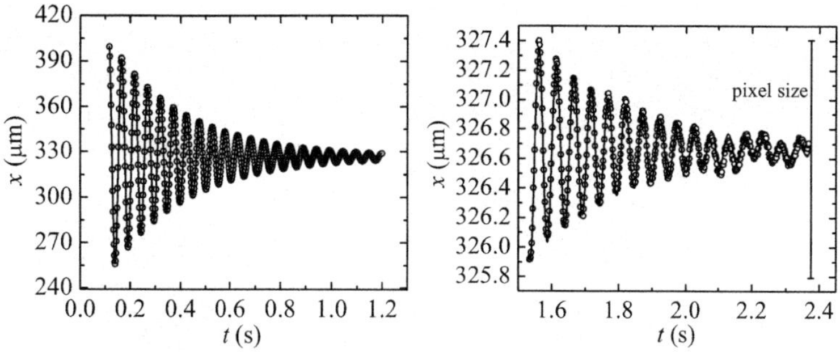

Fig. 4.11 Viscous damping of the lateral oscillation of a liquid bridge. Horizontal position $x(t)$ of a free surface point (symbols) and the fit of the dominant oscillation mode $x(t) = ae^{-\gamma t}\cos[\omega(t - t_0)] + a_0$ to the data (solid line). The mode fits the experimental data remarkably well over two orders of magnitude of $x(t)$ (reprinted with permission from Ref. [53])

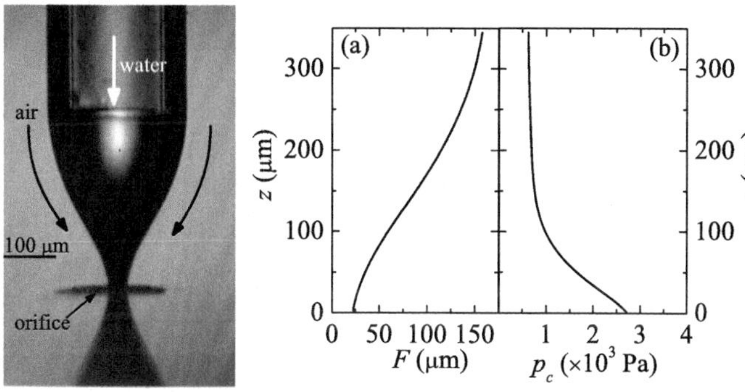

Fig. 4.12 (Left) Image of a water meniscus in gaseous flow focusing. (Right) Free surface distance $F(z)$ to the symmetry axis z (**a**) and capillary pressure $p_c(z)$ (**b**) (reprinted from Ref. [63] with permission of Elsevier)

The improved image processing technique combined with a smoothing algorithm provides accurate values of the free surface curvature even for relatively blurred images [60] and liquid shapes only a few pixels in width [61]. It is especially accurate when determining certain free surface motion features, like the oscillation amplitude and frequency. In this case, the detected point does not need to belong to the free surface but has to remain at a constant distance from that surface over its motion [62].

Figure 4.11 shows the subpixel accuracy of this method, which detects the free surface oscillation with an amplitude much smaller than the pixel size [53]. The improved image processing technique provides, for instance, the capillary pressure p_c over the free surface of the tapering meniscus in gaseous flow focusing (Fig. 4.12) [63]. Combined with the solution of the 1D model, the method allowed López-Herrera et al. [64] to "measure" the electric field over the interface in the cone-jet mode of electrospray.

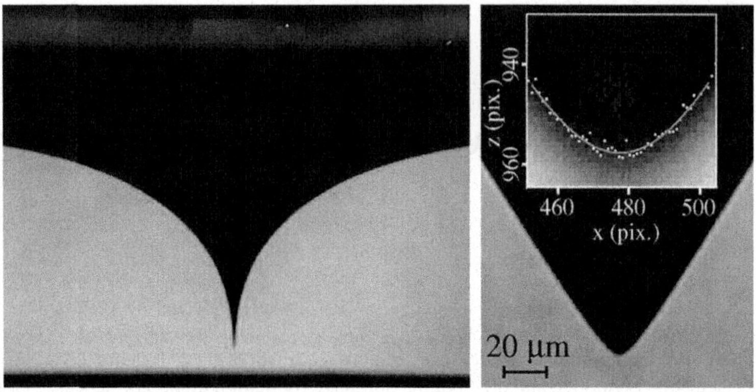

Fig. 4.13 Images of the hump tip in selective withdrawal. The solid white line is the Gaussian function fitted to the pixels corresponding to the interface (reprinted with permission from Ref. [65]. Copyright (2006) by the American Physical Society)

There are alternative methods to analyze the interface contour with pixel resolution. For instance, Courrech du Pont and Eggers [65] and Rubio and Montanero [17] measured the tip curvature in selective withdrawal by detecting the free surface contour with pixel accuracy and fitting a Gaussian function to the interface contour in that region (Fig. 4.13).

4.2.2 Flow Pattern Visualization

The measurement of the velocity field in tip streaming entails additional difficulties not present in other flows. Specifically, the optical distortion caused by the curved interface must be accounted for to provide accurate results. Time-resolved particle image velocimetry can be used to measure the velocity field in tip streaming. This system comprises a zoom lens mounted on a high-speed camera, a high-speed pulsed laser, and a PIV synchronizer. The liquid is seeded using glass microspheres of a diameter of the order of a few microns to visualize the flow field inside the tapering meniscus. Images are acquired at a few kfps. The images are processed using adaptive cross-correlation with an interrogation window with sides a few microns in length. Local velocity filters and interpolation of outliers can be used to obtain the final results. The flow can be easily visualized by taking images with long exposure times and low frame rates (Fig. 4.14-left) [66].

The flow structure in the tapering meniscus of, for instance, gaseous flow focusing can also be visualized by injecting a colored or opaque inner liquid into a transparent outer one. The fluid configuration is not illuminated using a backlight but with a cool white light beam at 45° azimuth from the camera axis (Fig. 4.14-right).

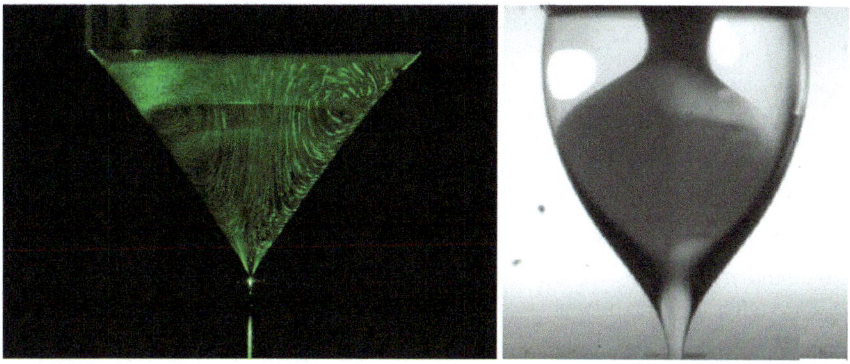

Fig. 4.14 Image of the tapering meniscus in electrospray (left) (reprinted from Ref. [66] with permission of AIP publishing) and gaseous flow focusing (right) (reprinted with permission from Ref. [67])

The spatiotemporal distribution of fluorescent surfactants on interfaces can be measured using high-speed laser-induced fluorescence. This technique has been applied, for instance, to the analysis of the coalescence of drops [68]. It might be applied to the study of tip streaming as well.

4.3 Searching for Scaling Laws

Many microfluidic applications involve complex phenomena for which theoretical analyses based on first principles may not lead to practical results. One of the most common elements of experimental analysis in microfluidics is the search for scaling laws to characterize the performance of a given configuration or device. Scaling laws allow one to unify the description of similar experiments and identify the physical mechanisms governing the associated applications. Here, we describe a simple method for finding the scaling law that best fits a set of experimental or simulation data.

Suppose the analyzed data involves n dimensional variables within the parameter region of interest. The Buckingham π theorem establishes that the sought empirical law can be written in terms of $n - j$ dimensionless groups, where j is the reduction of the problem. In most cases, j equals the number of fundamental dimensions of the problem. Hydrodynamic tip streaming involves the three fundamental dynamical dimensions (length, time, and mass). The charge has to be added in electrohydrodynamic tip streaming.

Consider, for instance, a hydrodynamic tip streaming problem involving $n = 6$ dimensional variables with $j = 3$. Taking into account the above comments, the sought empirical equation can be written as

$$\pi_1 = f(\pi_2, \pi_3), \tag{4.8}$$

where π_1, π_2, and π_3 are three independent dimensionless groups based on the six dimensional variables. The key element of this analysis is to assume that Eq. (4.8) is of the form

$$\pi_1 = C\pi_2^{\alpha}\pi_3^{\beta}, \tag{4.9}$$

where C is a dimensionless constant. This hypothesis is frequently referred to as *self-similarity* [69].

The values of the exponents α and β that yield the best fit (4.9) to the experimental data can be found using the following procedure [70]. We calculate the logarithmic errors

$$\varepsilon^{(k)} = \log\left[\left(\pi_2^{(k)}\right)^{\alpha}\left(\pi_3^{(k)}\right)^{\beta}\right] - \log\pi_1^{(k)} \quad (k = 1, 2, \ldots, K) \tag{4.10}$$

for fixed values of α and β. Here, $\{\pi_1^{(k)}, \pi_2^{(k)}, \pi_3^{(k)}\}$ represent the values of the corresponding dimensionless variables for the kth experimental realization. We assume that the probability density function (PDF) of the set of errors $\{\varepsilon^{(k)}\}$ is a normal distribution. Thus, one fits the function

$$h(x) = \frac{1}{\sqrt{2\pi}}\exp\left[-\frac{(x-\mu_0)^2}{2\sigma_0^2}\right] \tag{4.11}$$

to the PDF of $\{\varepsilon^{(k)}\}$. This process is repeated for different values of α and β: for each pair of values (α, β), we calculate $\{\varepsilon^{(k)}\}$ and fit (4.11) to the corresponding PDF. The optimum values α^* and β^* are those leading to the least standard deviation σ_0. It can be seen that one obtains the best grouping around a straight line when $\{\pi_1^{(k)}\}$ are plotted versus $\{\pi_2^{(k)\alpha^*}\pi_3^{(k)\beta^*}\}$ in a log-log graph.

Here, we consider an example to illustrate the method described above. For a fixed geometry, the transonic flow focusing (see Chap. 11) is governed by the set of dimensional parameters [71]

$$\{D_o;\, \mu_o, c_{vo}, c_{po}, \kappa_o;\, T_0, p_0, p_e;\, \rho_i, \mu_i, c_i, \kappa_i, \sigma;\, Q_i\}, \tag{4.12}$$

where D_o is the discharge orifice diameter, μ_o, c_{vo}, c_{po}, and κ_o are the gas viscosity, specific heat coefficients, and thermal conductivity, respectively, T_0 and p_0 are the upstream stagnation temperature and pressure, respectively, p_e is the exit pressure, ρ_i, μ_i, c_i, and κ_i are the liquid density, viscosity, specific heat coefficient, and thermal conductivity, respectively, σ is the surface tension, and Q_i the liquid flow rate.

The dimensionless jet diameter $d_j^* = d_j/D_o$ is a function of $Q = Q_i/Q_D$ ($Q_D = D_o\mu_i/\rho_i$), $p_r = p_0/p_e$, and eight additional nondimensional parameters. In the numerical simulations of Rubio et al. [71], these additional parameters were fixed,

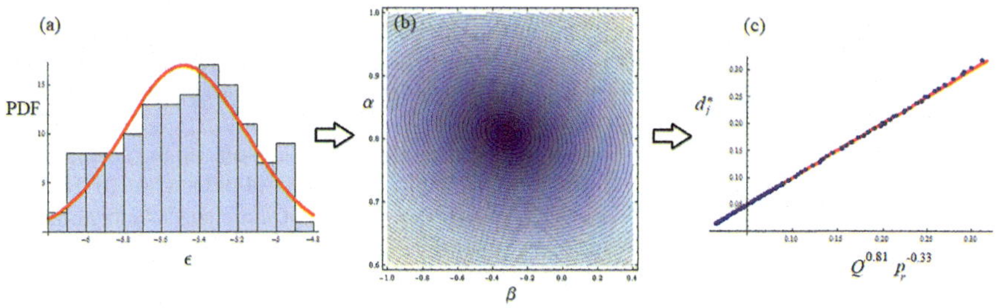

Fig. 4.15 Steps in the method used to determine the scaling law: **a** fitting of the normal distribution (red line) to the PDF of the logarithmic errors, **b** standard deviation of the normal distribution as a function of α and β, and **c** the resulting scaling law

and the values of Q and p_r were changed. We look for a scaling law of the form $d_j^* = C Q^\alpha p_r^\beta$. Figure 4.15 shows the steps to determine the optimum values of α and β. We fit the normal distribution to the PDF of the logarithmic errors calculated for given values of α and β (Fig. 4.15a). The optimum values of these exponents are those leading to the minimum standard deviation of the PDF (Fig. 4.15b). Figure 4.15c shows how the data are grouped around the corresponding law.

References

1. Harkins WD, Jordan HF (1930) Surface tension by the ring method. Science 72:73–75
2. Rotenberg Y, Boruvka L, Neumann AW (1983) Determination of surface tension and contact angle from the shapes of axisymmetric fluid interfaces. J Colloid Interface Sci 93.169–183
3. Anastasiadis SH, Chen JK, Koberstein JT, Siegel AF, Sohn JE, Emerson JA (1987) The determination of interfacial tension by video image processing of pendant fluid drops. J Colloid Interface Sci 119:55–66
4. Hoorfar M, Neumann AW (2004) Axisymmetric drop shape analysis (ADSA) for the determination of surface tension and contact angle. J Adhes 80:727–743
5. Zuo YY, Ding M, Bateni A, Hoorfar H, Neumann AW (2004) Improvement of interfacial tension measurement using a captive bubble in conjunction with axisymmetric drop shape analysis (ADSA). Colloids Surf 62:233–246
6. Cabezas MG, Bateni A, Montanero JM, Neumann AW (2006) Determination of surface tension and contact angle from the shapes of axisymmetric fluid interfaces without use of apex coordinates. Langmuir 22:10,053–10,060
7. Cabezas MG, Bateni A, Montanero JM, Neumann AW (2005) A new method of image processing in the analysis of axisymmetric drop shapes. Colloids Surf A 255:193–200
8. Passerone A, Liggieri L, Rando N, Ravera F, Ricci E (1991) A new experimental method for the measurement of the interfacial tension between immiscible fluids at zero bond number. J Colloid Interface Sci 146:152–162
9. Peters F, Arabali D (2013) Interfacial tension between oil and water measured with a modified contour method. Colloids Surf A Physicochem Eng Asp 426:1–5
10. Neeson MJ, Chan DYC, Tabor RF (2014) Compound pendant drop tensiometry for interfacial tension measurement at zero Bond number. Lamgmuir 30:15,388–15,391
11. Bateni A, Susnar SS, Amirfazli A, Neumann AW (2004) Development of a new methodology to study drop shape and surface tension in electric fields. Langmuir 20:7589–7597

12. Muñoz-Sánchez BN, Cabezas MG, Ferrera C, Herrada MA, Montanero J (2020) A method for measuring the interfacial tension for density-matched liquids. J Colloid Interface Sci 566:90–97
13. Henderson DM, Miles JW (1994) Surface-wave damping in a circular cylinder with a fixed contact line. J Fluid Mech 275:285–299
14. Lunkenheimer K, Czichocki G, Hirte R, Barzyk W (1995) Novel results on the adsorption of ionic surfactants at the air/water interface-sodium-n-alkyl sulfates. Colloid Surf A 101:187–197
15. Prosser AJ, Franses EI (2001) Adsorption and surface tension of ionic surfactants at the air-water interface: review and evaluation of equilibrium models. Colloids Surf A 178:1–40
16. Ponce-Torres A, Vega EJ, Montanero JM (2016) Effects of surface-active impurities on the liquid bridge dynamics. Exp Fluids 57:67
17. Rubio M, Montanero JM (2023) Influence of a soluble surfactant on the transition to tip streaming. Exp Therm Fluid Sci 141(110):776
18. Tajima K, Muramatsu M, Sasaki T (1970) Radiotracer studies on adsorption of surface active substance at aqueous surface. I. Accurate measurement of adsorption of Tritiated Sodium Dodecylsulfate. Bull Chem Soc Japan 43:1991–1998
19. Hines JD, Thomas RK, Garrett PR, Rennie GK, Penfold J (1997) Investigation of mixing inbinary surfactant solutions by surface tension and neutron reflection: anionic/nonionic and zwitterionic/nonionic mixtures. J Phys Chem B 101:9215–9223
20. Martínez-Balbuena L, Arteaga-Jiménez A, Hernández-Zapata E, Márquez-Beltrán C (2017) Applicability of the Gibbs adsorption isotherm to the analysis of experimental surface-tension data for ionic and nonionic surfactants. Adv Colloid Interface Sci 247:178–184
21. Qazi MJ, Schlegel SJ, Backus EHG, Bonn M, Bonn D, Shahidzadeh N (2020) Dynamic surface tension of surfactants in the presence of high salt concentrations. Langmuir 36:7956–7964
22. Eastoe J, Dalton JS (2000) Dynamic surface tension and adsorption mechanisms of surfactants at the air-water interface. Adv Colloid Interface Sci 85:103–144
23. Lin SY, McKeigue K, Maldarelli C (1990) Diffusion-controlled surfactant adsorption studied by pendant drop digitization. AIChE J 36:1785–1795
24. Joos P, Rillaerts E (1981) Theory on the determination of the dynamic surface tension with the drop volume and maximum bubble pressure methods. J Coll Interf Sci 79:96–100
25. Tricot YM (1997) Surfactants: static and dynamic surface tension, liquid film coating, vol 1. Chapman and Hall, pp 100–136
26. Herrada MA, Ponce-Torres A, Kaneelil PR, Pahlavan AA, Stone HA, Montanero JM (2022) Effect of a soluble surfactant on the linear stability of two-phase flows in a finite-length channel. Phys Rev Fluids 7(114):003
27. Herrada MA, Ponce-Torres A, Rubio M, Eggers J, Montanero JM (2022b) Stability and tip streaming of a surfactant-loaded drop in an extensional flow influence of surface viscosity. J Fluid Mech 934:A26
28. Zell ZA, Nowbahar A, Mansard V, Leal LG, Deshmukh SS, Mecca JM, Tucker CJ, Squires TM (2014) Surface shear inviscidity of soluble surfactants. Proc Natl Acad Sci 111:3677–3682
29. Ponce-Torres A, Rubio M, Herrada MA, Eggers J, Montanero JM (2020) Influence of the surface viscous stress on the pinch-off of free surfaces loaded with nearly-inviscid surfactants. Sci Rep 10(16):065
30. Rubio A, Galindo F, Vega EJ, Montanero JM, Cabezas MG (2022) Viscoelastic transition in transonic flow focusing. Phys Rev Fluids 7(074):201
31. Goldin M, Yerushalmi J, Pfeffer R, Shinnar R (1969) Breakup of a laminar capillary jet of a viscoelastic fluid. J Fluid Mech 38:689–711
32. Bazilevsky AV, Entov VM, Rozhkov AN (1990) Liquid filament microrheometer and some of its applications. In: Oliver DR (ed) Proocedings of third European rheology conference and golden jubilee meeting of the British society of rheology, vol 1, pp 41–43
33. Entov VM, Hinch EJ (1997) Effect of a spectrum of relaxation times on the capillary thinning of a filament of elastic liquid. J Non-Newtonian Fluid Mech 72:31–53
34. Sousa PC, Vega EJ, Sousa RG, Montanero JM, Alves MA (2017) Measurement of relaxation times in extensional flow of weakly viscoelastic polymer solutions. Rheol Acta 56:11–20

35. Anna SL, McKinley GH (2001) Elasto-capillary thinning and breakup of model elastic liquids. J Rheol 45:115–138
36. McKinley GH, Tripathi A (2000) How to extract the Newtonian viscosity from capillary breakup measurements in a filament rheometer. J Rheol 44:653–670
37. Bazilevsky AV, Voronkov SI, Entov VM, Rozhkov AN (1981) Orientational effects in capillary breakup of jets and threads of dilute polymer solutions (English version in, vol 26). Doklady Akademii Nauk SSSR 257:336–339
38. Schultz WW, Davis SH (1982) One-dimensional liquid fibers. J Rheol 26:331–345
39. Rodd LE, Scott TP, Cooper-White JJ, Mckinley GH (2005) Capillary break-up rheometry of low-viscosity elastic fluids. Appl Rheol 15:12–27
40. Dinic J, Biagioli M, Sharma V (2017) Pinch-off dynamics and extensional relaxation times of intrinsically semi-dilute polymer solutions characterized by dripping-onto-substrate rheometry. J Polym Sci B 55:1692–1704
41. Sharma V, Haward SJ, Serdy J, Keshavarz B, Soderlund A, Threlfall-Holmes P, McKinley GH (2015) The rheology of aqueous solutions of ethyl hydroxy-ethyl cellulose (EHEC) and its hydrophobically modified analogue (hmEHEC): extensional flow response in capillary break-up, jetting (ROJER) and in a cross-slot extensional rheometer. Soft Matter 11:3251–3270
42. Dinic J, Zhang Y, Jimenez LN, Sharma V (2015) Extensional relaxation times of dilute, aqueous polymer solutions. ACS Macro Lett 4:804–808
43. Sur S, Rothstein J (2018) Drop breakup dynamics of dilute polymer solutions: effect of molecular weight, concentration, and viscosity. J Rheol 62:1245–1259
44. Campo-Deaño L, Clasen C (2010) The slow retraction method (SRM) for the determination of ultra-short relaxation times in capillary breakup extensional rheometry experiments. J Non-Newtonian Fluid Mech 165:1688–1699
45. Vadillo DC, Mathues W, Clasen C (2012) Microsecond relaxation processes in shear and extensional flows of weakly elastic polymer solutions. Rheol Acta 51:755–769
46. Rubio M, Ponce-Torres A, Vega EJ, Montanero JM (2020) Experimental analysis of the extensional flow of very weakly viscoelastic polymer solutions. Materials 13:192
47. Rubio M, Vega EJ, Herrada MA, Montanero JM, Galindo-Rosales FJ (2020) Breakup of an electrified viscoelastic liquid bridge. Phys Rev E 101(033):103
48. Rubio M, Sadek S, Vega EJ, Gañán-Calvo AM, Montanero JM (2021) Electrical conductivity of a stretching viscoelastic filament. Materials 14:1–12
49. Agilent Technologies Ltd (2000) Agilent 16452A liquid test fixture-operation and service manual Japan
50. Sengwa RJ, Sankhla S (2007) Dielectric dispersion study of coexisting phases of aqueous polymeric solution: poly(vinyl alcohol) + poly(vinyl pyrrolidone) two-phase systems. Polymer 48:2737–2744
51. Herrada MA, López-Herrera JM, Gañán-Calvo AM, Vega EJ, Montanero JM, Popinet S, (2012) Numerical simulation of electrospray in the cone-jet mode. Phys Rev E 86(026):305
52. López M, Cabezas MG, Montanero JM, Herrada MA (2022) On the hydrodynamic focusing for producing microemulsions via tip streaming. J Fluid Mech 934:A47
53. Vega EJ, Montanero JM, Fernández J (2009) On the precision of optical imaging to study free surface dynamics at high frame rates. Exp Fluids 47:251–261
54. Canny J (1986) A computational approach to edge-detection. IEEE Trans Pattern Anal Mach Intell 8:679–698
55. Hoorfar M, Neumann AW (2006) Recent progress in axisymmetric drop shape analysis (ADSA). Adv Colloid Interface Sci 121:25–49
56. Ferrera C, Cabezas MG, Montanero JM (2006) An experimental analysis of the linear vibration of axisymmetric liquid bridges. Phys Fluids 18(082):105
57. Lowry BJ (1996) Pressure and stress measurement via image analysis (P-SIA) of axisymmetric drops and liquid bridges. J Colloid Interface Sci 176:284–297
58. Song B, Springer J (1996) Determination of interfacial tension from the profile of a pendant drop using computer-aided image processing: 2 experimental. J Colloid Interface Sci 184:77–91

59. Savitzky A, Golay MJE (1964) Smoothing and differentiation of data by simplified least squares procedures. Anal Chem 36:1627–1639
60. Vega EJ, Montanero JM, Ferrera C (2011) Exploring the precision of backlight optical imaging in microfluidics close to the diffraction limit. Measurement 44:1300–1311
61. Vega EJ, Montanero JM, Herrada MA, Ferrera C (2014) Dynamics of an axisymmetric liquid bridge close to the minimum-volume stability limit. Phys Rev E 90(013):015
62. Montanero JM, Ponce-Torres A (2020) Review on the dynamics of isothermal liquid bridges. Appl Mech Rev 72(010):803
63. Montanero JM, Vega EJ, Ferrera C (2009) Sub-micrometer precision of optical imaging to locate the free surface of a micrometer fluid shape. J Colloid Interface Sci 339:271–274
64. López-Herrera JM, Herrada MA, Montanero JM, Rebollo-Muñoz N, Gañán-Calvo AM (2013) On the validity and applicability of the one-dimensional approximation in cone-jet electrospray. J Aerosol Sci 61:60–69
65. Courrech du Pont S, Eggers J (2006) Sink flow deforms the interface between a viscous liquid and air into a tip singularity. Phys Rev Lett 96(034):501
66. Gupta A, Mishra BK, Panigrahi PK (2021) Internal and external hydrodynamics of Taylor cone under constant and alternating voltage actuation. Phys Fluids 33(117):118
67. Gañán-Calvo AM, Ferrera C, Torregrosa M, Herrada MA, Marchand M (2011) Experimental and numerical study of the recirculation flow inside a liquid meniscus focused by air. Microfluid Nanofluid 11:65–74
68. Dong T, Weheliye W, Angeli P (2019) Laser induced fluorescence studies on the distribution of surfactants during drop/interface coalescence. Phys Fluids 31(012):106
69. Barenblatt GI (2003) Scaling. Cambridge University Press, Cambridge
70. Montanero JM, Gañán-Calvo AM (2020) Dripping, jetting and tip streaming. Rep Prog Phys 83(097):001
71. Rubio M, Rubio A, Cabezas MG, Herrada MA, Gañán-Calvo AM, Montanero JM (2021) Transonic flow focusing: stability analysis and jet diameter. Int J Multiphase Flow 142(103):720

Chapter 5
Electrohydrodynamic Transient Tip Streaming

Abstract Tip streaming can be classified into two fundamentally different categories: transient tip streaming and periodic microdripping/steady microjetting. We devote this chapter and Chap. 6 to analyzing the first category.

Electrohydrodynamic and hydrodynamic forces can power transient tip streaming. This chapter presents some results for transient tip streaming driven by electrohydrodynamic forces. Special attention will be paid to a low-conductivity droplet subject to a strong electric field. We will consider the mechanism leading to the onset of tip streaming, examining the role of charge relaxation. The scaling laws for the size and charge of the first-emitted droplet are reviewed.

Keywords Electrohydrodynamic tip streaming · Charge relaxation · Electrospray

5.1 Transient Tip Streaming

Submillimeter droplets can be formed against the action of the surface tension and viscosity forces by different procedures. In tip streaming, the dispersed phase accelerates in the tip of a mother drop or film powered by the hydrostatic pressure and shear stress to the extent of beating the intense resistance offered by the capillary and viscous forces. A new interface is violently created by the fast extensional flow produced in the drop tip. The interface forms a fluid thread, which eventually pinches off, giving rise to tiny drops.

In the *transient* tip streaming realization described above, there is no control over the flow rate at which the dispersed phase is ejected. On the contrary, the system autonomously selects the time-dependent flow rate characterizing the ejection, which depends essentially on the magnitude of the applied external force and the fluid properties. The disintegration of the electrified pendant drop shown in Fig. 5.1 constitutes an excellent example of this. Under certain conditions, the system temporally adopts a state similar to the steady cone-jet mode of electrospray [1] (see chap. 8).

J. M. Montanero, *Tip Streaming of Simple and Complex Fluids*, Fluid Mechanics and Its Applications 137, https://doi.org/10.1007/978-3-031-52768-5_5

However, there are substantial differences between this transient state and the steady mode. If the dispersed phase is injected into the pendant drop at the appropriate flow rate, then the system eventually adopts the cone-jet mode of electrospray.

5.2 Electrohydrodynamic Transient Tip Streaming in Different Configurations

In his pioneering work, Rayleigh [3] predicted that a charged liquid droplet becomes unstable due to the growth of the quadrupole oscillation mode when the disruptive Coulomb force equals the attractive surface tension force (the so-called Rayleigh limit). He also claimed that higher multipole oscillations could cause the ejection of very fine jets from the poles of the electrified droplet. This probably constitutes the first example of tip streaming. The predictions of Rayleigh [3] were subsequently verified experimentally. The first verification was conducted by Taylor [4], who showed the ejection of very thin liquid jets from the conical tips of electrified films (Fig. 5.2). Duft et al. [5] observed the emission of Rayleigh jets from the tip streaming arising in the poles of levitated droplets (Fig. 5.2).

The transient ejection in electrohydrodynamic tip streaming has also been considered by Fontelos et al. [6], who studied theoretically the unsteady elementary disintegration of charged and neutral conducting drops under applied electric fields. The onset of tip streaming in low-conductivity drops after a step-change in the electric field magnitude has been examined both experimentally [7–10] and theoretically [10–15] in several works. In the latter case, the analysis has been conducted both assuming perfect volumetric charge relaxation (the leaky-dielectric model) [10–12, 16] and considering certain charge relaxation phenomena along the process [13–15, 17].

As can be observed in Figs. 5.1 and 5.2, tip streaming has been produced not only in levitated droplets [18–20] but also in films and pendant droplets in contact with an electrode. Electrohydrodynamic tip streaming from an initially spherical droplet can be triggered by an electric field producing either oblate or prolate deformations

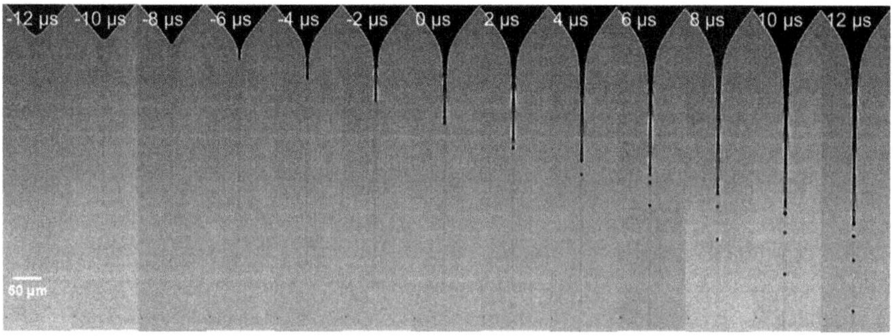

Fig. 5.1 Experimental images of the disintegration of a pendant octanol droplet suspended between two electrodes. The labels indicate the time to the first droplet formation (reprinted from Ref. [2] with permission of AIP Publishing)

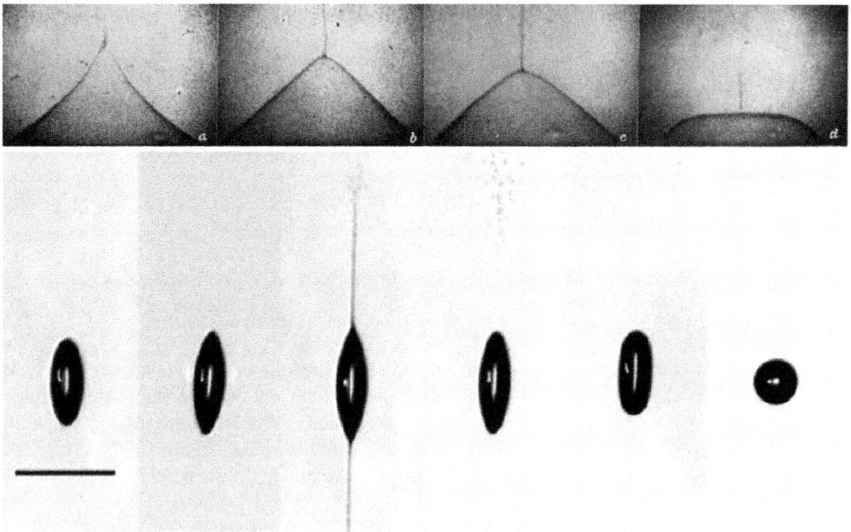

Fig. 5.2 (Upper images) Jet formation and subsequent collapse of a soap film (reprinted with permission from Ref. [4]). (Lower images) Disintegration of a levitated droplet of ethylene glycol charged to the Rayleigh limit (scale bar, 100 μm) (reprinted with permission from Ref. [5])

on the droplet in the direction of the applied field. Brosseau and Vlahovska [21] demonstrated that an oblate deformation could eventually lead to tip streaming from the equatorial rim generated by the sustained nonlinear deformation of the drop (Fig. 5.3). This phenomenon leads to the formation of a thin sheet that gives rise to thin toroidal rings by a 2D capillary breakup. These rings eventually break up in the azimuthal direction [21]. This process generates beautiful arrangements of monodisperse droplets in the equatorial plane of the parent drop. The 2D breakup can produce main and satellite rings, leading to a finite variety of droplet sizes.

The disparity between the system size and the scale of the tip streaming flow allows us to neglect the secondary role played by the boundary conditions. Therefore, all the experiments show essentially the same local electrohydrodynamic phenomenon. Thus, one can compare results obtained for different droplet volumes and fluid configurations. Figure 5.4 shows how the shapes of threads ejected by droplets of different volumes collapse when they are properly scaled.

As explained above, it is essential to distinguish the electrohydrodynamic tip streaming phenomenon described here and the classical steady cone-jet mode of electrospray [1]. In the first case, the drop ejects a thin liquid thread, which issues a train of tiny droplets. The liquid properties determine the masses and charges of those droplets, which reflect the electrohydrodynamic history leading to them. When this first phase of the ejection process ends, the system spontaneously reaches a quasi-steady regime characterized by a natural (intrinsic) flow rate Q^*, electric current, and droplet diameter, which essentially depend upon the liquid properties too [22]. This quasi-steady process gives rise to the steady cone-jet mode of electrospray if a

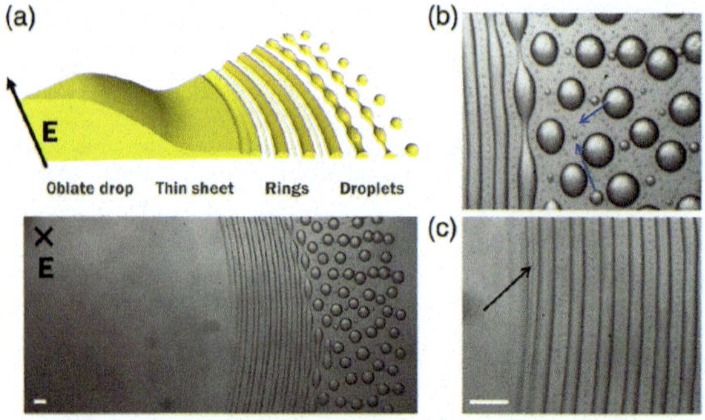

Fig. 5.3 Electrohydrodynamic equatorial tip streaming. **a** Three-dimensional rendering of the phenomenon. The thin sheet is the slightly brighter region right in front of the edge where the ring is forming. **b** The ring breakup produced two generations of satellite droplets (blue arrows). **c** Satellite ring that breaks up in droplets (black arrow). The scale bars in (**a**) and (**c**) are 100 μm (reprinted with permission from Ref. [21]. Copyright (2017) by the American Physical Society)

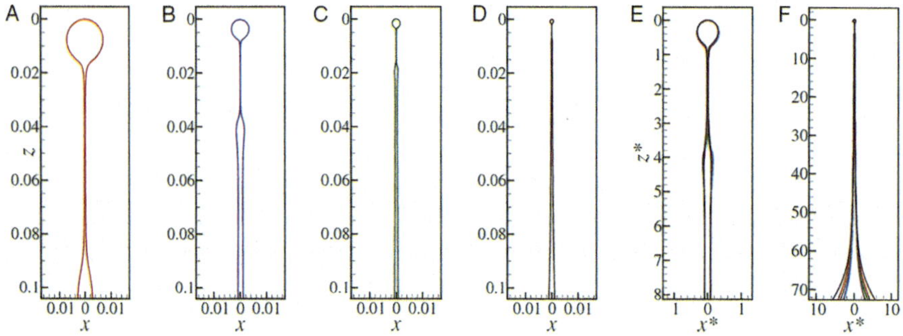

Fig. 5.4 Numerical simulations of the leaky-dielectric model for a droplet subject to a uniform electric field. Graphs (**a**)–(**d**) correspond to different droplet volumes, while (**e**) and (**f**) show the shapes scaled as explained by Collins et al. [12] (reprinted with permission from Ref. [12])

flow rate $Q_i \geq Q^*$ is prescribed, for instance, by injecting liquid into the tapering meniscus hanging on a feeding capillary. In this case, the injected flow rate Q_i becomes a control parameter. The scaling laws derived for the steady cone-jet mode of electrospray [1] cannot be applied to the transient electrohydrodynamic tip streaming phenomenon described here, a common mistake in the literature.

5.3 The Onset of Electrohydrodynamic Tip Streaming

The simplest way of investigating the critical conditions for the onset of electrohydrodynamic tip streaming is to increase progressively the volume $\mathcal{V}$ of a sessile (pendant) droplet sitting on (hanging from) an electrode while keeping constant the

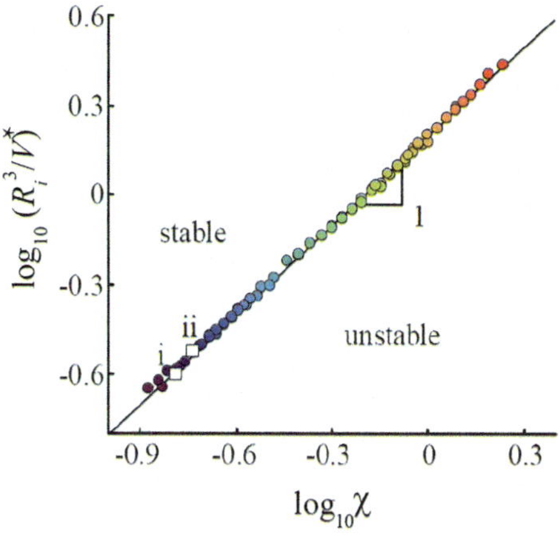

Fig. 5.5 Critical value of the dimensionless drop volume $\mathcal{V}^*/R_i^3$ as a function of the electrical Bond number χ. The symbols and the line correspond to the experimental values and the prediction (5.1), respectively. The squares are the results of the experiments conducted by Wilson and Taylor [24] (i) and Basaran and Scriven [25] (ii) (reprinted with permission from Ref. [23]. Copyright (2019) by the American Physical Society)

voltage V applied between the two electrodes. During the liquid injection, the electric relaxation time is much smaller than any characteristic hydrodynamic time, allowing the net free charge to screen the electric field perfectly. The droplet progressively stretches so that the capillary pressure balances the electrostatic pressure at any time. In this way, both the velocity and the inner electric fields vanish in the droplet before the instability arises. The critical volume $\mathcal{V}^*$ for the onset of tip streaming is not affected by the liquid properties except for the surface tension.

If the liquid volume is small enough, gravity plays a minor role. For a sufficiently large distance H between the two electrodes, $\mathcal{V}^*$ depends only on the characteristic electric field $E_c = V/H$, the surface tension σ, the triple contact line radius R_i, and the outer dielectric medium permittivity ε_o. The Buckingham theorem of dimensional analysis dictates that the dimensionless critical volume $\mathcal{V}^*/R_i^3$ must be a function of the electric Bond number $\chi = \varepsilon_o E_c^2 R_i/\sigma$. Beroz et al. [23] have determined that this function is (Fig. 5.5)

$$\frac{R_i^3}{\mathcal{V}^*} = \frac{\pi\,\chi}{2}. \tag{5.1}$$

The key insight of Beroz et al. [23] for deriving this simple formula was to reformulate the problem in terms of a droplet volume rather than its height, which is an output variable of the problem.

5.4 Global Stability Analysis

In the global linear stability analysis (see Chap. 3), one assumes the temporal dependence $\Phi = \Phi_0 + \phi\, e^{-i\omega t} + \text{c.c.}$, where Φ stands for any flow quantity, Φ_0 is its value in the base flow, ϕ is the spatial dependence of the eigenmode, and $\omega = \omega_r + i\omega_i$ is the eigenfrequency. The base flow becomes unstable when the growth rate ω_i of the dominant eigenmode (i.e., that with the largest ω_i) becomes positive.

Rubio et al. [26] calculated the eigenmodes for the problem described above by integrating the leaky-dielectric model. For a critical volume $\mathcal{V}^*/R_i^3$, a non-oscillatory instability ($\omega_r = \omega_i = 0$) arises. As expected, the stability limit perfectly agrees with (5.1) for arbitrary values of the applied voltage, the distance between electrodes, and the liquid properties.

The interface deformation caused by the critical eigenmode (Fig. 5.6) is practically independent of the liquid conductivity. This deformation is not localized at the droplet tip, which implies that the small local scale characterizing tip streaming is set during the nonlinear droplet deformation. Similar behavior has been observed in a surfactant-loaded drop in an extensional flow [27]. The tip curvature sharply increases at the onset of the tip streaming, suggesting that nonlinear effects become important at the early stage of the liquid ejection process.

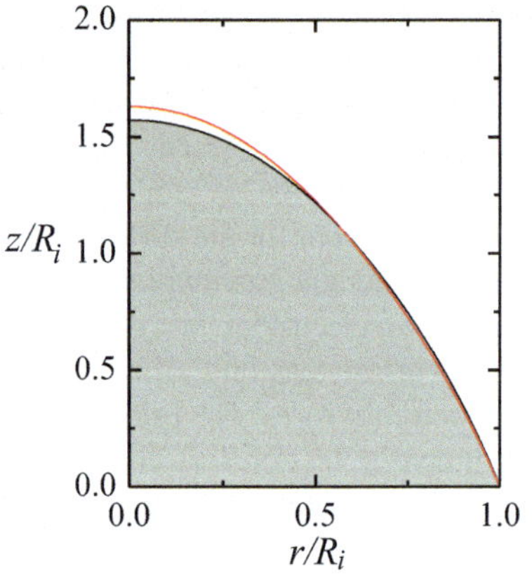

Fig. 5.6 Droplet shape in the base flow (shaded area) and interface displacement due to the critical eigenmode (red line) for glycerine at the marginal stability. The results for glycerine and glycerine+LiCl 0.01M overlap despite their considerably different conductivities. The magnitude of the perturbation has been chosen arbitrarily to appreciate the interface deformation (reprinted from Ref. [26] with permission of AIP Publishing)

5.5 The Role of the Charge Relaxation

In the experiments conducted by Beroz et al. [23] and Rubio et al. [26] (see Fig. 5.7), the apex of the sessile droplet moved very slowly towards the counter-electrode as the liquid was injected into the droplet [phase (I)]. Then, the droplet volume reached the stability limit, and the apex stretched more rapidly until reaching a quasi-conical shape [phase (II)]. This stretching led to tip streaming, in which a very thin liquid thread was ejected [phase (III)].

The droplet behavior following the instability drastically differed in the glycerine case, characterized by a much lower conductivity [26]. For the appropriate range of the applied voltage, the droplet apex oscillated once before ejecting the liquid thread (Fig. 5.8). As in the previous case, the apex moved slowly towards the counter-electrode as the liquid was injected [phase (I)]. Then, the apex stretched more rapidly until the maximum deformation was reached [phase (II)]. In the next stage, the apex contracted [phase (III)] and bounced [phase (IV)]. This rebound finally led to jet emission [phase (V)].

The glycerine droplet oscillation described above can be explained as follows. During the droplet quasi-static growth, the surface charge perfectly screens the inner electric field, and the liquid velocity vanishes. In this case, the electrostatic pressure

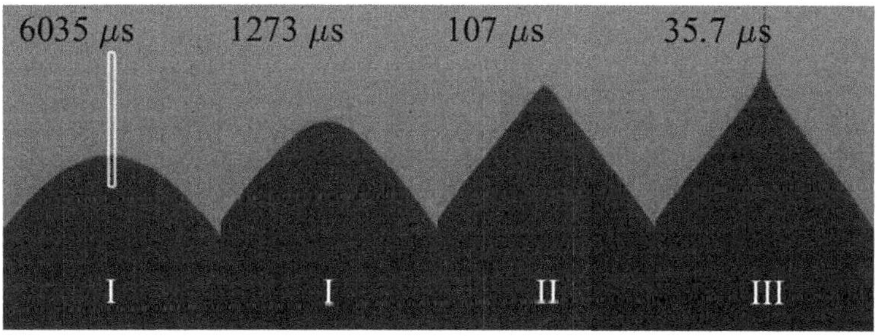

Fig. 5.7 Images of the apex of the glycerine+LiCl 0.01M droplet corresponding to the three phases described in the text. The labels indicate the time to the instant the ejected liquid thread touches the counter-electrode for the first time (reprinted from Ref. [26] with permission of AIP Publishing)

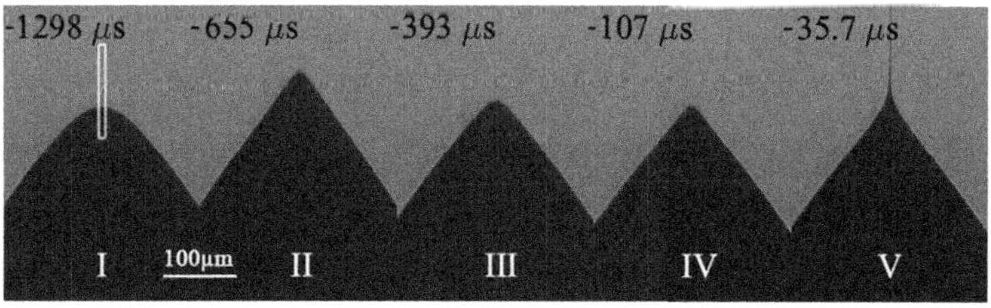

Fig. 5.8 Images of the glycerine droplet apex corresponding to the five phases described in the text [26]. The labels indicate the time to the instant the ejected liquid thread touches the counter-electrode for the first time (reprinted from Ref. [26] with permission of AIP Publishing)

is balanced by the capillary stress. The surface charge density and, therefore, the electrostatic pressure increase in the droplet apex as the volume increases. The curvature and, therefore, the capillary pressure also increase to balance the electrostatic pressure. This balance eventually breaks up, and the liquid moves towards the apex.

The flow at this instability stretches the interface and increases the curvature at the apex. If the charges were supplied at the required rate, this process would follow a self-similar path [14] in which the driver (electrostatic pressure) is balanced by inertia, surface tension, and viscosity. However, the surface charge is not restored at the necessary rate in the low-conductivity glycerine case, and the surface charge density (electrostatic pressure) decreases while the capillary pressure increases. The electrostatic pressure is insufficient to overcome the resistant forces, and the apex retracts. In this phase, the surface charge does not perfectly screen the external electric field. The resulting tangential electric field gives rise to electric shear stress that powers the liquid ejection after the apex rebound. The high liquid viscosity is a favorable factor in this part of the process because it ensures the transfer of momentum from the interface toward the bulk.

The dimensionless conductivity $\alpha^* = K\left[\rho R_i^3/(\sigma\varepsilon^2\varepsilon_o^2)\right]^{1/2}$ (ε is the liquid relative permittivity) between the inertio-capillary time $t_{\rho\sigma} = (\rho R_i^3/\sigma)^{1/2}$ and the electric relaxation time $t_e = \varepsilon\varepsilon_o/K$ for glycerine is considerably smaller than for the rest of the liquids analyzed by Beroz et al. [23] and Rubio et al. [26]. This indicates that charge is not transferred to the interface at the necessary speed in that case, which explains the anomalous behavior of this liquid.

As mentioned above, the interface deformation given by the dominant eigenmode is non-oscillatory and hardly depends on the dimensionless conductivity (Fig. 5.6). This indicates that the oscillation of the glycerine droplet is a nonlinear effect.

The onset of the electrohydrodynamic tip streaming has also been studied by solving the charge-conservative model (see Chap. 2) for a low-conductivity pendant droplet subject to a step change in the electric field magnitude. The numerical results satisfactorily agreed with the experiments (Fig. 2.2) [10]. The simulations showed that the flow at the onset of the ejection is driven by the electric shear stress exerted on the interface, as previously demonstrated by Collins et al. [11] by integrating the leaky-dielectric model (see Chap. 2) for an electrified liquid film.

5.6 Scaling Laws for the First-Emitted Droplet

Studies have focused not only on the onset of tip streaming but also on the characteristics of the first-emitted droplet (Figs. 5.4 and 5.9). As a first approximation, the ejection of the first droplet in electrohydrodynamic tip streaming is supposed to result from a local process taking place in a region much smaller than the parent drop. In this case, one can assume that the geometrical variables do not significantly affect the phenomenon (Fig. 5.4) [12]. The applied electric field magnitude is expected to have little or, at most, moderate influence because it must lie within a relatively narrow range for the ejection to occur.

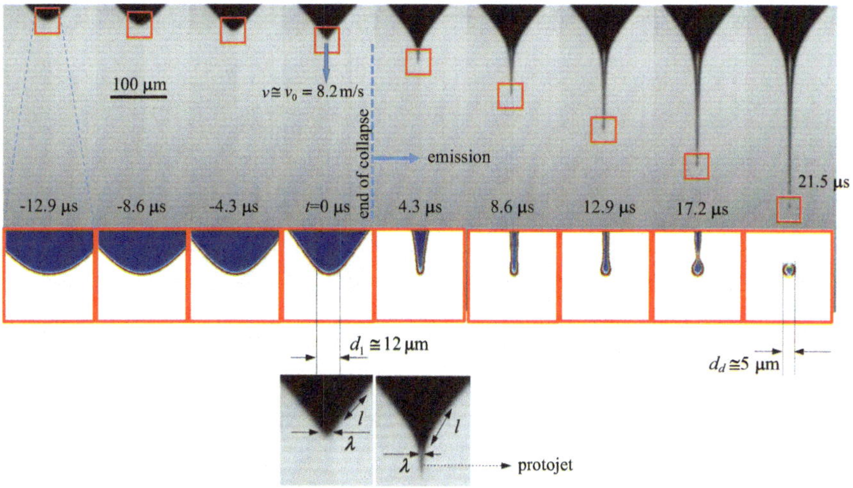

Fig. 5.9 First-droplet ejection from electrohydrodynamic tip streaming in a 1-octanol droplet [14]. The upper images and central graphs correspond to an experiment and numerical simulation of the charge-conservative model, respectively. The colors indicate the dimensionless volume charge density from 0 (blue) to 1 (red). The two lower experimental images show the magnification of the meniscus tip for $t = 0$ and $t = 4.3\,\mu$s. The process is divided into two stages: the droplet tip collapse for $t < 0$ and the jet emission for $t > 0$

Under the above approximations, the parameters that essentially characterize the problem are the liquid properties (density ρ, viscosity μ, surface tension σ, electrical conductivity K, and electrical permittivity ε_i) and the electrical permittivity ε_o of the dielectric environment (vacuum or gas). The intrinsic characteristic length $d_o = (\sigma\varepsilon_o^2/\rho K^2)^{1/3}$, time $t_o = \varepsilon_o/K$, velocity $v_o = (\sigma K/\rho\varepsilon_o)^{1/3}$, electric field magnitude $E_o = (\sigma/d_o\varepsilon_o)^{1/2}$, and the electric charge $q_o = \varepsilon_o E_o d_o^2$ are defined from those physical properties.

If one uses the above characteristic quantities, the problem is governed by the electrohydrodynamic Reynolds number $\delta_\mu = \rho d_o v_o/\mu$ and relative permittivity $\varepsilon = \varepsilon_i/\varepsilon_o$ [28]. The dimensionless diameter d/d_o and electric charge q/q_R of the first-ejected droplet are expected to depend on those two numbers exclusively. Here, $q_R = (8\pi^2\varepsilon_o\sigma d^3)^{1/2}$ is the Rayleigh limit of charge [3].

Collins et al. [12] analyzed the problem mentioned above by numerically solving the leaky-dielectric model. They concluded that $q/q_R = 0.44$. Gañán-Calvo et al. [14] rationalized their numerical and experimental results in terms of the self-similar collapse of the drop's tip, the lack of full charge relaxation from the bulk to the free surface, and the balance between inertia, axial viscous stresses, surface tension, and electrostatic suction during the jet's evolution. This analysis led to the scaling laws $d/d_o = 0.6\,\zeta_d$ and $q/q_R = 0.48\,\zeta_R$, where $\zeta_d = \delta_\mu^{-1/3}\varepsilon^{5/12}$ and $\zeta_R = \delta_\mu^{-1/6}\varepsilon^{-1/24}$ (Fig. 5.10).

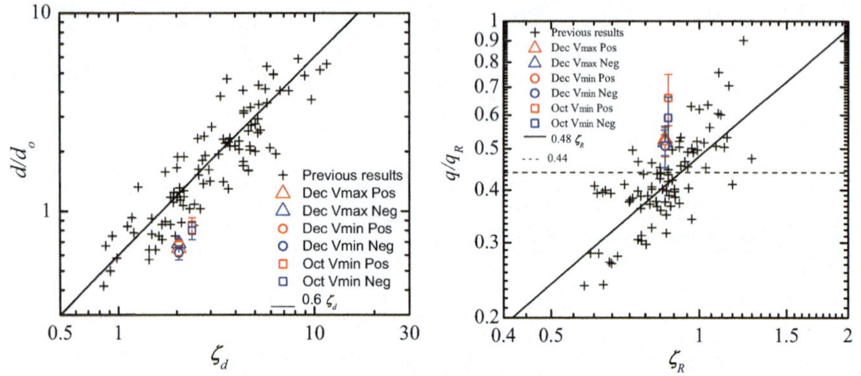

Fig. 5.10 $d/d_o(\zeta_d)$ (left) and $q/q_R(\zeta_R)$ (right). The triangles, circles, and squares are the experimental results in Ref. [2]. The meaning of the labels can be found in that reference. The crosses in the left-hand graph correspond to numerical and experimental results for d/d_0 [14]. The crosses in the right-hand graph are numerical results for q_R/q_0 [14]. The solid lines are the scaling laws $d/d_o = 0.6\,\zeta_d$ and $q/q_R = 0.48\,\zeta_R$ [14]. The dashed line in the right-hand graph corresponds to the prediction $q/q_R = 0.44$ [12] (reprinted from Ref. [2] with permission of AIP Publishing)

Rubio et al. [2] measured the diameter and charge of the first-emitted droplet with unprecedented accuracy. The diameters were compared with previous experimental and numerical data [14]. The droplet diameters measured by Rubio et al. [2] are consistent with the exponent of the scaling law derived by Gañán-Calvo et al. [14]. However, the prefactor of that scaling law leads to an overestimation of the droplet diameter. It must be noted that this scaling law does not account for the influence of the voltage magnitude and polarity, which played a non-negligible role in the experiments of Rubio et al. [2].

The numerical results of both Collins et al. [12] and Gañán-Calvo et al. [14] slightly underestimate the ratio q/q_R measured by Rubio et al. [2]. During the ultra-fast ejection of the first-emitted droplet, the mother drop apex suddenly stretches, giving rise to a large fresh free surface per unit volume. This process occurs with a characteristic time comparable to the electric relaxation time $t_e = \varepsilon_i/K$ (see Sect. 2.2). For this reason, the constant (ohmic) conductivity considered in those theoretical predictions may not accurately describe the conduction of charge from the bulk to the newly created surface.

When the polarity is inverted, the chemical composition of the ions transferred to the free surface changes, and so do the mobilities of those ions. This effect is not accounted for by the ohmic model, in which the conductivity is assumed independent of the polarity. The polarity effect is expected to become more noticeable for liquids with anions and cations of very different sizes (mobilities), such as ionic liquids. These deviations from the ohmic approximation are expected to leave their mark on the diameter and charge of the first-emitted droplet.

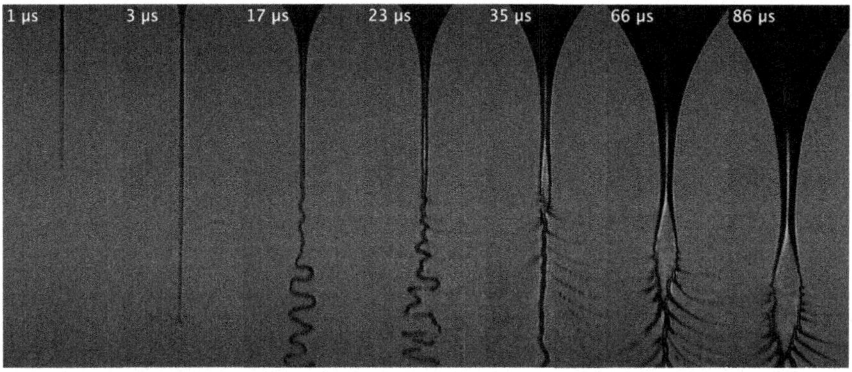

Fig. 5.11 Sequence of images of the electrohydrodynamic response of a propanol pendant drop following voltage switching. The instant $t = 0$ approximately corresponds to the formation of the first (practically invisible) emitted droplet. The varicose, kink, ramified, and multiple-branch modes [29] grow during the experiment (reprinted from Ref. [2] with permission of AIP Publishing)

Unfortunately, it is difficult to extend the experimental study conducted by Rubio et al. [2] to other liquids, which would allow one to explore the dependency of d/d_o and q/q_R on their corresponding scaled variables ζ_d and ζ_R. This probably requires increasing the electrode-to-electrode distance and the applied voltage to unreachable values with standard experimental setups. To avoid electrohydrodynamic instabilities, one must select the liquid conductivity and viscosity appropriately within a narrow range.

5.7 Other Instabilities

Different complex electrohydrodynamic phenomena arise when a pendant droplet is subjected to a sharp step change in the electric field magnitude. For the sake of illustration, Fig. 5.11 shows the electrohydrodynamic response of a pendant propanol droplet. A stream of tiny droplets is ejected from the apex of the mother parent drop for times in the interval 0–3 μs. These droplets are produced by the growth of the varicose instability in the emitted liquid thread. The surface charge in the thread increases over time, which gives rise to a sequence of non-axisymmetric instabilities. The whipping, kink, ramified, and multiple-branch instabilities, frequently observed in electrospray experiments as the electric Bond number increases [29], arise during this experiment.

References

1. Gañán-Calvo AM, López-Herrera JM, Herrada MA, Ramos A, Montanero JM (2018) Review on the physics of electrospray: from electrokinetics to the operating conditions of single and coaxial Taylor cone-jets, and AC electrospray. J Aerosol Sci 125:32–56
2. Rubio M, Sadek SH, Gañán-Calvo AM, Montanero JM (2021) Diameter and charge of the first droplet emitted in electrospray. Phys Fluids 33(032):002
3. Rayleigh JWS (1881) On the equilibrium of liquid conducting masses charged with electricity. Proc Roy Soc 5:110
4. Taylor G (1964) Disintegration of water drops in electric field. Proc R Soc Lond A 280:383–397
5. Duft D, Achtzehn T, Muller R, Huber BA, Leisner T (2003) Coulomb fission: Rayleigh jets from levitated microdroplets. Nature 421:128
6. Fontelos MA, Kindelán U, Vantzos O (2008) Evolution of neutral and charged droplets in an electric field. Phys Fluids 20(092):110
7. Oddershede L, Nagel SR (2000) Singularity during the onset of an electrohydrodynamic spout. Phys Rev Lett 85:1234–1237
8. Stachewicz U, Dijksman JF, Yurteri CU, Marijnissen JC (2007) Experiments on single event electrospraying. Appl Phys Lett 91(254):109
9. Paine MD (2009) Transient electrospray behaviour following high voltage switching. Microfluid Nanofluid 6:775–783
10. Ferrera C, López-Herrera JM, Herrada MA, Montanero JM, Acero AJ (2013) Dynamical behavior of electrified pendant drops. Phys Fluids 25(012):104
11. Collins RT, Jones JJ, Harris MT, Basaran OA (2008) Electrohydrodynamic tip streaming and emission of charged drops from liquid cones. Nat Phys 4:149–154
12. Collins RT, Sambath K, Harris MT, Basaran OA (2013) Universal scaling laws for the disintegration of electrified drops. Proc Nat Acad Sci 110:4905–4910
13. Pillai R, Berry JD, Harvie DJE, Davidson MR (2016) Electrokinetics of isolated electrified drops. Soft Matter 12:3310–3325
14. Gañán-Calvo AM, López-Herrera JM, Rebollo-Muñoz N, Montanero JM (2016) The onset of electrospray: the universal scaling laws of the first ejection. Sci Rep 6(32):357
15. Dastourani H, Jahannama MR, Eslami-Majd A (2018) A physical insight into electrospray process in cone-jet mode: Role of operating parameters. Int J Heat Fluid Flow 70:315–335
16. Gawande N, Mayya YS, Thaokar R (2020) Jet and progeny formation in the Rayleigh breakup of a charged viscous drop. J Fluid Mech 884:A31
17. Mohammadi K, Movahhedy MR, Khodaygan S (2019) A multiphysics model for analysis of droplet formation in electrohydrodynamic 3D printing process. J Aerosol Sci 135:72–85
18. Garton CG, Krasucki Z (1964) Bubbles in insulating liquids: stability in an electric field. Proc Roy Soc Lon A 280:211–226
19. Achtzehn TR, Muller TA, Duft D, Leisner T (2005) The Coulomb instability of charged microdroplets: dynamics and scaling. Eur Phys J D 34:311–313
20. Giglio E, Gervais B, Rangama J, Manil B, Huber BA, Duft D, Muller R, Leisner T, Guet C (2008) Shape deformations of surface-charged microdroplets. Phys Rev E 77(036):319
21. Brosseau Q, Vlahovska PM (2017) Streaming from the equator of a drop in an external electric field. Phys Rev Lett 119(034):501
22. Gañán-Calvo AM, Rebollo-Muñoz N, Montanero JM (2013) Physical symmetries and scaling laws for the minimum or natural rate of flow and droplet size ejected by Taylor cone-jets. New J Phys 15(033):035
23. Beroz J, Hart AJ, Bush JM (2019) Stability limit of electrified droplets. Phys Rev Lett 122(244):501
24. Wilson CTR, Taylor GI (1925) The bursting of soap-bubbles in a uniform electric field. Math Proc Cambridge Philos Soc 22:728–730
25. Basaran OA, Scriven LE (1990) Axisymmetric shapes and stability of pendant and sessile drops in an electric field. J Colloid Interface Sci 140:10–30

26. Rubio M, Rodríguez-Díaz P, López-Herrera JM, Herrada MA, Gañán-Calvo AM, Montanero JM (2023) The role of charge relaxation in electrified tip streaming. Phys Fluids 35(017):131
27. Herrada MA, Ponce-Torres A, Rubio M, Eggers J, Montanero JM (2022) Stability and tip streaming of a surfactant-loaded drop in an extensional flow influence of surface viscosity. J Fluid Mech 934:A26
28. Gañán-Calvo AM (1999) The surface charge in electrospraying: its nature and its universal scaling laws. J Aerosol Sci 30:863–872
29. Cloupeau M, Prunet-Foch B (1994) Electrohydrodynamic spraying functioning modes: a critical review. J Aerosol Sci 25:1021–1036

Chapter 6
Hydrodynamic Transient Tip Streaming

Abstract As mentioned in the previous chapter, tip streaming can be categorized into two different classes: transient tip streaming and microdripping/microjetting. This chapter reviews some transient tip streaming flows produced by hydrodynamics means. Specifically, we consider a surfactant-loaded droplet in a linear extensional flow, the viscous entrainment of selective withdrawal, and bubble bursting. The chapter closes by mentioning other examples that have received less attention.

We consider both the subcritical steady flow and the onset of tip streaming in a droplet submerged in a linear extensional flow, paying attention to the effect of a surfactant monolayer. We present the same analysis for the viscous entrainment of selective withdrawal. With respect to bubble bursting, we review the major results for simple liquids and summarize recent studies for liquids containing polymers and surfactants.

Keywords Linear flows · Viscous entrainment · Selective withdrawal · Bubble bursting

6.1 Tip Streaming in Linear Flows

6.1.1 The Subcritical Steady Flow

The deformation and stability of a small drop immersed in a viscous flow is a paradigmatic problem for evaluating the role of different factors involved in interfacial dynamics. Suppose the droplet size is much smaller than the scale of variation of the imposed external flow. In that case, the droplet deforms and bursts due to the local velocity gradient at the droplet location; thus, the outer flow can be regarded as linear. Hyperbolic, simple shear, and extensional (straining) flows are examples of this type of fluid motion.

J. M. Montanero, *Tip Streaming of Simple and Complex Fluids*, Fluid Mechanics and Its Applications 137, https://doi.org/10.1007/978-3-031-52768-5_6

Taylor was one of the first to analyze the dynamics of a clean droplet in a linear flow both theoretically [1, 2] and experimentally [3], showing how the droplet can develop conical ends before its breakup. Subsequent studies have examined the deformation and instability of bubbles and drops in both shear and extensional flows (see, e.g., [4, 5]).

According to the slender-body theory of Taylor [2], a drop with vanishing viscosity in a uniaxial extensional flow adopts a conical shape. Buckmaster [6] demonstrated that the conical end is not a solution to the Stokes equations. Numerical simulations have shown that the tip is rounded, with a curvature increasing exponentially with the unperturbed external velocity at the tip [7]. For a finite droplet viscosity, the pressure buildup near the apex leads to a saddle-node bifurcation at

$$Ca_G = \frac{5^{1/2}2^{2/3}}{24}\hat{\lambda}^{-1/6}, \tag{6.1}$$

where $Ca_G = \mu_o Ga/\sigma$ is the capillary number, G is the intensity of the extensional flow, a is the droplet radius, and $\hat{\lambda} = \mu_i/\mu_o$ [7]. The flow mode adopted by the droplet for supercritical capillary numbers can be predicted neither from the steady solution nor from its linear stability. We will come back to this point in the next section.

Courrech du Pont and Eggers [8] studied the effect of an outer axisymmetric Stokes flow on an interface with vanishing viscosity on the inner side of the interface. They concluded that the interface near the tip exhibits a universal self-similar shape (Fig. 6.1) whose curvature at the tip, κ_{tip}, obeys the exponential law

$$\kappa_{\text{tip}} \propto L^{-1} \exp(4Ca_{\text{tip}}^2), \tag{6.2}$$

where L is a characteristic length of the system (for instance, the drop radius), $Ca_{\text{tip}} = v_{\text{tip}}^o \mu_o/\sigma$ is the capillary number, v_{tip}^o is the velocity of the outer flow imposed at the tip position, and μ_o is the outer fluid viscosity. The above result does not depend on the configuration used to produce the outer flow and ca be applied to, for instance, selective withdrawal.

6.1.2 The Effect of Surfactants: Tip Streaming

Tip streaming is one of the most exciting phenomena arising when a droplet is submerged in a linear flow. In his pioneering work, De Bruijn [9] described the tip streaming in a surfactant-laden droplet submerged in a simple shear flow (Fig. 6.2). In this case, reducing the interfacial tension in the drop pole resulted in the ejection of a fluid thread much smaller than the drop size. This was one of the first mechanisms used to produce tip streaming in a clear and reproducible way. De Bruijn [9] concluded that surfactants trigger tip streaming, which disappears as surfactants are convected away from the tip. In fact, there is a widespread belief that a surfactant

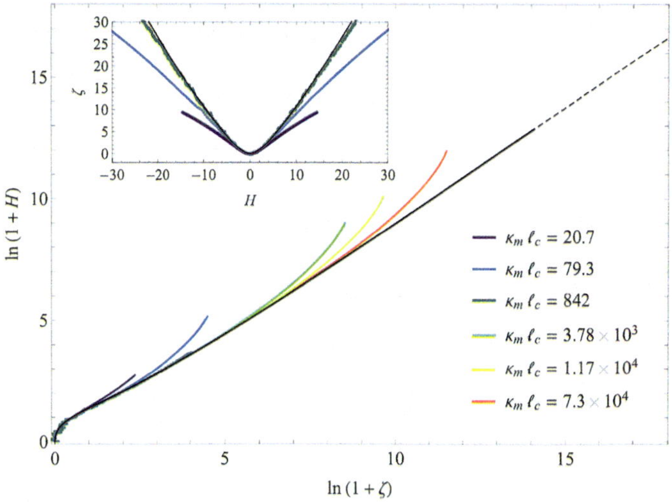

Fig. 6.1 Comparison between the similarity solution $H(\xi)$ (black solid line) and shapes obtained from both selective withdrawal experiments and full numerical simulations of a droplet in an extensional flow for different values of the tip curvature ($\kappa_m \equiv \kappa_{\text{tip}}$) (colored lines). The capillary length and the scaled variables H and ξ were defined by [8] (reprinted with permission from Ref. [8])

Fig. 6.2 Tip streaming produced in a Couette device at a shear rate of 19.5 s^{-1}. The image shows a drop of water submerged in 1000 cSt silicone oil. A surfactant was dissolved in the water droplet. The image height is 0.5 mm (reprinted with permission from Ref. [10]. Copyright (1997) by the American Physical Society)

monolayer is necessary to produce tip streaming in unbounded shear and extensional flows. The numerical simulations presented in Sect. 6.1.3 show this is false.

While a simple shear flow can be easily produced experimentally [11], a uniaxial extensional flow facilitates the theoretical analysis owing to the existence of a symmetry axis [4, 5]. Consider a droplet of radius a, density ρ_i and viscosity μ_i placed in the center of a flow of a liquid of density ρ_o and viscosity μ_o (Fig. 6.3). The imposed (unperturbed) external flow obeys the equations

$$u^{(\text{ext})} = -Gr/2, \quad w^{(\text{ext})} = G\,z, \quad p^{(\text{ext})} = \text{const.}, \tag{6.3}$$

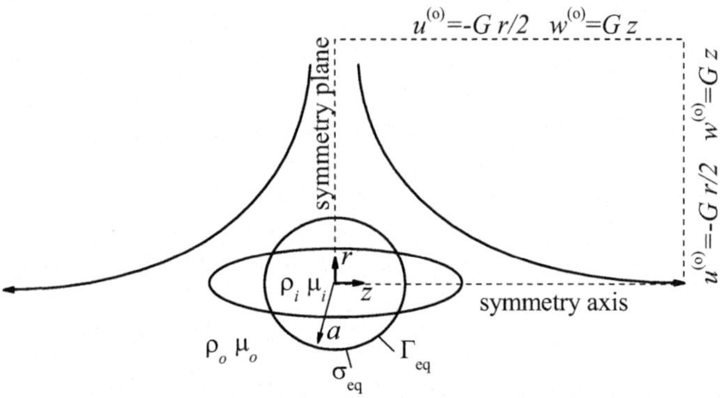

Fig. 6.3 Sketch of the fluid domain. The dashed line indicates the computational domain

where $u^{(\mathrm{ext})}$ and $w^{(\mathrm{ext})}$ are the radial and axial components of the velocity field, G is the intensity of the extensional flow, and $p^{(\mathrm{ext})}$ is the outer pressure field. Sufficiently far from the droplet, the outer flow approximately equals (6.3), i.e., $u^{(\mathrm{o})} \simeq u^{(\mathrm{ext})}$, $w^{(\mathrm{o})} \simeq w^{(\mathrm{ext})}$, and $p^{(\mathrm{o})} \simeq p^{(\mathrm{ext})}$.

The droplet and outer fluid interface is loaded with an insoluble surfactant. The surfactant surface concentration (surface coverage) and the interfacial tension at equilibrium are Γ_{eq} and σ_{eq}, respectively. The outer flow deforms the droplet and produces an inhomogeneous surfactant distribution over the interface. The Langmuir equation of state (2.41) approximately gives the variation of the interfacial tension with the surfactant surface concentration.

The simulations of Eggleton [12] for negligible inertia showed how the outer flow drags the surfactant molecules toward the poles of the droplet. This effect causes the reduction of the interfacial tension in the droplet apex, which can result in tip streaming, as occurs in a simple shear flow [9]. When the Capillary number

$$Ca_G = \frac{\mu_o\, G\, a}{\sigma_{\mathrm{eq}}} \tag{6.4}$$

exceeds a critical value, the droplet ejects a thin liquid thread from its poles (Fig. 6.4). The diameter of the ejected thread for the critical capillary number increases with the initial surfactant surface concentration Γ_{eq} [12].

Wang et al. [13] studied the tip streaming numerically in a droplet covered with a soluble surfactant and placed in a uniaxial extension flow. They showed that the adsorption of a small amount of surfactant onto the interface produces tip streaming. The size of the emitted droplet decreases as the Biot number (the ratio of the flow time scale to that for surfactant desorption) increases.

Herrada et al. [14] analyzed both the flow stability for subcritical capillary numbers and the droplet breakup mode beyond the stability limit. As in clean droplets, the instability corresponds to a saddle-node bifurcation (Fig. 6.5). The Marangoni stress balances the viscous shear stress exerted by the outer flow, while the inner shear stress

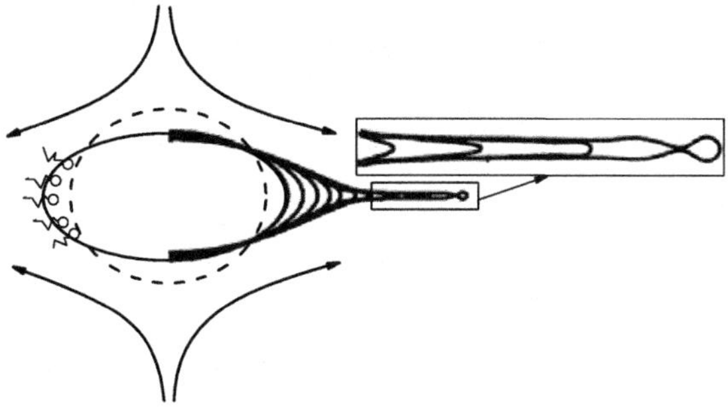

Fig. 6.4 Breakup of a droplet loaded with an insoluble surfactant and submerged in a linear extensional flow. The contours on the right side were calculated from the numerical simulation of the Stokes flow for $Ca_G = 0.065$ and $\hat{\lambda} = \mu_i/\mu_o = 0.05$. The values of the rest of the governing parameters can be found in Ref. [12] (reprinted with permission from Ref. [12]. Copyright (2001) by the American Physical Society)

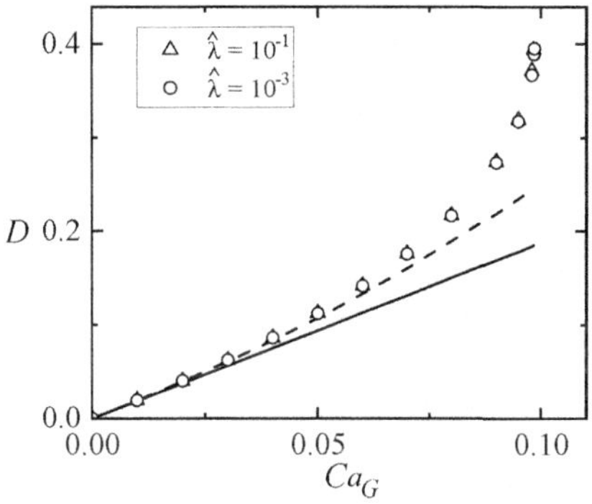

Fig. 6.5 Deformation $D = (\hat{a} - \hat{b})/(\hat{a} + \hat{b})$ ($\hat{a}$ and $\hat{b}$ are the half-length and half-breadth of the cross-sectional shape, respectively) of a surfactant-covered droplet as a function of the capillary number Ca_G for $\hat{\lambda} = \mu_i/\mu_o = 0.1$ and 10^{-3}. The values of the rest of the governing parameters can be found in Ref. [14]. The symbols are the simulation results, while the solid and dashed lines correspond to the first-order and third order perturbation theory of Vlahovska et al. [15]. The perturbation theory results are insensitive to the viscosity ratio value

becomes negligible. The surfactant monolayer practically immobilizes the interface. Therefore, the flow inside the droplet vanishes. This implies that the droplet viscosity significantly affects neither the droplet deformation nor the flow stability (Fig. 6.5). This resembles what occurs in a droplet subject to an external electric field. In this case, electrical charges are distributed over the interface to screen the external electric field. Then, the shear Maxwell stress vanishes, and so does the flow inside the droplet (Fig. 6.6).

The critical capillary number considerably decreases in the presence of a surfactant monolayer. The transient state resulting from the instability of the extensional flow is fundamentally different for drops with and without surfactant [14]. Tip streaming occurs only in the presence of surfactants, while the central pinching mode is

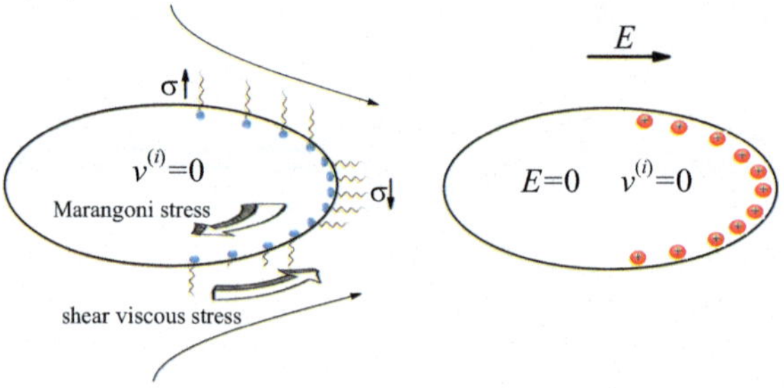

Fig. 6.6 Surfactant-loaded droplet submerged in an extensional flow (left) and droplet subjected to an external electric field (right). In the two cases, the flow inside the droplet practically vanishes

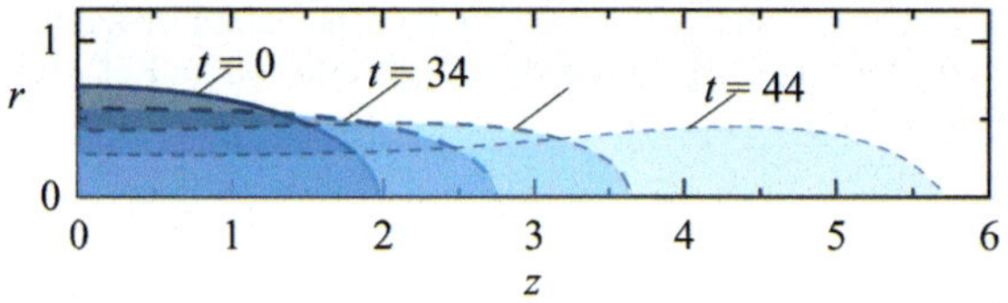

Fig. 6.7 Shape of an inertialess droplet without surfactant at different instants for $\hat{\lambda} = \mu_i/\mu_o = 0.1$ and $Ca_G = 0.175$ [14]

consistently observed for a clean interface. This mode thins the central part of the drop, which splits into two droplets whose sizes are commensurate with that of the mother drop (Fig. 6.7).

The critical eigenmode leading to tip streaming in the presence of surfactant is qualitatively the same as that yielding the central pinching mode for a clean interface [14]. This indicates that the small local scale characterizing tip streaming is set during the nonlinear droplet deformation. In an extensional flow, the force exerted by the outer stream increases with the distance to the droplet center. As the droplet stretches, the apex experiences more intense viscous forces than the rest of the droplet. This explains why the deformation progressively localizes around the apex, and tip streaming arises in the nonlinear deformation.

Herrada et al. [14] also incorporated in their simulations the effect of the shear and dilatational surface viscosities μ^s and κ^s. The viscous surface stress was given by the Boussinesq-Scriven constitutive equation [16, 17], and the surface viscosity depended exponentially on the surface pressure [18, 19] (see Sect. 2.4.4). Interestingly, shear viscous surface stress considerably alters the tip streaming arising in the supercritical regime, even for very small surface viscosities. The viscous surface stress alters the balance of normal interfacial stresses and the surfactant transport over the stretched interface. Specifically, the shear viscosity slightly reduces the surfactant in the droplet tip, which may entail a sharp increase in the interfacial tension. For this reason, the size of the ejected droplet significantly increases with the shear surface viscosity (Fig. 6.8).

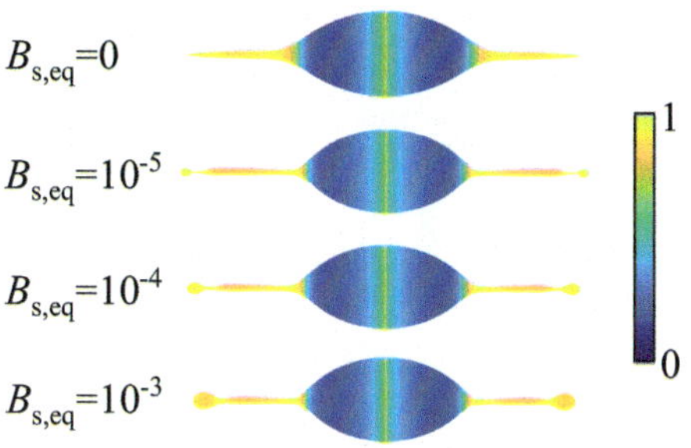

Fig. 6.8 Transient simulation for $B_{s,\text{eq}} = 0$, 10^{-5}, 10^{-4}, and 10^{-3} [14]. Here, $B_{s,\text{eq}} = B^s_\mu = \mu^s_{\text{eq}}/(\mu_o a)$ is the value of the Boussinesq number for the surfactant concentration at equilibrium ($t = 0$). The values of the rest of the governing parameters can be found in Ref. [14]. The colors indicate the surfactant surface concentration

Interestingly, the above phenomenon occurs even for very small surface viscosities, such as that of SDS, a surfactant commonly used in experiments. The shear surface viscosity of SDS takes values as small as 10^{-10} Pa s m [20, 21]. If we consider a droplet $a = 1$ mm in radius submerged in a liquid bath with viscosity $\mu_0 = 10$ mPa s, the shear surface viscosity value mentioned above leads to Boussinesq numbers of the order of 10^{-5}.

6.1.3 *The Role of the Triple Contact Line Anchorage*

To our knowledge, tip streaming in a surfactant-free droplet has not yet been observed in the shear and extensional flows. This occurs because the droplet breaks up following the central pinching mode at capillary numbers smaller than those leading to tip streaming. However, if the droplet equator is pinned, the central pinching is suppressed, the capillary number can be increased beyond the critical value for the unpinned case, and tip streaming is produced.

Figure 6.9 shows the last instant of the transient simulations of the Stokes flow when a droplet is placed in an extensional flow for $\hat{\lambda} = 0.1$ and 0.0125. The capillary numbers $Ca_G = 0.31$ and 0.56 are the corresponding critical values. The values leading to central pinching instability for the unpinned case are $Ca_G = 0.18$ and $Ca_G = 0.26$ for $\hat{\lambda} = 0.1$ and 0.0125, respectively. The simulations show the critical role played by the equator anchorage condition. The size of the emitted droplet decreases as the viscosity ratio decreases. A tiny droplet is ejected for $\hat{\lambda} = 0.0125$. We conclude that the extensional flow does not require the help of surfactants to produce tip streaming from a droplet.

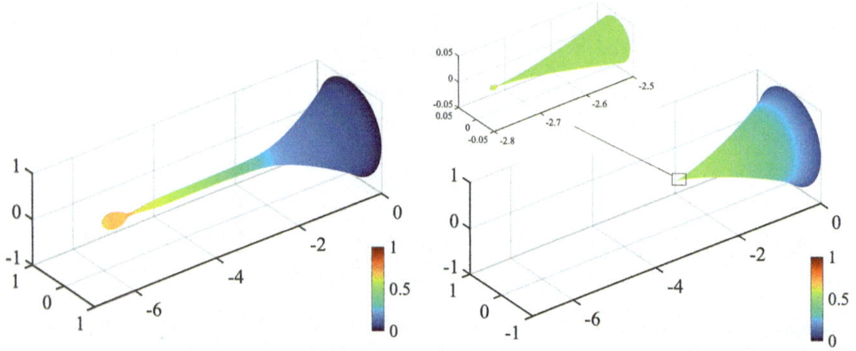

Fig. 6.9 Droplet shape at the end of the transient simulation for ($\hat{\lambda} = 0.1, Ca_G = 0.31$) (left) and ($\hat{\lambda} = 0.0125$, $Ca_G = 0.56$) (right) [22] The color scale indicates the magnitude of the interface velocity relative to the magnitude of the imposed external flow at that point

6.2 Viscous Entrainment of Selective Withdrawal

6.2.1 *Formulation of the Problem*

The so-called viscous entrainment mode of the selective withdrawal configuration is probably the simplest example of hydrodynamic tip streaming without surfactant. In selective withdrawal, a liquid layer is withdrawn across a cylindrical capillary placed in front of the interface between that layer and another fluid [23–26] (Fig. 6.10). The interface forms a hump in front of the capillary due to the viscous stresses exerted by the withdrawn liquid. When the flow rate exceeds a critical value, those stresses overcome the resistance of the interfacial tension, and the hump emits a jet much thinner than the collector capillary (viscous entrainment).

The flow analyzed in this section is sketched in Fig. 6.10. A liquid with density ρ_1 and viscosity μ_1 is located below another liquid immiscible with the former and with density ρ_2 ($\rho_2 < \rho_1$) and viscosity μ_2. An interface with surface tension σ separates the liquids. Phase (2) is suctioned through a cylindrical capillary of diameter D at a constant flow rate Q. The capillary is located at a distance H_0 from the unperturbed (horizontal) interface.

The problem can be formulated in terms of the dimensionless distance H_0/D, the density and viscosity ratios, $\hat{\rho} = \rho_1/\rho_2$ and $\hat{\lambda} = \mu_1/\mu_2$, as well as the Ohnesorge and capillary numbers,

$$\text{Oh} = \frac{\mu_2}{\rho_2 U_\sigma D} = \frac{\mu_2}{(\rho_2 \sigma D)^{1/2}} \quad \text{and} \quad Ca = \frac{\mu_2 U}{\sigma}, \tag{6.5}$$

where $U_\sigma = (\sigma/\rho_2 D)^{1/2}$ is the inertio-capillary velocity and $U = 4Q/(\pi D^2)$ is the mean velocity in the collector tube. In a typical experimental run, the flow rate Q is varied while keeping the rest of the dimensional parameters constant. We use the

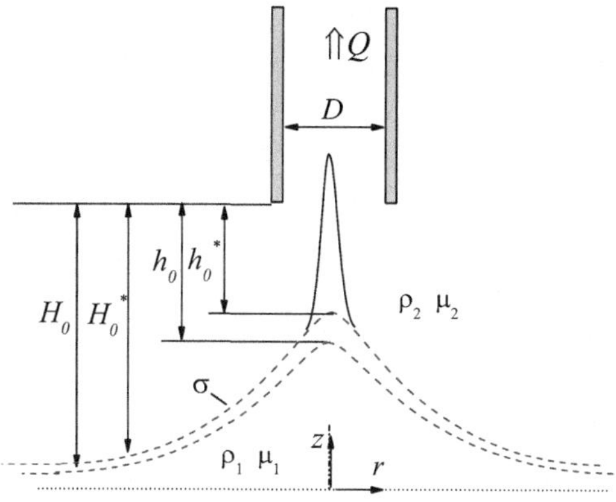

Fig. 6.10 Sketch of the onset of viscous entrainment mode in selective withdrawal. The dashed line represents the interface of subcritical/marginally stable steady flows

inertio-capillary velocity U_σ (instead of the convective velocity U) in the definition of the Ohnesorge number for the experimental run to correspond to fixed values of all the dimensionless numbers except for the capillary number Ca. This allows one to determine the critical capillary number Ca^* above which viscous entrainment is found as a function of $\{H_0/D, \hat{\rho}, \hat{\lambda}, \text{Oh}\}$. If the flow is dominated by viscosity, then $Ca^* = f(H_0/D, \hat{\lambda})$.

In selective withdrawal, the dispersed phase is suctioned from a big reservoir (measured in terms of the capillary diameter) without directly controlling its flow rate. In fact, the dispersed phase flow rate is a function of the above-mentioned governing parameters. In the confined version of the selective withdrawal geometry [27], the value of this parameter is prescribed by injecting the dispersed phase through a capillary in front of the collecting tube (see Chap. 1).

6.2.2 The Subcritical Steady Flow

The subcritical steady flow of selective withdrawal has received attention over the last three decades. Both experimental and numerical works have focused on the shape of the hump and the flow stability [23, 26, 28–34].

In liquid-liquid systems, the steady-state profiles can be scaled onto a single similarity profile [28, 29]. The curvature κ_{tip} of the hump tip displays the power-law scaling behavior (Fig. 6.11)

$$\kappa_{\text{tip}}^{-1} \propto \left(\frac{Q_c - Q}{Q}\right)^{\alpha}, \tag{6.6}$$

where Q_c is the critical flow rate leading to viscous entrainment and $\alpha > 0$. The law (6.6) is cut-off by a finite value of κ_{tip}^{-1}, indicating a first-order transition to viscous entrainment. The capillary length sets the scale for the interface curvature right before the transition to viscous entrainment [29]. The hump height is proportional to the logarithm of the curvature at the hump tip [31].

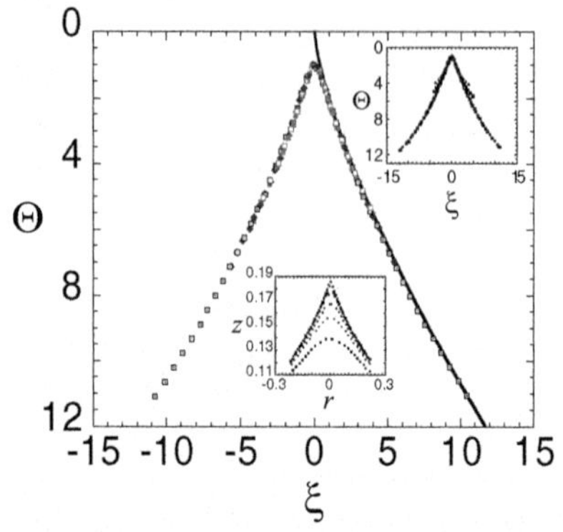

Fig. 6.11 The scaled hump profiles for a given distance H_0. The lower inset shows the non-scaled hump profiles. The upper inset compares the universal curves for different values of H_0. The scaled variables Θ and ξ were defined by Cohen and Nagel [28] (reprinted with permission from Ref. [28]. Copyright (2002) by the American Physical Society)

Lister [23] studied the linear stability of this liquid-liquid steady flow numerically, describing the stabilizing effect of the interfacial tension. The transition to viscous entrainment corresponds to a saddle-node bifurcation for the hump solutions [31], as occurs in a droplet submerged in an extensional flow [14]. Berkenbusch et al. [31] claimed that a steady-state singularity does not control the transition to viscous entrainment.

Surprisingly, Cohen [29] found no dependence of the scalings and similarity profiles on the lower fluid viscosity in his experiments. The experiments of Blanchette and Zhang [26] and Rubio and Montanero [35] showed that the critical condition at the transition to viscous entrainment is not significantly affected by the dispersed-phase viscosity μ_1, which suggests that the lower liquid dynamics hardly influence the flow stability.

The system behavior drastically changes when a gas rests on a liquid phase withdrawn from below. In the experiments of Courrech du Pont and Eggers [30], the curvature of the air-liquid interface seemed to diverge as the distance from the capillary approached its critical value. No evidence of air entrainment was found unless the tip was inside the capillary. As mentioned above, Courrech du Pont and Eggers [8] analytically showed in their subsequent work that the interface shape near the hump tip for zero viscosity ratio is universal, and the tip curvature always remains finite, although it grows exponentially with the square of the capillary number [Eq. (6.2)].

6.2.3 The Transient Viscous Entrainment Mode

The flow sketched in Fig. 6.10 deforms the interface producing a hump located at a distance h_0 from the tube end. In the subcritical steady mode, the liquid (1) recirculates in the hump, driven by the viscous stress exerted by phase (2). Consider an experiment in which h_0 progressively decreases as the liquid (2) is withdrawn. For the appropriate experimental conditions, the interface suddenly stretches when h_0 reaches its critical value h_0^*, producing a thin liquid thread that penetrates the capillary (viscous entrainment). This thread gives rise to a jet that breaks into tiny droplets due to the Rayleigh capillary instability.

This tip streaming phenomenon is supposed to arise on a time scale much shorter than that characterizing the decrease of h_0 in the subcritical regime. Therefore, the flow right before the viscous entrainment can be considered quasi-steady. A true steady jetting regime can be established by replacing the withdrawn flow rate Q. The vertical distance H_0 between the unperturbed (horizontal) interface and the capillary becomes a control parameter. It is worth mentioning that the periodic microdripping mode of tip streaming has not as yet been observed in selective withdrawal.

Rubio and Montanero [35] conducted experiments turning the system upside down. They considered that tip streaming occurred if the diameter of the first droplet emitted by viscous entrainment was smaller than $0.2D$. Using this criterion, the critical capillary number Ca^* took practically the same value, $Ca^* \simeq 0.19$, despite the noticeable variations of the viscosity ratio $\hat{\lambda}$ and Ohnesorge number Oh. However, the shape of the ejected thread depended significantly on $\hat{\lambda}$.

6.2.4 *The Effect of Surfactants*

The surfactant effects have been studied by dissolving SDS at the concentration $\hat{c}$ in the dispersed phase [35]. These effects were characterized by the dimensionless surfactant concentration $c = \hat{c}/\hat{c}_{\text{cmc}}$, measured in terms of the critical micelle concentration $\hat{c}_{\text{cmc}}$. The most apparent consequence of adding a surfactant is the global reduction of the interfacial tension. To eliminate this effect from the analysis, the equilibrium interfacial tension (instead of that of the clean interface) is used to define the Ohnesorge and capillary numbers (6.5).

Let us explain this in more detail. Consider a fluid-fluid system with a clean interface and another with the same densities and viscosities but with an interface loaded with a surfactant monolayer producing the same equilibrium interfacial tension. They are characterized by the same values of the dimensionless numbers except for c. These two systems may behave differently only due to the in-homogeneity of the interfacial tension when the surfactant monolayer is present. This in-homogeneity produces *local* soluto-capillarity and Marangoni convection. Therefore, the surfactant concentration c quantifies only these two effects, not the global reduction of the interfacial tension.

The experiments of Rubio and Montanero [35] showed that the critical capillary number for tip streaming hardly depends on the surfactant concentration for the viscous case $\hat{\lambda} = 0.002$. This indicates that Marangoni convection and surfactant surface diffusion counterbalance the weak surfactant convection towards the hump tip, and the surfactant distribution over the interface becomes practically homogeneous. For

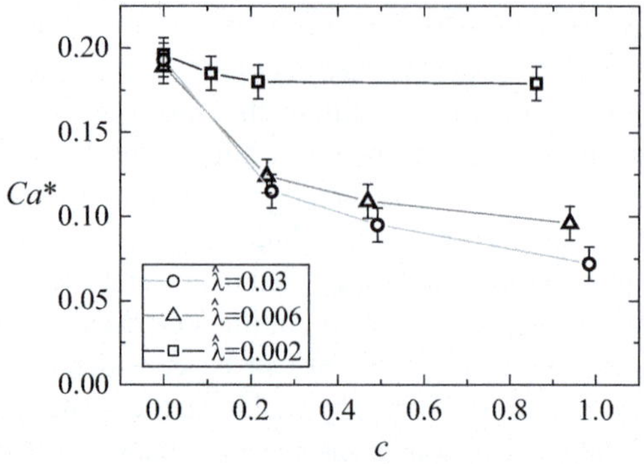

Fig. 6.12 Critical capillary number Ca^* versus the surfactant concentration c for $\{\hat{\rho} = 0.57, \hat{\lambda} = 0.03\}$ (circles), $\{\hat{\rho} = 0.55, \hat{\lambda} = 0.006\}$ (triangles), and $\{\hat{\rho} = 0.53, \hat{\lambda} = 0.002\}$ (squares) (reprinted from Ref. [35] with permission of Elsevier)

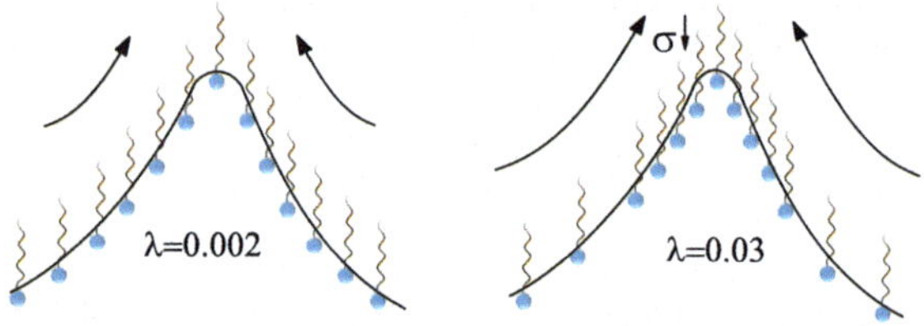

Fig. 6.13 Sketch of the monolayer behavior in the more (left) and less (right) viscous cases

this reason, the critical condition is practically the same as that of a surfactant-free system with the same equilibrium interfacial tension (Fig. 6.12).

The behavior of the less viscous (more convective) systems $\hat{\lambda} = 0.006$ and 0.03 considerably differs from that described above. In these systems, the withdrawal flow rate takes higher values for the same capillary numbers, enhancing the surfactant convection towards the steady hump tip (Fig. 6.13). This means that the interfacial tension at that point is significantly smaller than the corresponding equilibrium value, favoring the tip streaming mode. For this reason, this mode is reached for significantly smaller capillary numbers (Fig. 6.12).

For $c \gtrsim 1$, the interface at the steady hump tip saturates, the surfactant concentration is practically homogeneous, and the interface is expected to behave as a clean surface with the same interfacial tension. In this case, the reduction of the capillary number for $\hat{\lambda} = 0.006$ and 0.03 can be explained as follows. When the steady flow is perturbed, interface is created at the hump tip on a time scale much smaller than the surfactant adsorption time. Consequently, the interfacial tension increases, producing Marangoni convection toward the tip of the incipient thread (Fig. 6.14). This convection helps the liquid overcome the adverse pressure gradient in the tip, favoring the growth of a thin thread characteristic of tip streaming in viscous entrainment. This mechanism is absent in a clean interface.

An effect similar to that described above has been reported in the prejetting regime of electrified tip streaming in a leaky-dielectric liquid, where the flow is driven by

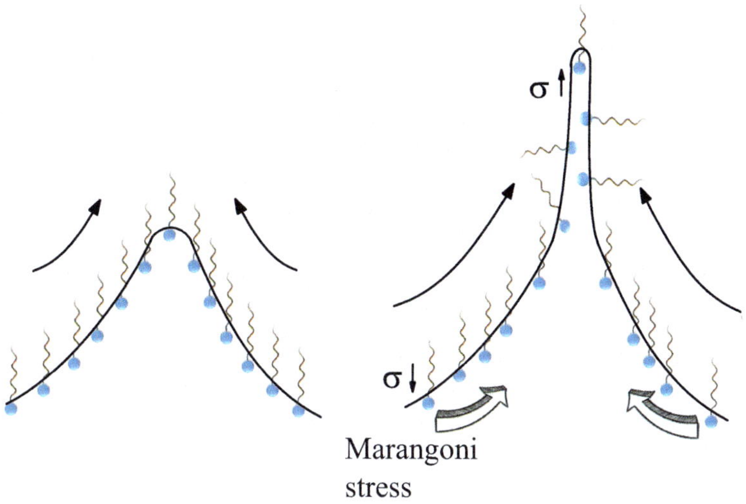

Fig. 6.14 Sketch of the monolayer behavior for $c \gtrsim 1$

the electric shear stress exerted on the interface due to the tangential component of the electric field [36, 37].

Viscous entrainment in the presence of a surfactant monolayer is an excellent example of the effect of local soluto-capillarity and interfacial elasticity (Marangoni stress). The combined action of these two factors explains the experimental results shown in Fig. 6.12. The effects of local soluto-capillarity and interfacial elasticity can be unveiled only if the dimensionless numbers are defined in terms of the equilibrium interfacial tension and not the value corresponding to the clean interface.

6.3 Bubble Bursting

When a bubble rising in a liquid of density ρ_l and viscosity μ_l reaches the free surface, a thin film (the bubble cap) forms. This film drains and breaks up under the action of the surface tension σ and gravity g. The film shattering expels tens of small drops [38] (Fig. 6.15). Jiang et al. [39] have explained the complex relations between bubble size and the number of drops produced per bubble through a mechanism involving the flapping shear instability of the cap with the outer environment. Bursting bubbles with radii $R \lesssim 0.4\ell_c$ ($\ell_c = (2\sigma/\rho_l g)^{1/2}$ is the capillary length) are mainly responsible for submicron film drop production.

In addition to the mechanism described above, capillary waves travel down over the free surface in the first moments after rupture. They collide at the bubble bottom, reverse the free surface curvature, and create a region of high curvature which ejects a Worthington-like jet upwards [41]. Under certain conditions, the jet breaks up into

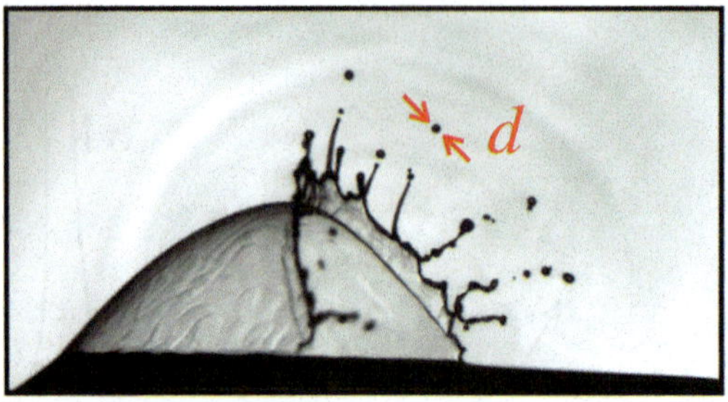

Fig. 6.15 Film drops [40]. https://creativecommons.org/licenses

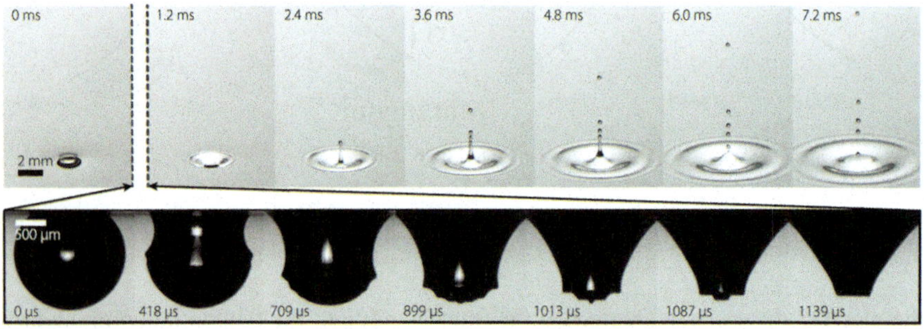

Fig. 6.16 Time sequence of bubble bursting in water (reprinted from Ref. [42] with permission of AIP Publishing)

one or more droplets (Fig. 6.16). The jet breakage is not driven by the Rayleigh capillary instability but by the end pinching mechanism. Bubble bursting frequently exhibits a strong focusing effect owing to the nearly cylindrical collapse of the driving capillary wave at the bottom of the parent bubble. This phenomenon is a remarkable example of transient tip streaming.

Droplet production from bursting bubbles is of great relevance in fields ranging from disease transfer [43] to earth science [44, 45]. It has been estimated that between 10^{18} and 10^{20} bubbles burst every second across the oceans [46], which entails a global emission of about 10^{12}–10^{14} kg per year of sea salt (Fig. 6.17). Most of the water droplets expelled by the film shattering are less than 1 μm in radius, while the radii of the jet droplets span a range from 2 to 500 μm. It is a question of debate whether primary aerosols in the marine environment are produced by film or jet drops.

6.3.1 Production of Jet Drops in Simple Liquids

The production of jet drops in simple liquids has been extensively studied over the last two decades. Here, we summarize some relevant results. Experiments have

Fig. 6.17 Production of primary aerosols in the marine environment. Illustration of the turbulent fragmentation (spume drops) and bubble bursting (film and jet droplets) responsible for spray formation (reprinted from Ref. [47] with permission of AIP Publishing)

shown that viscosity and gravity oppose the emission of jet droplets [48]. As a first approximation, the effect of the gas dynamics is neglected, and the phenomenon can be described in terms of the Ohnesorge and gravitational Bond numbers

$$\mathrm{Oh} = \frac{\mu_l}{(\rho_l R_c \sigma)^{1/2}} \quad \text{and} \quad B = \frac{\rho_l g R_c^2}{\sigma}, \tag{6.7}$$

based on the cap radius R_c. Bubble bursting can be regarded as the parametrically simplest transient tip streaming.

The liquid viscosity hinders jet formation in two ways: (i) through the damping of the precursor capillary waves that merge at the bubble base and (ii) through the energy dissipation during the jet evolution. Experiments have shown that jet droplets form if Oh and B take values approximately smaller than 0.037 and 3, respectively [49]. According to this criterion, jet droplets of water are produced for 4.2 μm $\lesssim R_c \lesssim$ 8 mm. Walls et al. [49] have shown that viscosity and gravity collaborate in reducing the parameter window where the jet droplet is emitted so that emission does not occur even if Oh$\lesssim$ 0.037 and $B \lesssim$ 3. A surface microlayer on the liquid bath surface results in a multi-phase jet and affects the size of the emitted droplet [50].

The numerical simulations of Duchemin et al. [41] showed an optimal liquid viscosity for which the focusing of the jet is most efficient and produces very tiny droplets. Ghabache et al. [42] observed that the jet speed drastically increases with

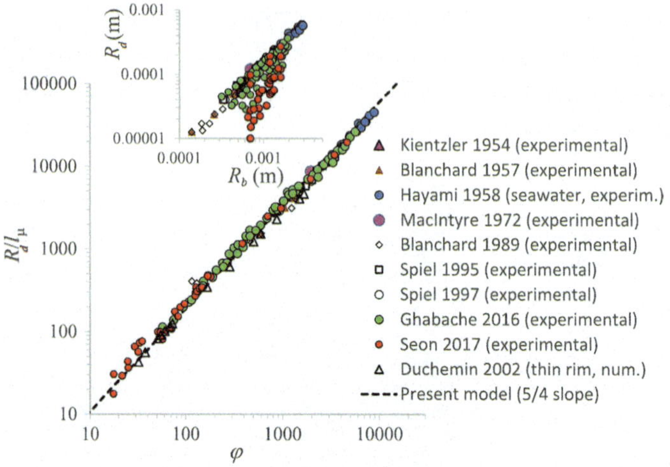

Fig. 6.18 Experimental validation of the scaling law (6.8) for the jet droplet radius R_d (reprinted with permission from Ref. [52]. Copyright (2017) by the American Physical Society)

the liquid viscosity for values less than 6 mPa s. In addition, the jet drop shrank and seemed to reach a minimum radius for $\mu_l \simeq 6 - 7$ mPa s.

The bubble radius R_b and jet droplet radius R_d approximately verify the relationship $R_d \propto R_b^{6/5}$ [51], which suggests a scaling law of the form $B_d = B_b^{6/5} H(\text{Mo})$, where $B_d = \rho_l g R_d^2/\sigma$ and $B_b = \rho_l g R_b^2/\sigma$ are the droplet and bubble Bond numbers, respectively, while $\text{Mo} = g\mu_l^4/(\rho_l\sigma^3)$ is the Morton number. The scaling law $H = \mathcal{A}\text{Mo}^{-1/3}$ ($\mathcal{A}$ is a constant) accurately fits the experimental data [51].

Gañán-Calvo [52] deduced scaling laws for the jet droplet velocity V_d and radius R_d by assuming that (i) the terms in the axial momentum equation are balanced, an argument similar to that used in electrohydrodynamic tip streaming [53], and (ii) the bubble surface energy feeds viscous dissipation in the capillary waves and the jet kinetic energy. These two hypotheses lead to

$$\frac{V_d}{V_\mu} = \varphi^{-3/4}, \quad \frac{R_d}{\ell_\mu} = k_d \varphi^{5/4}, \tag{6.8}$$

where $V_\mu = \sigma/\mu_l$ and $\ell_\mu = \mu_l^2/(\rho_l\sigma)$ are the characteristic (intrinsic) velocity and length, $\varphi = [(\text{Oh}^*/\text{Oh}) - 1]/\text{Oh}$ is related to the excess of mechanical energy necessary to form the jet, and Oh^* and k_d are two constants. The comparison with experimental data showed a remarkable agreement (Fig. 6.18) [45].

The scaling laws (6.8) were refined to include the gravity effect and the cases with very small Ohnesorge numbers [54]. For $\text{Oh} = \text{Oh}^*$, $\varphi = 0$, and the droplet velocity diverges while its radius vanishes. This critical Ohnesorge number corresponds to the optimal liquid viscosity mentioned above. For $B \simeq 0$, simulations [55, 56] indicate

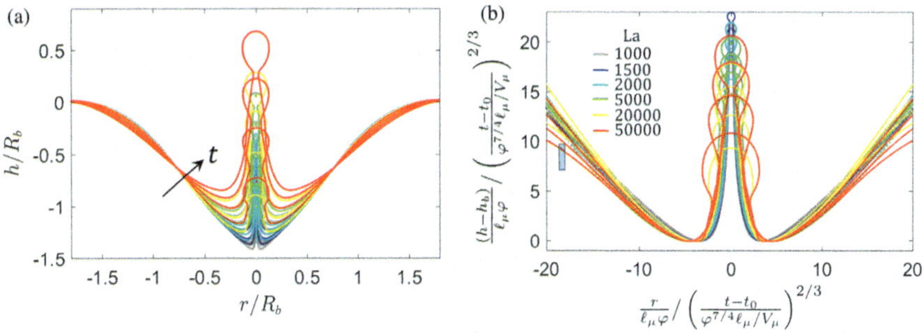

Fig. 6.19 Scaling of the profiles $h(r)$ of cavity collapse (reprinted with permission from Ref. [60]. Copyright (2018) by the American Physical Society)

that $\text{Oh}^* \simeq 0.03$, while Gañán-Calvo and López-Herrera [57] have suggested the value 0.034.

The existence of an optimum value $\text{Oh} = \text{Oh}^*$ for the jet formation has been attributed to an energy-focusing effect [55] during the bubble collapse. Gañán-Calvo and López-Herrera [57] have argued that the mechanical energy per unit volume focused on the ejected ligament is maximized for that optimum value of the Ohnesorge number. Yang et al. [58] proposed the following mechanism for oil-coated bubble bursting: the precursor capillary waves ahead of the most energetic one (the last wave) are completely damped before reaching the oil-coated cavity bottom for the optimum viscosity so that the self-similar collapse of the dominant wave evolves unaffected by the preceding short-wavelength capillary ripples.

Deike et al. [55] examined the bubble bursting both experimentally and numerically. They proposed the scaling law for the jet velocity

$$Ca = k_v(B_b)\text{La}^{-3/4}(\text{La}_*^{-1/2} - \text{La}^{-1/2})^{-3/4}, \tag{6.9}$$

where $Ca = V_d \mu_l / \sigma$ is the capillary number, $k_v = k_{v0}\text{La}_*^{3/8}(1 + \alpha B_b)^{-3/4}$, k_{v0} and La_* are two constants, and $\text{La} = \text{Oh}^{-1}$ is the Laplace number. More recently, Berny et al. [56] have proposed a simple scaling law for the total vertical momentum of the ejected droplets. They have also shown that the jet droplets significantly contribute to the amount of evaporated water. The size and speed of primary jet drops are robust to initial perturbations, while the generation of secondary drops is influenced by those conditions [59].

Lai et al. [60] presented a universal scaling law to describe the liquid-gas interface as a function of time, the liquid properties, and the initial bubble radius. Specifically, the time-dependent profiles of the cavity collapse obey the inertio-capillary (inviscid) scaling $(t - t_0)^{2/3}$ both before and after the curvature reversal at $t = t_0$ (Fig. 6.19).

Gordillo and Rodriguez-Rodriguez [61] concluded that the dynamics near the jet ejection are determined by the capillary wave with the smallest wavelength created during the coalescence of the bubble and not attenuated by viscosity. This theory predicts the non-monotonous dependence of the jet velocity upon the liquid viscosity observed both experimentally and numerically. It has been extended to obtain the scaling laws $R_d/R_b = K_b[1 - (\mathrm{Oh}/\mathrm{Oh}')^{1/2}]$ for $\mathrm{Oh} < \mathrm{Oh}'$ and $R_d \simeq 18\mu_l^2/(\rho_l\sigma)$ otherwise, where $K_b \simeq 0.2$ and $\mathrm{Oh}' \simeq 0.03$ [62].

Yang et al. [58] have shown that the bursting of a millimeter-sized bubble contaminated by an oil coating can generate much smaller and faster drops than its clean counterpart. The oil layer becomes thicker as the oil is swept toward the cavity bottom, which separates the capillary waves on the oil-air interface from those traveling over the oil-water interface. The former propagates faster than the latter, finally collapsing at the cavity bottom. A cone-shaped cavity is formed in the oil domain, which generates an upward oily jet. Bubble bursting in the presence of compound [63] interfaces has been examined experimentally.

As mentioned above, jet drops are critical to health and the environment because they can carry viruses, bacteria, and pollutants, such as microplastics, into the air. Particles sufficiently small to fit within the thin microlayer surrounding the bubble are assumed to collect into the top jet drop. It is believed that the particles extending beyond this microlayer end up elsewhere. Dubitsky et al. [64] have shown that particles with diameters significantly larger than the microlayer thickness can also be transported into the top jet drop and still exhibit an enrichment peak.

6.3.2 *Production of Jet Drops in Complex Liquids*

The non-Newtonian effects on the production of jet drops have been considered in very few works. Sanjay et al. [65] have described how viscoplasticity alters the capillary waves that trigger the jet ejection and their interaction at the bottom of the cavity. They determined the parameter region where bubble bursting can be found in the plane defined by the Ohnesorge and plastocapillary numbers.

Surfactant is expected to play a fundamental role in the bubble bursting across the oceans due to the diversity of the surfactant molecules present in them. The collective dynamics and statistics of air bubbles evolving freely at the surface of water in the presence of a surfactant monolayer have been investigated experimentally in a few works. Coalescence at the surface is prevented, and bubble lifetime is increased for high surfactant concentrations [66]. The number of supermicron droplets does not depend monotonically on the surfactant concentration [67].

The influence of the surfactant monolayer on the ejection of jet droplets has been studied numerically by Constante-Amores et al. [68]. Their simulations show that the Marangoni stress caused by a soluble surfactant drives motion from high to low surfactant concentration regions. As occurs in a droplet in an extensional flow [14], this stress immobilizes the interface, which delays the cavity collapse and affects the jet breakup process. The numerical simulations of Boulton-Stone [69] showed that the surface dilatational viscosity drastically reduces the amount of surface compression and slows and even prevents the jet formation, while shear surface viscosity has a smaller influence on the interface dynamics but affects the surfactant surface distribution.

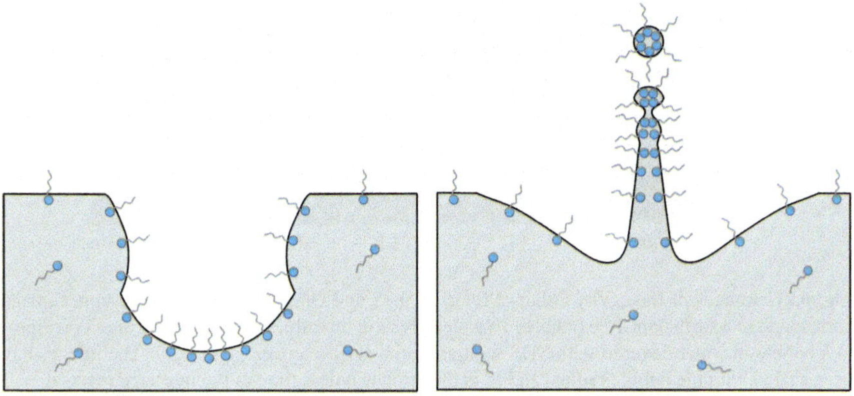

Fig. 6.20 Sketch of the surfactant convection towards the first-emitted droplet

The entire process leading to the formation of the jet droplets takes a time much shorter than the characteristic adsorption and desorption times [70–72]. This implies that most surfactants can be regarded as insoluble even though they are dissolved in the liquid phase. This is the case of SDS, used in bubble-bursting experiments.

Pierre et al. [73] have experimentally shown how Marangoni stress produced by an SDS monolayer alters the bubble collapse, leading to fewer, smaller, and faster jet drops. This stress can even suppress the drop production for a particular concentration. Vega and Montanero [74] have studied the bursting of a bubble also covered with SDS, comparing their experiments with those of surfactant-free bubbles characterized by the same Bond and Ohnesorge numbers values to determine the surfactant monolayer's effect. Their results confirm the theoretical predictions of Constante-Amores et al. [68]. Specifically, the bursting of a surfactant-covered bubble takes longer than the bursting of a surfactant-free bubble with the same equilibrium surface tension due to the interfacial elasticity. In addition, the surfactant is dragged by the flow toward the jet tip so that the monolayer covering the first-emitted droplet practically saturates (Fig. 6.20).

A tiny bubble is formed at the bottom of a quasi-saturated bubble for values of B and Oh for which this phenomenon does not occur without surfactant (Fig. 6.21). In addition, the jet escapes from the end-pinching mechanism for specific conditions where no jet droplet is emitted. These two phenomena are caused by the intense Marangoni stress arising during the cavity collapse and the jet emission, respectively. For a sufficiently large surfactant concentration, a diminutive droplet manages to escape from the jet before the end-pinching mechanism is suppressed (Fig. 6.22). The droplet is much smaller than its surfactant-free counterpart but moves at a similar speed [74].

Rodríguez-Díaz et al. [75] studied the bursting of bubbles in aqueous solutions of low-molecular-weight PEO at sufficiently small concentrations for the mixtures to behave as weakly-viscoelastic (quasi-Newtonian) liquids (Fig. 6.23). The experiments showed that the cavity collapse is affected by the interfacial effects of the PEO

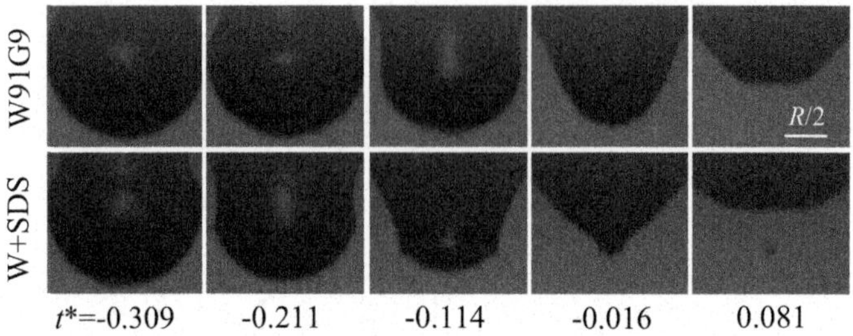

Fig. 6.21 Images of the cavity collapse for $B = 0.15$ and $\mathrm{Oh} = 0.0049$ [74]. The upper images correspond to a surfactant-free bubble in a water-glycerol mixture, while the lower ones correspond to a bubble in water covered with SDS with an initial surface coverage of 0.99. The images have been scaled with the bubble radius, and t^* is the dimensionless time to the interface reversal

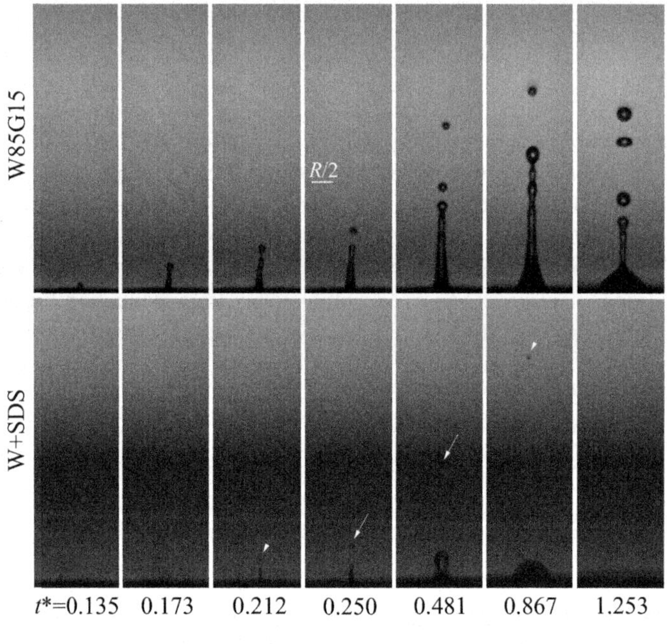

Fig. 6.22 Images of the jet and droplet ejection for $B = 0.13$ and $\mathrm{Oh} = 0.0064$ [74]. The upper images correspond to a surfactant-free bubble in a water-glycerol mixture, while the lower ones correspond to a bubble in water covered with SDS with an initial surface coverage of 0.99. The images have been scaled with the bubble radius, and t^* is the dimensionless time to the interface reversal. The arrows point at the ejected diminutive droplet

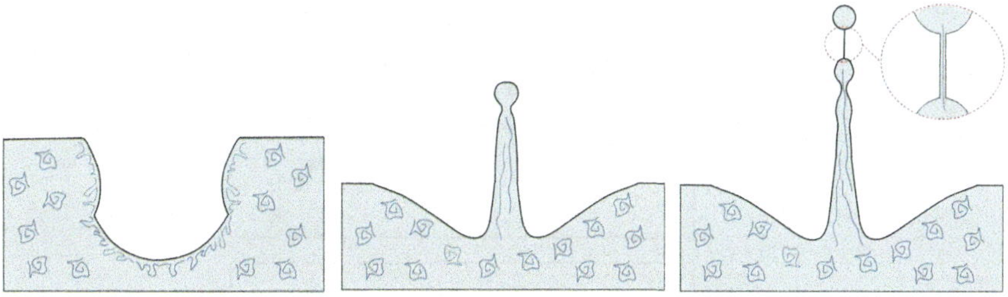

Fig. 6.23 Cavity shape evolution from film breakage and jet emission. Polymers form a monolayer at the free surface that affects the cavity collapse. The bulk viscoelasticity influences the jet emission due to the polymer stretching

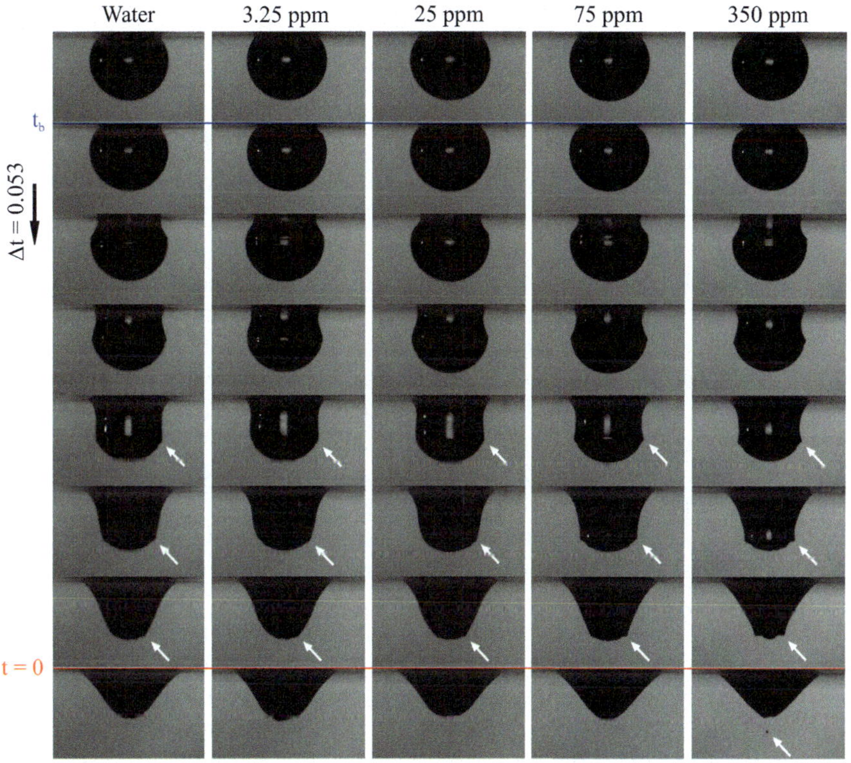

Fig. 6.24 Cavity shape evolution from film breakage to interface reversal. The size and time have been properly scaled for comparison, as explained in Ref. [75]. The arrows point at the short-wavelength capillary waves and the tiny bubble formed at the cavity bottom for $c = 350$ ppm (reprinted from Ref. [75] with permission of AIP publishing)

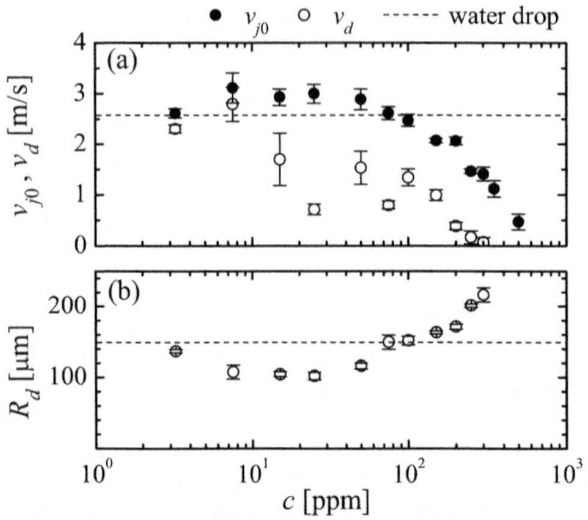

Fig. 6.25 **a** Velocity v_{j0} of the jet tip (solid symbols) when its vertical position coincides with that of the free surface of the tank. Velocity v_d of the first-emitted droplet (open symbols) right after the droplet is released. **b** Radius R_d of the first-emitted droplet. The dashed line is the first-emitted droplet velocity measured for water. The bubble radius is $R \simeq 1$ mm (reprinted from Ref. [75] with permission of AIP publishing)

monolayer, while the ejection and breakage of the jet are influenced by viscoelasticity in the bulk.

For a small concentration interval, the interfacial effects dampen short-wavelength waves during the cavity collapse (Fig. 6.24). This enhances the energy focusing that leads to the jet ejection, which results in a significant increase (decrease) in the first-emitted droplet velocity (radius) (Fig. 6.25). A tiny bubble is formed at the cavity bottom for $c = 350$ ppm. This effect is observed in Newtonian liquids for higher values of the liquid viscosity (Ohnesorge number) [57], which suggests that extensional thickening around the bubble is responsible for this phenomenon.

The intense straining flow driving the jet emission increases the extensional viscosity, which reduces (increases) the first-emitted droplet velocity (radius). A thin elasto-capillary filament is formed in the nonlinear phase of the jet breakage, even for very small polymer concentrations (Fig. 6.26). The increase in the extensional viscosity and the downward force exerted by the elasto-capillary filament on the first-emitted droplet slows down the droplet. For sufficiently large polymer concentrations, the first-emitted droplet and those ejected subsequently coalescence. As a result, the emitted droplet is much larger and slower than its Newtonian counterpart. For sufficiently large concentrations, the beads on the elasto-capillary filament coalescence with the bath before they are released (Fig. 6.26). Thus, viscoelasticity suppresses the emission of jet droplets even for weakly-viscoelastic liquids.

As explained in this section, both surfactants and polymers can eliminate the ejection of jet droplets. While surfactants produce Marangoni stresses that enable the escape from the end-pinching mechanism, the polymer stretching slows down the jet and creates a tensile force that prevents the ejection of the viscoelastic droplet.

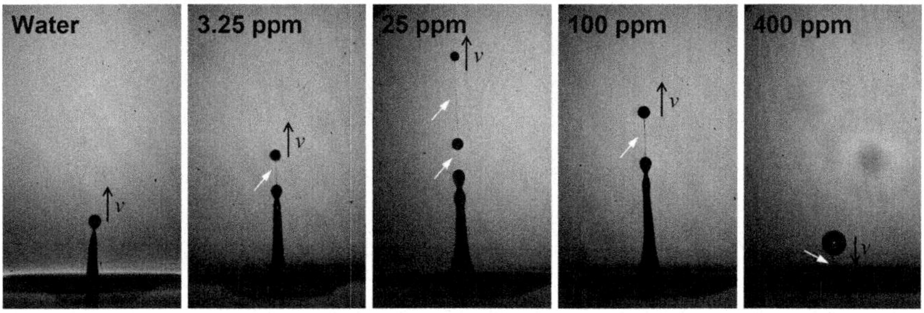

Fig. 6.26 Images of the emitted jet for different PEO concentrations. The white arrows point at the elasto-capillary filament. For water and $c = 3.25$ and 100 ppm, the image corresponds to the instant right before the droplet is released. For $c = 25$ ppm, the filament remains longer and breaks when the droplet is out of the vision field. For $c = 400$ ppm, the droplet falls back and coalesces with the bath. The bubble radius is around 1 mm (reprinted from Ref. [75] with permission of AIP publishing)

The presence of SDS and PEO molecules in water allows one to observe a bubble bursting regime where a tiny daughter bubble is entrapped at the cavity bottom and the jet drop ejection is suppressed (Figs. 6.21 and 6.23). This behavior has been observed in surfactant-free Newtonian liquids only for Ohnesorge numbers close to the critical one ($\text{Oh}^* \simeq 0.03$), much higher than those characterizing the experiments with SDS [74] and PEO [75]. Ji et al. [76] have also observed this bubble entrainment regime when the interface was covered with an adsorbed protein layer. Since the bubble bursting time was much smaller than the surface stress relaxation time, the adsorbed layer responded as a solidlike elastic surface. Ji et al. [76] claimed that the bubble entrainment was triggered by the dilatational surface stress rather than the local soluto-capillarity effect and Marangoni stress produced by the protein layer.

6.4 Other Tip Streaming Realizations

Tip streaming can be produced by the so-called inertial focusing, a phenomenon closely related to bubble bursting. In inertial focusing, the free surface deformation produced by standing (Faraday) waves can entrain an air bubble beneath the free surface, drastically changing the surface topology. The resulting inertial collapse focuses the fluid kinetic energy along the central axis and produces a thin, high-speed vertical jet (Fig. 6.27).

The thread and ligaments from the impact of liquid pools and drops may also be categorized as tip streaming under certain conditions (Fig. 6.28). In this case, the curvature of liquid menisci and cavities also reverses violently, which gives rise to highly focused jets.

Gravity alone does not seem capable of producing true tip streaming, probably because of its inability to focus the flow toward the tip. However, gravity combined

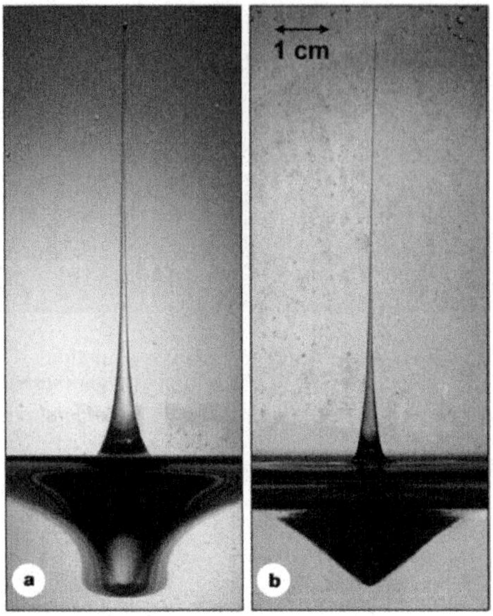

Fig. 6.27 Jet produced by inertial focusing. Image **a** is a four-exposure image to show four time-steps during the jet formation (reprinted with permission from Ref. [77])

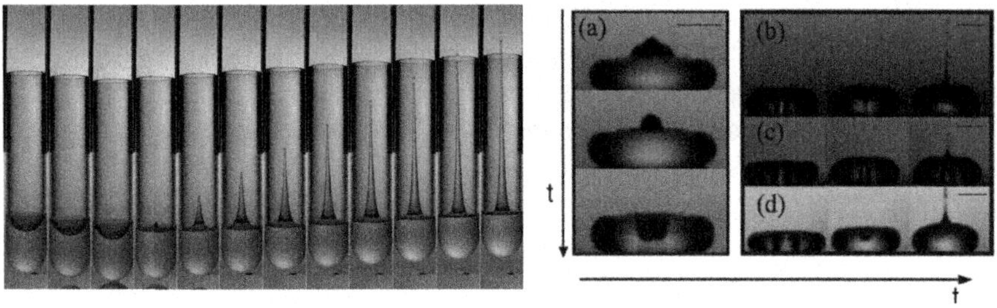

Fig. 6.28 (Left) Birth of a rapidly ascending jet following the impact on a rigid floor of a cylindrical tube. (Ref. [78], reproduced with permission). (Right) Ejection of a jet following the impact of a millimetric water drop on a super-hydrophobic surface (reprinted with permission from Ref. [79]. Copyright (2006) by the American Physical Society)

with the centrifugal force arising in a rotating cylindrical container produces a "bath-tub vortex" as the liquid drains out through a small hole [80, 81] (Fig. 6.29). Bubbling from tip streaming takes place at the tip of the needle-like surface depression for sufficiently high rotation speeds. The frequency of bubble formation increases, and the bubble size decreases as the rotation speed increases.

A laser beam crossing a near-critical phase-separated liquid, which experiences large fluctuations in its index of refraction, produces a large-scale deformation of the interface, which can emit a liquid jet [82] (Fig. 6.30).

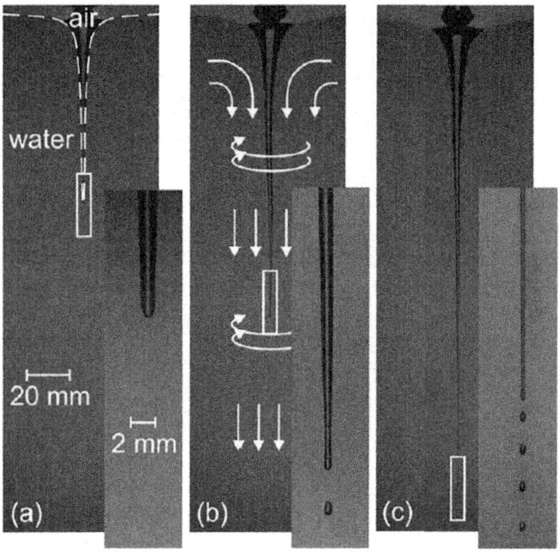

Fig. 6.29 Bathtub vortex formed when the container rotates at 13.5 rpm (**a**), 24 rpm (**b**), and 30.5 rpm (**c**) (reprinted with permission from Ref. [81]. Copyright (2009) by the American Physical Society)

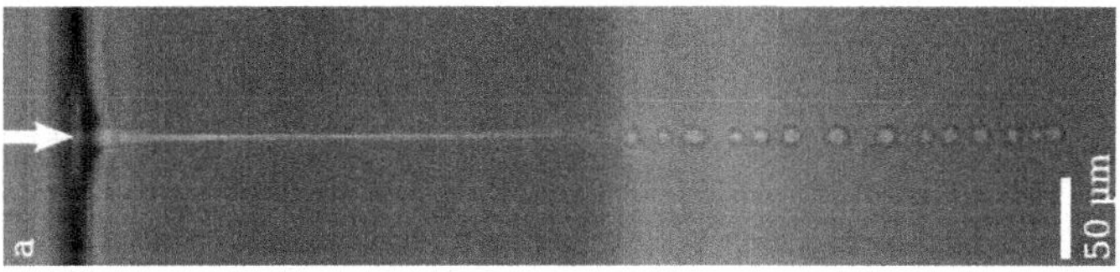

Fig. 6.30 Jet formed by a laser beam perpendicular to the interface (reprinted with permission from Ref. [82]. Copyright (2007) by the American Physical Society)

References

1. Taylor GI (1932) The viscosity of a fluid containing small drops of another fluid. Proc R Soc Lond A 138:41–48
2. Taylor GI (1964) Conical free surfaces and fluid interfaces. In: Gortler H (ed) Proceedings of the 11th international congress of applied mathematics. Springer, Heidelberg, pp 790–796
3. Taylor GI (1934) The formation of emulsions in definable fields of flow. Proc R Soc Lond Ser A 146:501–523
4. Rallison JM (1984) The deformation of small viscous drops and bubbles in shear flows. Annu Rev Fluid Mech 16:45–66
5. Stone HA (1994) Dynamics of drop deformation and breakup in viscous fluids. Ann Rev Fluid Mech 26:65–102
6. Buckmaster JD (1972) Pointed bubbles in slow viscous flow. J Fluid Mech 55:385–400
7. Eggers J, Courrech du Pont S (2009) Numerical analysis of tips in viscous flow. Phys Rev E 79(066):311
8. Courrech du Pont S, Eggers J (2020) Fluid interfaces with very sharp tips in viscous flow. Proc Natl Acad Sci 117:32,238–32,243
9. De Bruijn RA (1993) Tipstreaming of drops in simple shear flows. Chem Eng Sci 48:277–284

10. Eggers J (1997) Nonlinear dynamics and breakup of free-surface flows. Rev Mod Phys 69:865–929
11. Bentley BJ, Leal LG (1986) An experimental investigation of drop deformation and breakup in steady, two-dimensional linear flows. J Fluid Mech 167:241–283
12. Eggleton CD, Tsai TM, Stebe KJ (2001) Tip streaming from a drop in the presence of surfactants. Phys Rev Lett 87(048):302
13. Wang Q, Siegel M, Booty MR (2014) Numerical simulation of drop and bubble dynamics with soluble surfactant. Phys Fluids 26(052):102
14. Herrada MA, Ponce-Torres A, Rubio M, Eggers J, Montanero JM (2022) Stability and tip streaming of a surfactant-loaded drop in an extensional flow influence of surface viscosity. J Fluid Mech 934:A26
15. Vlahovska PM, Lawzdziewicz JB, Loewenberg M (2009) Small-deformation theory for a surfactant-covered drop in linear flows. J Fluid Mech 624:293–337
16. Scriven LE (1960) Dynamics of a fluid interface equation of motion for Newtonian surface fluids. Chem Eng Sci 12:98–108
17. Langevin D (2014) Rheology of adsorbed surfactant monolayers at fluid surfaces. Annu Rev Fluid Mech 46:4765
18. Kim K, Choi SQ, Zell ZA, Squires TM, Zasadzinski JA (2013) Effect of cholesterol nanodomains on monolayer morphology and dynamics. Proc Natl Acad Sci 110:E3054–E3060
19. Samaniuk JR, Mermant J (2014) Micro and macrorheology at fluid-fluid interfaces. Soft Matt 10:7023–7033
20. Zell ZA, Nowbahar A, Mansard V, Leal LG, Deshmukh SS, Mecca JM, Tucker CJ, Squires TM (2014) Surface shear inviscidity of soluble surfactants. Proc Natl Acad Sci 111:3677–3682
21. Ponce-Torres A, Rubio M, Herrada MA, Eggers J, Montanero JM (2020) Influence of the surface viscous stress on the pinch-off of free surfaces loaded with nearly-inviscid surfactants. Sci Rep 10(16):065
22. Rubio M, Montanero JM, Eggers J, Herrada MA (2024) Stable production of fluid jets with vanishing diameters via tip streaming. J Flui Mech 893: A4
23. Lister JR (1989) Selective withdrawal from a viscous two-layer system. J Fluid Mech 198:231–254
24. Cohen I, Li H, Hougland JL, Mrksich M, Nagel SR (2001) Using selective withdrawal to coat microparticles. Science 292:265–267
25. Case SC, Nagel SR (2007) Spout states in the selective withdrawal of immiscible fluids through a nozzle suspended above a two-fluid interface. Phys Rev Lett 98(114):501
26. Blanchette F, Zhang WW (2009) Force balance at the transition from selective withdrawal to viscous entrainment. Phys Rev Lett 102(144):501
27. Evangelio A, Campo-Cortés F, Gordillo JM (2015) Pressure gradient induced generation of microbubbles. J Fluid Mech 778:653–668
28. Cohen I, Nagel SR (2002) Scaling at the selective withdrawal transition through a tube suspended above the fluid surface. Phys Rev Lett 88(074):501
29. Cohen I (2004) Scaling and transition structure dependence on the fluid viscosity ratio in the selective withdrawal transition. Phys Rev E 70(026):302
30. Courrech du Pont S, Eggers J (2006) Sink flow deforms the interface between a viscous liquid and air into a tip singularity. Phys Rev Lett 96(034):501
31. Berkenbusch MK, Cohen I, Zhang WW (2008) Liquid interfaces in viscous straining flows: numerical studies of the selective withdrawal transition. J Fluid Mech 613:171–203
32. Eggers J, Courrech du Pont S (2010) Comment on force balance at the transition from selective withdrawal to viscous entrainment. Phys Rev Lett 105(089):401
33. Zhoua D, Feng JJ (2010) Selective withdrawal of polymer solutions: experiments. J Non-Newtonian Fluid Mech 165:829–838
34. Zhoua D, Feng JJ (2010) Selective withdrawal of polymer solutions: computations. J Non-Newtonian Fluid Mech 165:839–851
35. Rubio M, Montanero JM (2023) Influence of a soluble surfactant on the transition to tip streaming. Exp Therm Fluid Sci 141(110):776

36. Collins RT, Jones JJ, Harris MT, Basaran OA (2008) Electrohydrodynamic tip streaming and emission of charged drops from liquid cones. Nat Phys 4:149–154
37. Ferrera C, López-Herrera JM, Herrada MA, Montanero JM, Acero AJ (2013) Dynamical behavior of electrified pendant drops. Phys Fluids 25(012):104
38. Lhuissier H, Villermaux E (2012) Bursting bubble aerosols. J Fluid Mech 696:5–44
39. Jiang X, Rotily L, Villermaux E, Wang X (2022) Submicron drops from flapping bursting bubbles. Proc Natl Acad Sci 19:34
40. Villermaux E, Wang X, Deike L (2023) Bubbles spray aerosols: certitudes and mysteries. Proc Natl Acad Sci Nexus (in Press)
41. Duchemin L, Popinet S, Josserand C, Zaleski S (2002) Jet formation in bubbles bursting at a free surface. Phys Fluids 14:3000–3008
42. Ghabache E, Antkowiak A, Josserand C, Seon T (2014) On the physics of fizziness: How bubble bursting controls droplets ejection. Phys Fluids 26(121):701
43. Blanchard D, Syzdek L (1970) Mechanism for the water-to-air transfer and concentration of bacteria. Science 170:626–628
44. Boyce SG (1951) Source of atmospheric salts. Science 113:620–621
45. Deike L (2022) Mass transfer at the ocean-atmosphere interface: the role of wave breaking, droplets, and bubbles. Annu Rev Fluid Mech 54:191–224
46. MacIntyre F (1972) Flow patterns in breaking bubbles. J Geophys Res 77:5211–5228
47. Gañán-Calvo AM (2023) The ocean fine spray. Phys Fluids 35(023):317
48. Lee JS, Weon BM, Park SJ, Je JH, Fezzaa K, Lee WK (2011) Size limits the formation of liquid jets during bubble bursting. Nat Commun 2:367
49. Walls PLL, Henaux L, Bird JC (2015) Jet drops from bursting bubbles: how gravity and viscosity couple to inhibit droplet production. Phys Rev E 92(021):002(R)
50. Ji B, Yang Z, Feng J (2021) Compound jetting from bubble bursting at an air-oil-water interface. Nat Commun 12:6305
51. Ghabache E, Seon T (2016) Size of the top jet drop produced by bubble bursting. Phys Rev Fluids 1(051):901(R)
52. Gañán-Calvo AM (2017) Revision of bubble bursting: universal scaling laws of top jet drop size and speed. Phys Rev Lett 119(204):502
53. Gañán-Calvo AM, López-Herrera JM, Rebollo-Muñoz N, Montanero JM (2016) The onset of electrospray: the universal scaling laws of the first ejection. Sci Rep 6(32):357
54. Gañán-Calvo AM (2018) Scaling laws of top jet drop size and speed from bubble bursting including gravity and inviscid limit. Phys Rev Fluids 3(091):601(R)
55. Deike L, Ghabache E, Liger-Belair G, Das AK, Zaleski S, Popinet S, Seon T (2018) Dynamics of jets produced by bursting bubbles. Phys Rev Fluids 3(013):603
56. Berny A, Deike L, Seon T, Popinet S (2020) Role of all jet drops in mass transfer from bursting bubbles. Phys Rev Fluids 5(033):605
57. Gañán-Calvo AM, López-Herrera JM (2021) On the physics of transient ejection from bubble bursting. J Fluid Mech 929:A12
58. Yang Z, Ji B, Ault JT, Feng J (2023) Enhanced singular jet formation in oil-coated bubble bursting. Nat Phys 19:884–890
59. Berny A, Deike L, Seon T, Popinet S (2022) Size and speed of jet drops are robust to initial perturbations. Phys Rev Fluids 7(013):602
60. Lai CY, Eggers J, Deike L (2018) Bubble bursting: universal cavity and jet profiles. Phys Rev Lett 121(144):501
61. Gordillo JM, Rodriguez-Rodriguez J (2019) Capillary waves control the ejection of bubble bursting jets. J Fluid Mech 867:556–571
62. Blanco-Rodríguez FJ, Gordillo JM (2020) On the sea spray aerosol originated from bubble bursting jets. J Fluid Mech 886:R2
63. Feng J, Roché M, Vigolo D, Arnaudov LN, Stoyanov SD, Gurkov TD, Tsutsumanova GG, Stone HA (2014) Nanoemulsions obtained via bubble-bursting at a compound interface. Nat Phys 10:606–612

64. Dubitsky L, McRae O, Bird JC (2023) Enrichment of scavenged particles in jet drops determined by bubble size and particle position. Phys Rev Lett 130(054):001
65. Sanjay V, Lohse D, Jalaal M (2021) Bursting bubble in a viscoplastic medium. J Fluid Mech 922:A2
66. Neel B, Deike L (2021) Collective bursting of free-surface bubbles, and the role of surface contamination. J Fluid Mech 917:A46
67. Neel B, Erinin MA, Deike L (2021) Role of contamination in optimal droplet production by collective bubble bursting. Geophys Res Lett 49:e2021GL096,740
68. Constante-Amores CR, Kahouadji L, Batchvarov A, Shin S, Chergui J, Juric D, Matar O (2021) Dynamics of a surfactant-laden bubble bursting through an interface. J Fluid Mech 911:A57
69. Boulton-Stone JM (1995) The effect of surfactant on bursting gas bubbles. J Fluid Mech 302:231–257
70. Roche M, Aytouna M, Bonn D, Kellay H (2009) Effect of surface tension variations on the pinch-off behavior of small fluid drops in the presence of surfactants. Phys Rev Lett 103(264):501
71. Mayer HC, Krechetnikov R (2012) Landau-Levich flow visualization: revealing the flow topology responsible for the film thickening phenomena. Phys Fluids 24(052):103
72. Kamat PM, Wagoner BW, Castrejón-Pita AA, Castrejón-Pita JR, Anthony CR, Basaran OA (2020) Surfactant-driven escape from endpinching during contraction of nearly inviscid filaments. J Fluid Mech 899:A28
73. Pierre J, Poujol M, Seon T (2022) Influence of surfactant concentration on drop production by bubble bursting. Phys Rev Fluids 7(073):602
74. Vega E, Montanero J (2024) Influence of a surfactant on bubble bursting. Exp Therm Fluid Sci 151(111):097
75. Rodríguez-Díaz P, Rubio A, Montanero JM, Gañán-Calvo A, Cabezas MG (2023) Bubble bursting in a weakly-viscoelastic liquid. Phys Fluids 35(102):107
76. Ji B, Yang Z, Wang Z, Ewoldt RH, Feng J (2023) Secondary bubble entrainment via primary bubble bursting at a viscoelastic surface. Phys Rev Lett 131(104):002
77. Zeff BW, Kleber B, Fineberg J, Lathrop DP (2000) Singularity dynamics in curvature collapse and jet eruption on a fluid surface. Nature 403:401–404
78. Antokowiak A, Bremond N, Le Dices S, Villermaux E (2007) Short-term dynamics of a density interface following an impact. J Fluid Mech 577:241–250
79. Bartolo D, Josserand C, Bonn D (2006) Singular jets and bubbles in drop impact. Phys Rev Lett 96(124):501
80. Andersen A, Bohr T, Stenum B, Rasmussen JJ, Lautrup B (2003) Anatomy of a bathtub vortex. Phys Rev Lett 91(104):502
81. Bergmann R, Andersen A, van der Meer D, Bohr T (2009) Bubble pinch-off in a rotating flow. Phys Rev Lett 102(104):502
82. Schroll RD, Wunenburger R, Casner A, Zhang WW, Delville JP (2007) Liquid transport due to light scattering. Phys Rev Lett 98(133):601

Chapter 7
Microfluidic Configurations for Producing Tip Streaming

Abstract As described in the previous chapter, tip streaming can be produced by accelerating the fluid in the tip of a mother drop. Stresses of hydrodynamic or electrohydrodynamic nature drive this acceleration to the extent of overcoming the resistance offered by the capillary and viscous forces. This phenomenon has modest practical (technological) applications due to the intrinsic unsteady character of the process. Besides, the experimenter has limited control over the flow outcome (for instance, the size of the emitted droplets), which can be achieved only by fixing the flow rate at which the dispersed phase is ejected.

When the droplet is attached to a capillary and fed at the appropriate flow rate, the system eventually adopts either the microdripping or microjetting mode. Microdripping periodically emits tiny droplets of almost equal size from the droplet tip, while microjetting steadily ejects a thin, long, fluid thread. Microdripping and microjetting can produce droplets with the desired morphology, size, and electrical charge.

This chapter describes the *axisymmetric* microfluidic configurations used to produce microdripping and microjetting. Specifically, we consider the cone-jet mode of electrospray and the coflowing, flow focusing, and confined selective withdrawal configurations. The characteristics of the selective withdrawal and electrified films are also discussed. We introduce and explain the meaning of the dimensionless numbers characterizing the corresponding flows. The numerical and experimental results will be discussed in the following chapters.

Keywords Electrospray · Coflowing · Flow focusing · Selective withdrawal

J. M. Montanero, *Tip Streaming of Simple and Complex Fluids*, Fluid Mechanics and Its Applications 137, https://doi.org/10.1007/978-3-031-52768-5_7

7.1 Introduction

7.1.1 Energy Focusing and Recirculation Patterns

Viscosity and surface tension dominate at the small scale relevant to microfluidics. In standard dripping and jetting, the energy necessary to overcome those forces is injected into the dispersed phase upstream in the feeding system. For instance, the pressure applied to the fluid reservoir converts into kinetic energy in the nozzle orifice, beating the opposition of viscosity and surface tension. In many inkjet printer heads, the upstream pressure is increased by some piezoelectric or thermal actuator in a short time interval to produce drops on demand. Standard jetting is produced if the ratio between the kinetic and interfacial energies (the Weber number) is sufficiently high.

In tip streaming, an external agent provides energy that gently pushes the fluid toward the tip of the stretched and sharpened meniscus attached to a feeding capillary. In this way, the energy transmitted to the dispersed phase is focused on the critical region where the droplet/jet forms. The fluid accelerates in the meniscus tip to overcome the intense resistance offered by the viscous stress and capillary pressure, which allows the formation of droplets, bubbles, and jets much smaller than the discharge orifice. This constitutes the essential advantage of tip streaming over the rest of microfluidic techniques. Matter can be fragmented or deposited in the fluid phase at the micro- and nanometer scales using millimeter devices, simplifying the manufacturing of those devices and avoiding clogging effects. Figure 7.1 illustrates how hydrodynamic and electrohydrodynamic forces can increase the resolution in non-contact printing in microelectronic devices.

The microdripping and microjetting modes of tip streaming are sustained by complex flows even if Newtonian fluids are used without surfactants. As mentioned above, the external forces stretch the droplet attached to the feeding capillary, forming an acute tapering meniscus. This fluidic structure results from the delicate balance between inertia, surface tension, and viscosity, depending on the specific situation. For high dispersed-phase viscosities and injected flow rates, the fluid generally moves towards the meniscus tip following roughly a 1D flow. However, for sufficiently small viscosities and flow rates, the shear stress caused by the driving force typically induces a recirculation pattern with a stagnation point next to the region where the droplet/bubble/jet is emitted [2, 3]. Recirculation can also appear for small dispersed-phase Reynolds numbers, even in Stokes flows. As explained in the next chapters, the recirculation cells growing in the tapering meniscus may play a substantial role in the tip streaming stability, limiting the range of injected flow rates for which a steady jetting regime can be obtained.

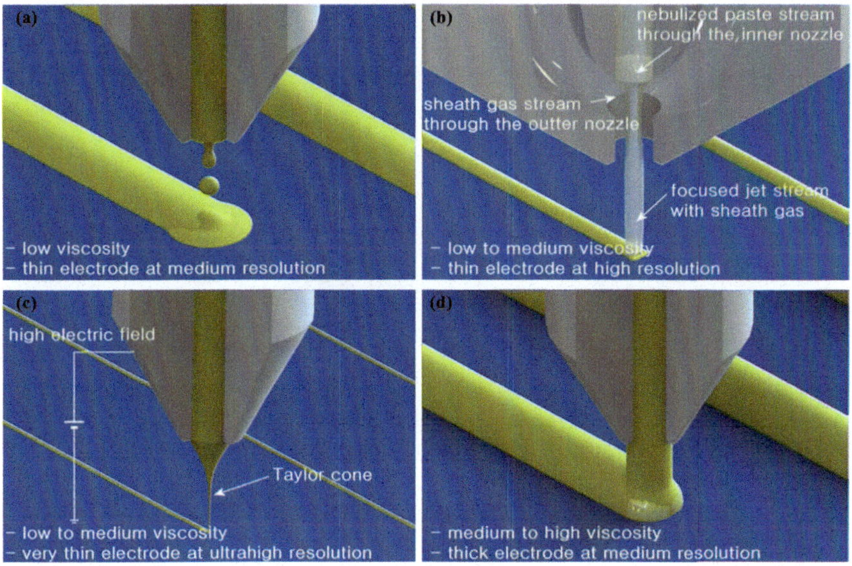

Fig. 7.1 Schematic illustration of non-contact printing in microelectronic devices: dripping (**a**), gaseous flow focusing (**b**), the cone-jet mode of electrospray (**c**), and conventional dispensing printing (**d**) [1]

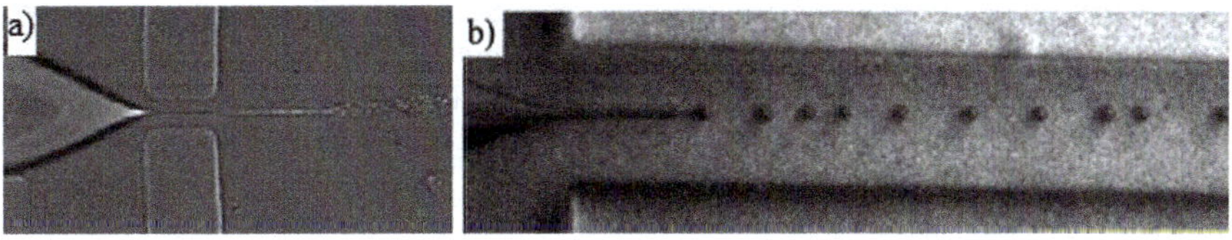

Fig. 7.2 **a** Liquid-liquid flow focusing for producing microemulsions (reprinted from Ref. [4] with permission of AIP Publishing). **b** Gas-liquid flow focusing for producing microbubbles (reprinted with permission from Ref. [5])

7.1.2 Two-Dimensional and Axisymmetric Topologies

Hydrodynamic focusing is a commonly used method to produce tip streaming both in liquid-liquid (Fig. 7.2a) and liquid-gas systems (Fig. 7.2b). In this case, the external agent is an outer fluid stream forced to coflow with the dispersed phase across the discharge orifice. This technique is frequently implemented in rectangular microfluidic devices. This geometry is referred to as the planar (two-dimensional, 2D) topology to distinguish it from the axisymmetric one.

The major advantage of the 2D topology is that the devices can be manufactured in essentially one single step. Two procedures are used for this purpose: (i) photolithography and etching in substrates of silicon, and (ii) soft lithography in substrates of

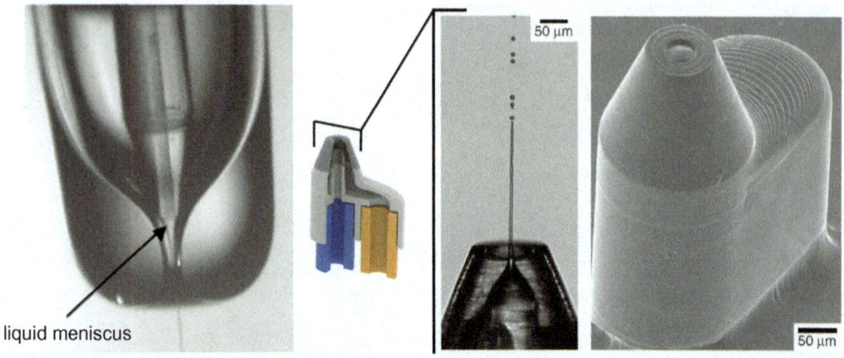

Fig. 7.3 (Left) Gaseous flow focusing device fabricated from a flame polished glass capillary (reprinted from Ref. [10] with permission of IOP Publishing). (Right) Ultracompact nozzle printed with two-photon polymerization (Nanoscribe GmbH, Karlsruhe, Germany) [16]

polymer materials (PDMS) [6, 7]. These procedures also allow the manufacturing of microchannels for sorting and manipulating the produced drops.

The planar topology presents important drawbacks. In these devices, the emerging droplet frequently touches the walls of the channel, which can damage fragile particles and cause problems associated with the competitive wetting between immiscible liquids [8]. Many applications require specific coatings and dedicated surfactants to avoid these problems. In addition, PDMS channels can deform with large applied pressures and swell in strong organic solvents and siloxane-based compounds.

The problems mentioned above are eliminated in an axisymmetric microfluidic device. In this topology, the continuous phase surrounds and shields the dispersed one in the circular cross-section of the outlet channel. Axisymmetric devices are more robust and resistant to aggressive chemical conditions than their 2D counterparts. The major advantage of the axisymmetric topology over the 2D one is its higher throughput. The considerable increase in the droplet production rate in the 3D geometry is crucial for the mass production of droplets, emulsions, and capsules in the pharmaceutical and food industries.

The axisymmetric devices are typically fabricated with borosilicate glass [9, 10], a smooth and transparent material. Glass nozzles with the desired converging or diverging rate can be produced in a controlled manner with the fire shaping procedure (Fig. 7.3) [11, 12]. They can also be made of metal, which does not swell and can easily be functionalized to control surface properties. Needles manufactured by a laser beam welding process have been assembled to fabricate triaxial flow focusing devices [13]. Alternative fabrication processes of axisymmetric devices include soft-lithography [14], ceramic [15] or polymer (Cellena®, OneNeb®, Ingeniatrics Tec. S.L.) microinjection molding.

The complex fabrication process of an axisymmetric device constitutes the major drawback of this topology. Axisymmetric tip streaming is very sensitive to slight asymmetries or imperfections of the ejector. The ejector commonly consists of sev-

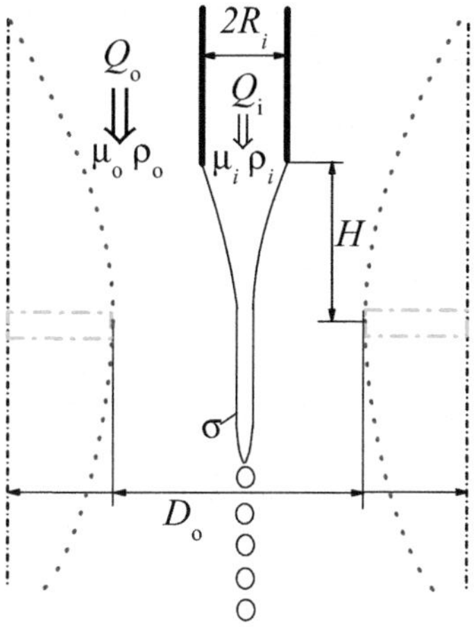

Fig. 7.4 Sketch of the injection system in an axisymmetric microfluidic configuration driven by hydrodynamic forces. The short-dashed (black), dotted (blue), and dashed (green) lines represent the shape of the outer channel in the coflowing, nozzle flow focusing, and plate-orifice flow focusing configurations, respectively

eral submillimeter pieces. Every piece must be axisymmetric, which constitutes a stringent requirement when manufacturing elements such as glass nozzles [12, 17]. In addition, all the elements must be carefully aligned. The large pressure applied during the experiment may misalign them, which forces the experimenter to disassemble and re-assemble the device. For these reasons, the fabrication techniques are art-dependent and not scalable in most cases. High-resolution 3D printing can manufacture devices with submicron resolution, avoiding the difficulties mentioned above. The 2-photon polymerization technique [18] is a remarkable example of this (Fig. 7.3). However, the high cost of these printers is not affordable for many laboratories.

A large body of literature shows the multiple capabilities of the different versions of 2D hydrodynamic focusing. Excellent reviews have summarized the significant achievements in this field [19–23]. This chapter describes the axisymmetric configurations used to produce microdripping and microjetting, which have been reviewed on much fewer occasions [24–28].

7.2 Axisymmetric Injection

The dispersed phase in an axisymmetric microfluidic device is injected at a constant flow rate Q_i across a feeding capillary (nozzle) of inner radius R_i. The fluid density and viscosity are ρ_i and μ_i, respectively. The surrounding medium can be a vacuum or a fluid bath in electrospray. In hydrodynamic focusing, the dispersed inner phase coflows with the continuous outer phase, which is immiscible with the former. The density and viscosity of the outer fluid are ρ_o and viscosity μ_o, respectively, while the interface surface tension is σ. The inner and outer streams cross an orifice of diameter D_o located at a distance H from the end of the feeding capillary (Fig. 7.4).

If the dispersed phase is a gas, the pressure upstream of the injection system can be fixed. In this case, the outcome may be different from that in which the flow rate is prescribed. It is worth mentioning that care must be taken when injecting gases in a microfluidic application because pressure fluctuations taking place both upstream and downstream of the injection circuit may lead to significant variations in the instantaneous flow rate. These pressure fluctuations can also trigger oscillations in liquid-liquid systems, especially close to the system stability limits. Pressure fluctuations can be caused by syringe pumps and can travel along the injection system.

In many experiments, the end of the feeding capillary is sharpened so that the triple contact line pins to the edge independently from the contact angle. If this does not occur, the contact angle on the capillary edge must remain within the interval between the advancing and receding contact angles (contact angle hysteresis) [29, 30] for the triple contact line to be pinned at any times (see Chap. 2). This is crucial for a successful tip streaming realization.

7.2.1 Stabilization Methods

Several techniques have been proposed to stabilize the microjetting mode of tip streaming. For instance, double-flow focusing nozzles have been used for improving the jet stability over the single nozzle ejector and reducing the amount of sample transported by the jet (Fig. 7.5a) [31]. As mentioned above, recirculation cells arise in the tapering meniscus for sufficiently small values of the dispersed phase viscosity μ_i and injected flow rate Q_i. The growth of these cells can destabilize the flow in the meniscus, limiting the range of the inner flow rate for which microjetting can be obtained. Different methods have been used to suppress this instability mechanism. Acero et al [32] introduced a sharpened rod coaxially into the feeding capillary of the gaseous flow focusing to enhance this configuration stability (Fig. 7.5b). This rod directs the flow in the meniscus tip, which prevents the formation of large recirculation cells even for small injected flow rates, reducing down to one fifth the minimum flow rate for microjetting.

The double nozzle configuration has also been used to stabilize the Taylor cone in electrospray (Fig. 7.6a). Morad et al. [33] installed a hemispherical cap above the nozzle to extend the Taylor cone-jet stability margin and produce a very high throughput (Fig. 7.6b). Sen et al. [34] introduced a very thin wire inside the liquid meniscus to stop the recirculation cell appearing for small flow rates in electrospray. Lozano and Martínez-Sánchez [35] used an externally wetted emitter made of a grooved tungsten wire electrochemically etched to electrospray liquids.

The methods mentioned above preserve flow symmetry. Nonaxisymmetric modifications of the injection system have been employed as well. One possibility is to replace the feeding capillary with a hypodermic needle (Fig. 7.7a). In this way, the tapering meniscus is substituted by a much more stable liquid film, which flows over the needle's outer surface toward the tip. The jet is emitted from a tiny cone formed in the needle tip. This configuration has been used both in electrospray [36–38] and

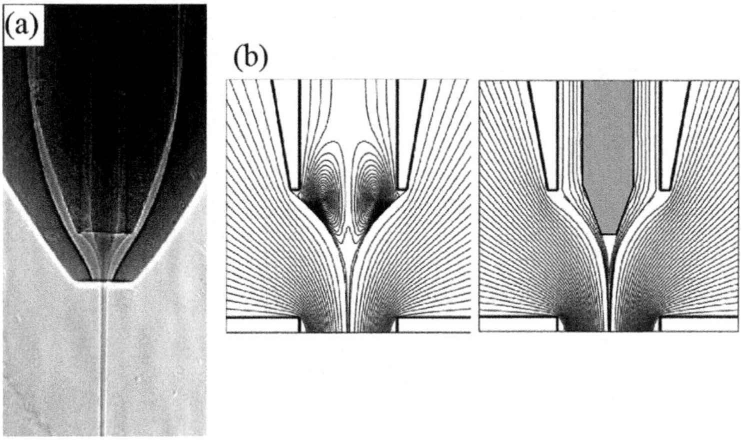

Fig. 7.5 Modifications of the injection system of gaseous flow focusing: **a** double-flow focusing nozzle [31] and **b** sharpened rod (reprinted from Ref. [32] with permission of IOP Publishing)

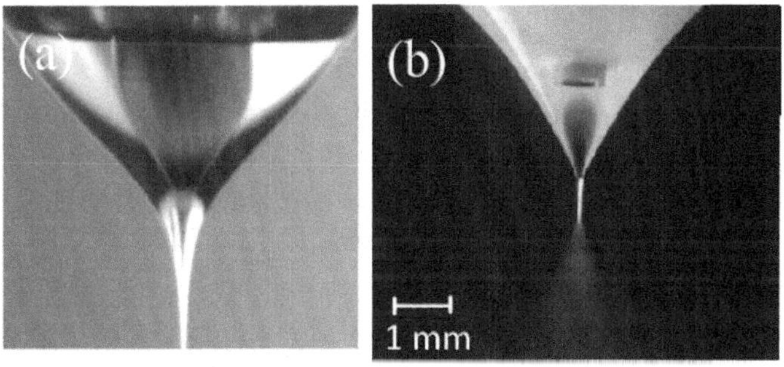

Fig. 7.6 Modifications of the injection system of electrospray: **a** double nozzle (reprinted with permission of YFLOW©) and **b** extender cap [33]

gaseous flow focusing [39, 40]. The idea proposed by Acero et al. [39] has been implemented in transonic flow focusing to reduce the jet's diameter when delivering the sample in Serial Femtosecond Crystallography (Fig. 7.7b) [41].

7.2.2 Dimensionless Numbers

The fluid injection system described above can be characterized in terms of the density and viscosity ratios

$$\rho = \frac{\rho_o}{\rho_i}, \quad \mu = \frac{\mu_o}{\mu_i}, \tag{7.1}$$

and the Weber and Reynolds numbers

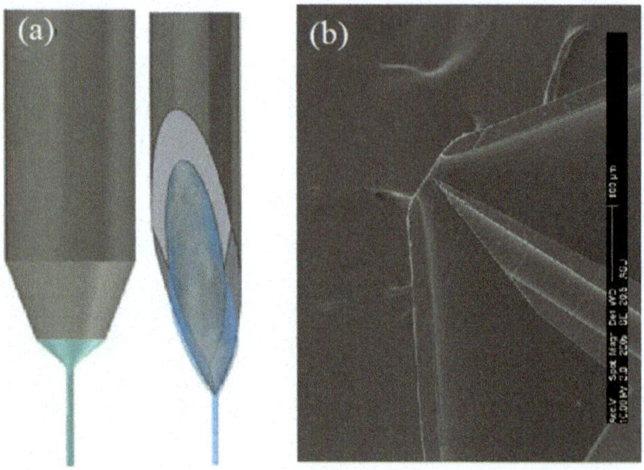

Fig. 7.7 Nonaxisymmetric modifications of the injection system of gaseous flow focusing: **a** hypodermic needle (reprinted from Ref. [39] with permission of IOP Publishing) and **b** beveled nozzle (reprinted with permission from Ref. [41]. Copyright by The Optical Society)

$$\mathrm{We}_i = \frac{\rho_i V_i^2 R_i}{\sigma}, \quad \mathrm{Re}_i = \frac{\rho_i V_i R_i}{\mu_i}, \tag{7.2}$$

where $V_i = Q_i/(\pi R_i^2)$ is the mean velocity in the feeding capillary. One can replace either We_i or Re_i with the Ohnesorge number

$$\mathrm{Oh}_i = \frac{\mathrm{We}_i^{1/2}}{\mathrm{Re}_i} = \frac{\mu_i}{(\rho_i \sigma R_i)^{1/2}} = \frac{\mu_i}{\rho_i U_\sigma R_i}, \tag{7.3}$$

where $U_\sigma = (\sigma/\rho_i R_i)^{1/2}$ is the inertio-capillary velocity. For $\mathrm{Re}_i \ll 1$, the inertia of the inner fluid becomes negligible, and the fluid injection can be described just in terms of the viscosity ratio μ and the capillary number

$$Ca_i = \frac{\mathrm{We}_i}{\mathrm{Re}_i} = \frac{\mu_i V_i}{\sigma}. \tag{7.4}$$

The use of the Ohnesorge number (7.3) entails a practical advantage. In a typical experimental run, we change the flow rate Q_i while keeping the rest of the dimensional parameters constant. In this case, the Ohnesorge number Oh_i is fixed, and only the other dimensionless number (either We_i or Re_i or Ca_i) varies. This allows us to determine, for instance, the critical capillary number for the onset of the microjetting instability as a function of Oh_i and the rest of the governing parameters.

For sufficiently large values of the Weber and Reynolds numbers, inertia overcomes surface tension and viscosity resistance during the fluid ejection, and the feeding capillary emits a jet (jetting mode). When this condition does not hold, the collaboration of some electrohydrodynamic or hydrodynamic driving force is required to produce that effect. Under the appropriate parameter conditions, tip streaming is obtained. The following sections introduce the dimensionless numbers characterizing the driving forces leading to tip streaming.

7.3 Electrospray

In electrospray and electrospinning, the interaction between electric, capillary, and viscous stresses on a liquid meniscus gives rise to the formation of charged jets and droplets. The steady cone-jet mode of electrospray and electrospinning is a classical example of microjetting induced by electric fields.

The effect of the electric field can be quantified through the electric Bond (Taylor) number

$$\chi = \frac{\varepsilon_o E^2 R_i}{\sigma}, \tag{7.5}$$

where E is a characteristic electric field (typically, the ratio of the applied voltage to the characteristic length R_i) and ε_o is the electrical permittivity of the outer insulator medium. The electric Bond number measures the electrostatic pressure $\varepsilon_o E^2$ in terms of the capillary pressure σ/R_i.

The electrical properties of the electrosprayed (electrospun) phase are taken into account through the relative permittivity ε and dimensionless electrical conductivity α:

$$\varepsilon = \frac{\varepsilon_i}{\varepsilon_o}, \qquad \alpha = K\left(\frac{\rho_i R_i^3}{\sigma \varepsilon_o^2}\right)^{1/2} = \frac{\varepsilon t_{\rho\sigma}}{t_e}. \tag{7.6}$$

The conductivity α is proportional to the ratio of the inertio-capillary time $t_{\rho\sigma} = (\rho_i R_i^3/\sigma)^{1/2}$ to the electric relaxation time $t_e = \varepsilon\varepsilon_o/K_i$.

The above interpretation of the dimensionless electrical conductivity helps analyze the onset of electrospray when an electric field is applied to a pendant drop (see Chap. 5). However, in the steady cone-jet mode of electrospray (electrospinning), the electric relaxation time t_e must be compared with the residence time t_r of a surface element in the cone-jet transition region (see Chap. 8). If $t_e \ll t_r$, the net free charge distributes over the interface to cancel the voltage variations throughout the electrosprayed phase, and the problem becomes independent of the conductivity. This does not occur for low-conductivity liquids and flow rates close to the minimum flow rate stability limit [42], where α becomes a relevant parameter.

The cone-jet mode of electrospray can be formulated in terms of χ, ε, and α, the parameters characterizing the fluid injection (e.g., Oh_i and We_i), and those defining the geometric boundary conditions [42]. As explained in Chap. 8, the device's geometrical features and associated lengths have little effect on the minimum flow rate stability limit due to the local character of the flow in the critical cone-jet transition region [43]. On this condition, one considers the characteristic length $d_o = [\sigma\varepsilon_o^2/(\rho_i K^2)]^{1/3}$ and velocity $v_o = [\sigma K/(\rho_i \varepsilon_o)]^{1/3}$, defined in terms of the properties of the fluids exclusively. Then, we define the electrohydrodynamic Reynolds number δ_μ and the relative flow rate Q as

$$\delta_\mu = \frac{\rho_i v_o d_o}{\mu_i} = \left(\frac{\sigma^2 \rho_i \varepsilon_o}{\mu_i^3 K}\right)^{1/3}, \qquad Q = \frac{Q_i}{Q_c}, \tag{7.7}$$

where $Q_c = v_o d_o^2 = \sigma \varepsilon_o/(\rho_i K)$. These parameters can be obtained as $\delta_\mu = \alpha^{-1/3}\text{Oh}_i^{1/2}$ and $Q = \pi\,\alpha\,\text{We}_i$. As a first approximation, the dimensionless minimum flow rate $Q_{\min}$ obeys the functional relationship $Q_{\min} = Q_{\min}(\beta, \delta_\mu)$ (see Chap. 8) [43].

7.4 Coflowing

Hydrodynamic forces can be employed to control the pinching of fluid-liquid interfaces and produce the microdripping and microjetting modes of tip streaming. Microfluidic devices designed for this purpose can be grouped into three main classes: cross-flowing systems, co-flowing streams, and flow focusing [22]. Only the last two classes can be realized in an axisymmetric configuration.

In a coflowing device, an outer fluid stream is coaxially injected across a cylindrical tube of radius R_o to produce jets, drops, and bubbles. The momentum transferred to the inner phase essentially comes from the tangential (shear) viscous stress exerted by the outer stream. For this reason, the intensity of the driving force is quantified through the outer capillary number

$$Ca_o = \frac{\mu_o V_o}{\sigma}, \tag{7.8}$$

where V_o is a characteristic velocity of the outer stream. The capillary number measures the characteristic shear stress $\mu_o V_o/R_i$ in terms of the capillary pressure σ/R_i.

The coflowing device is probably the simplest hydrodynamic configuration to produce tip streaming. If one neglects the thickness of the inner feeding capillary wall, the only geometrical parameter involved in the problem is the ratio of radii

$$R = R_i/R_o. \tag{7.9}$$

We can distinguish between coflowing realizations with R of order unity [44–46] and those with $R \ll 1$ [47, 48]. Their behaviors are substantially different owing to the stabilizing role of confinement [49] and the influence of this factor on the nonlinear phase of the jet breakup.

As mentioned above, the response of the dispersed phase for small Reynolds numbers can be characterized in terms of the inner capillary number (7.4) and the viscosity ratio μ. These dimensionless parameters can be combined to replace the outer capillary number (7.8) with the flow rate ratio Q_r [44]:

$$Q_r = \frac{Q_o}{Q_i} = \frac{R^2 - 1}{\mu} \frac{Ca_o}{Ca_i}. \tag{7.10}$$

If the inner phase inertia comes into play, the Reynolds, Ohnesorge, or Weber number must also be considered. The density ratio always takes values of the order of unity; therefore, it plays a secondary role in the analysis.

Due to the smallness of the Reynolds number, the flow fully develops in the tube at a relatively short distance from the inner feeding capillary. Under this condition, the jet diameter d_j is given by the expression

$$\frac{d_j}{2R_o} = \left(1 + Q_r + \sqrt{Q_r(Q_r + \mu)}\right)^{-1/2} \tag{7.11}$$

and the pressure gradient dp/dz obeys the formula

$$\frac{dp/dz}{\sigma/R_o^2} = -8\, Ca_o \mu^{-2} \left[2 + \mu\left(Q_r^{-1} + \mu - 2\right) + 2(\mu - 1)\sqrt{1 + \mu Q_r^{-1}}\right]. \tag{7.12}$$

These expressions also apply to the selectived withdrawal and confined selective withdrawal configurations described below.

7.5 Flow Focusing and Confined Selective Withdrawal

In flow focusing, the coflowing stream is forced to cross an orifice of diameter D_o located in front of the feeding capillary at a distance H from its end (Fig. 7.4). This originates a favorable gradient of hydrostatic pressure [50] not present in the coflowing configuration described above. The combination of this pressure gradient and both shear [51] and extensional viscous stresses enhances the stretching of the tapering meniscus attached to the feeding capillary, which favors the transition to tip streaming.

Two geometrical parameters enter the problem due to the existence of a discharge orifice in flow focusing: the orifice diameter D_o and the orifice-to-capillary distance H. The orifice diameter D_o is chosen as the characteristic length because it determines the size of the region where the focusing effect takes place. The ratio $\hat{H} = H/D_o$ must be considered because it significantly affects the meniscus slenderness and its stability [52].

The ratio $2R_i/D_0$ is expected to influence the flow behavior as well. However, it generally takes values around unity and is not typically taken into account. The role played by the orifice length has been neglected too. The ratio (7.9) between the radii of the feeding and outer capillaries has little influence on the flow focusing effect and is not considered in the analysis either.

In viscosity-dominated experiments of liquid-liquid flow focusing, we typically use the same non-geometrical dimensionless numbers as those of a coflowing device, i.e., the outer capillary number (7.8) and the viscosity ratio μ (or the flow rate ratio (7.10)). Inertia can become significant in flow focusing due to the larger speeds typically reached in the experiments. In this case, the Reynolds/Weber/Ohnesorge number enters the problem too. As mentioned above, the density ratio takes values of the order of unity, and its influence is frequently neglected. The dimensionless numbers describing the liquid-liquid flow focusing configuration can also be used for characterizing the confined selective withdrawal technique.

The focusing of a liquid stream by an outer high-speed gas current [50] constitutes a phenomenon considerably different from that in the liquid-liquid case. In the incompressible regime, the main force driving the liquid ejection is the hydrostatic pressure drop Δp caused by the gas acceleration across the discharge orifice. This quantity frequently substitutes the outer flow rate Q_o as a control parameter. In the microdripping mode [53], the pressure force is quantified by the ratio

$$\mathscr{P} = \frac{R_i \Delta p}{\sigma} = \frac{R_i}{d_\sigma} \quad (d_\sigma = \sigma/\Delta p) \tag{7.13}$$

of the pressure drop Δp to the capillary stress σ/R_i.

The incompressible microjetting regime for low-viscosity liquids is typically analyzed in terms of the jet's Reynolds and Weber numbers

$$\text{Re}_j^{\text{FF}} = \frac{\rho_i v_j^{\text{FF}} d_j^{\text{FF}}}{\mu_i}, \quad \text{We}_j^{\text{FF}} = \frac{\rho_i (v_j^{\text{FF}})^2 d_j^{\text{FF}}}{\sigma}, \tag{7.14}$$

where $v_j^{\text{FF}} = 4Q_i/[\pi(d_j^{\text{FF}})^2]$ and

$$d_j^{\text{FF}} = \left(\frac{8\rho Q_i^2}{\pi^2 \Delta p}\right)^{1/4} \tag{7.15}$$

are the jet's velocity and diameter calculated in terms of the injected flow rate Q_i and applied pressure drop Δp from mass and energy conservation (see Chap. 10) [51]. The dimensionless numbers (7.14) allow one to determine whether the jet is convectively or absolutely unstable (see Chap. 3) and, therefore, if flow focusing operates in the microjetting or microdripping mode.

As explained in Chap. 10, the formula (7.15) does not hold when focusing viscous liquids due to the considerable energy dissipation caused by the liquid viscosity. The formula (7.15) does not apply to viscoelastic liquids either. For this reason, we do not use the above parameter representation in those cases.

In transonic flow focusing [54], additional parameters must be considered due to the gas compressibility, specifically, the heat coefficients and thermal conductivities of the gas and liquid phases, as well as the upstream stagnation temperature and pressure. In this case, using dimensionless numbers leads to a less significant reduction of the number of governing parameters. For this reason, both numerical and experimental works have formulated the problem in terms of dimensional quantities.

Focusing a gaseous stream with an outer liquid current leads to a monodispersed microbubbling regime [55–58]. This phenomenon's strong oscillatory character suggests using the flow rate ratio Q_r as the major governing parameter. The flow is essentially inviscid, and the density ratio takes very small values. Therefore, neither the Reynolds number nor the density ratio comes into play.

7.6 Selective Withdrawal and Electrified Films

The steady microjetting mode for producing drops and bubbles does not necessarily require a feeding capillary through which the dispersed phase is injected. In the selective withdrawal technique, an upper liquid layer is suctioned across a cylindrical tube placed in front of the interface between that layer and a lower one. The interface forms a hump in front of the tube due to the viscous stresses exerted by the upper layer. Under certain parameter conditions, these stresses overcome the surface tension, and a very thin spout of the lower liquid is ejected across the collector tube. This is the so-called viscous entrainment mode of the selective withdrawal technique, an example of microjetting [59–62].

This visco-capillary phenomenon is governed by the capillary number [62]

$$Ca_o^{\mathrm{sw}} = \frac{\mu_o Q_o}{4\pi \widetilde{H} \sigma}, \tag{7.16}$$

where μ_o and Q_o are the upper-layer viscosity and flow rate, respectively, and $\widetilde{H}$ is the distance between the feeding capillary and the undisturbed (horizontal) interface. This dimensionless number expresses the competition between the viscous driving force and the surface tension opposing the ejection. The outer capillary number (7.8) characterizing the coflowing configuration can be used in selective withdrawal as well [63].

An electrohydrodynamic process analogous to the selective withdrawal occurs when an electrified collector tube is placed in front of a liquid film of finite conductivity [64, 65]. For small applied voltages, the free surface forms a hump in front of the tube owing to the action of the electric stresses. Above a critical voltage, the hump ejects a very thin liquid thread that moves into the tube. In this technique, the role of the capillary number (7.16) is played by the electric Bond number (7.5).

7.7 Surfactants and Viscoelasticity

7.7.1 Surfactants

The existence of a surfactant monolayer adsorbed onto the interface substantially modifies the tip streaming flow in the microfluidic configurations described above. As explained below, if the variation of the interfacial tension with the surface coverage is approximated by, e.g., the Langmuir equation of state (7.23), the Marangoni (elasticity) number and the dimensionless surface concentration,

$$\mathrm{Ma} = \frac{\Gamma_\infty R_g T}{\sigma_c} \quad \text{and} \quad \hat{\Gamma}_{\mathrm{eq}} = \frac{\Gamma_{\mathrm{eq}}}{\Gamma_\infty}, \tag{7.17}$$

characterize the effect of the monolayer. Here, Γ_∞ is the maximum packing density (which is taken as the reference value for the surface coverage), R_g the universal gas constant, T the temperature, and σ_c the characteristic value of the interfacial tension. In addition, Γ_{eq} is the surfactant surface concentration at equilibrium.

The surface Peclet number $\mathrm{Pe}_s = L\sigma_c/(\mu_i \mathscr{D}_s)$ is used to quantify the diffusion of the surfactant molecules over the interface. Here, $\mathscr{D}_s$ is the surface diffusion coefficient and L a characteristic length (e.g., the capillary radius R_i).

When the dilatational and shear surface viscosities, κ^s and μ^s, are accounted for, the respective Boussinesq numbers

$$B_\kappa^s = \frac{\kappa^s}{\mu_c L} \quad \text{and} \quad B_\mu^s = \frac{\mu^s}{\mu_c L} \tag{7.18}$$

enter the problem. Here, μ_c is a characteristic viscosity (either μ_i or μ_o). Alternatively, it may be useful to compare the surface viscosities with inertia and surface tension by defining the surface Ohnesorge numbers [66, 67],

$$\mathrm{Oh}_\kappa^s = \frac{\kappa^s}{(\rho_i \sigma_c L^3)^{1/2}} \quad \text{and} \quad \mathrm{Oh}_\mu^s = \frac{\mu^s}{(\rho_i \sigma_c L^3)^{1/2}}. \tag{7.19}$$

7.7.1.1 Soluble Surfactants

If the surfactant solubility plays a significant role, the adsorption/desorption process must be modeled. If one adopts, e.g., the approximation (2.38), the dimensionless adsorption and desorption coefficients

$$\hat{k}_a = \frac{k_a c_{\mathrm{cmc}} L \mu_c}{\Gamma_\infty \sigma_c} \quad \text{and} \quad \hat{k}_d = \frac{k_d L \mu_c}{\sigma_c} \tag{7.20}$$

are considered, where k_a and k_d are the adsorption and desorption coefficients, respectively, and c_{cmc} is the critical micelle concentration, which is taken as the reference concentration. Alternatively, it may be useful to compare adsorption/desorption with convection by defining the adsorption/desorption Biot number

$$\tilde{k}_a = \frac{k_a c_{\mathrm{cmc}} L}{\Gamma_\infty v_c}, \quad \text{and} \quad \tilde{k}_d = \frac{k_d L}{v_c}, \tag{7.21}$$

where the viscous-capillary velocity σ_c/μ_c has been replaced with the characteristic velocity v_c.

In addition, the volumetric Peclet number $\mathrm{Pe} = L\sigma_c/(\mu_c\mathscr{D})$ ($\mathscr{D}$ is the volumetric diffusion coefficient) is introduced to quantify the surfactant diffusion across the bulk. The dimensionless surface concentration $\hat{\Gamma}_{\mathrm{eq}}$ can be replaced with the corresponding dimensionless volumetric concentration $\hat{c}_\infty = c_\infty/c_{\mathrm{cmc}}$ because the Langmuir isotherm [Eq. (2.39)] links them at equilibrium:

$$\frac{\hat{c}_\infty}{\hat{k}_d/\hat{k}_a} = \frac{\hat{\Gamma}_{\mathrm{eq}}}{1-\hat{\Gamma}_{\mathrm{eq}}}. \tag{7.22}$$

To summarize, insoluble surfactants are typically characterized by the set of dimensionless numbers $\{\mathrm{Ma}, \hat{\Gamma}_{\mathrm{eq}}, B^s_\kappa, B^s_\mu, \mathrm{Pe}_s\}$. The parameters $\{\hat{k}_a, \hat{k}_d, \mathrm{Pe}\}$ must also be considered in the presence of solubility effects.

7.7.2 The Characteristic Surface Tension

The most apparent and trivial surfactant effect on the flow is the global reduction of the interfacial tension. This effect may mask the more interesting role played by local soluto-capillarity, Marangoni stress, and surface viscosity. To eliminate the global reduction of the interfacial tension from the analysis, we use the equilibrium interfacial tension (instead of the clean interface surface tension) as the characteristic surface tension σ_c in all the dimensionless numbers involving this quantity.

Consider a fluid-fluid system with a clean interface and another loaded with a surfactant monolayer producing the same equilibrium interfacial tension. These two systems may behave differently due to the inhomogeneity of the interfacial tension and the surface viscosity when the surfactant monolayer is present. With the choice $\sigma_c = \sigma_{\mathrm{eq}}$, these systems are characterized by the same values of the dimensionless numbers of the surfactant-free case (provided that the densities and bulk viscosities are the same in the two systems). Therefore, the surfactant's dimensionless numbers quantify the non-trivial effects of local soluto-capillarity, Marangoni convection, and surface viscosity (not the global reduction of the interfacial tension).

Using σ_{eq} as the reference value for σ leads to the dimensionless Langmuir equation of state

$$\hat{\sigma} = 1 + \mathrm{Ma} \ln\left(\frac{1-\hat{\Gamma}}{1-\hat{\Gamma}_{\mathrm{eq}}}\right), \tag{7.23}$$

where $\hat{\sigma} = \sigma/\sigma_{\rm eq}$, Ma$=\Gamma_\infty R_g T/\sigma_{\rm eq}$, and $\hat{\Gamma} = \Gamma/\Gamma_\infty$. The linearized equation of state for $\hat{\Gamma} \simeq \hat{\Gamma}_{\rm eq}$ is

$$\hat{\sigma} = 1 - \text{Ma}\left(\frac{\hat{\Gamma} - \hat{\Gamma}_{\rm eq}}{1 - \hat{\Gamma}_{\rm eq}}\right), \tag{7.24}$$

which shows that both the Marangoni number Ma and the dimensionless surface concentration $\hat{\Gamma}_{\rm eq}$ determine the surfactant strength. Ma$\gtrsim 0.1$ typically corresponds to strong surfactants, which are expected to produce noticeable effects.

It is worth noting that using $\sigma_{\rm eq}$ as the reference value for σ leads to a Marangoni number that depends on the surfactant concentration because $\sigma_{\rm eq}$ is a function of $\Gamma_{\rm eq}$. In this case, Ma is not an intrinsic parameter of the surfactant-liquid system but a measure of the strength of a surfactant monolayer with a given surface density. In other words, experiments with different surfactant concentrations are characterized by different values of the Marangoni number Ma$=\Gamma_\infty R_g T/\sigma_{\rm eq}$.

The use of $\sigma_{\rm eq}$ as the characteristic surface tension may not be appropriate in certain tip streaming configurations, such as flow focusing, in which the surfactant is swept away from the interface and the surface tension takes values much higher than $\sigma_{\rm eq}$.

7.7.3 Viscoelasticity

In the microjetting mode of tip streaming, the liquid sharply accelerates at the tip of the tapering meniscus, leading to high strain rates. This can induce the coil-stretch transition of polymer molecules even with low molecular weights [68]. In this case, the elastic stresses built up in the meniscus tip may substantially alter the flow stability and, therefore, the performance of the microjetting mode. For this reason, the most significant dimensionless number is the Weissenberg number Wi $= \lambda_e \langle\dot{\varepsilon}\rangle$, where λ_e is the extensional relaxation time (see Chap. 4), $\langle\dot{\varepsilon}\rangle = v_\ell/\ell_m$ is the effective (mean) strain rate along the meniscus axis, v_ℓ is the liquid velocity at the end of the meniscus, and ℓ_m is the meniscus length. For a sufficiently large value of Wi, the coil-stretch transition takes place at the tip of the tapering meniscus.

The effective strain rate $\langle\dot{\varepsilon}\rangle$ is not known a priori, and, therefore, the Weissenberg number cannot be strictly considered as a control parameter. The Deborah number De $= \lambda_e/t_{\rho\sigma}$ can be used for this purpose. Here, $t_{\rho\sigma} = (\rho_i L^3/\sigma)^{1/2}$ is the inertio-capillary time defined in terms of the characteristic length L (for instance, the orifice diameter D_o in flow focusing). However, it should be noted that the comparison between λ_e and $t_{\rho\sigma}$ may not be relevant for the local extensional flow occurring in the meniscus tip.

References

1. Shin D, Yoo S, Song H, Tak H, Byun D (2005) Electrostatic-force-assisted dispensing printing to construct high-aspect-ratio of 0.79 electrodes on a textured surface with improved adhesion and contact resistivity. Sci Rep 5:16,704
2. Herrada MA, Gañán-Calvo AM, Ojeda-Monge A, Bluth B, Riesco-Chueca P (2008) Liquid flow focused by a gas: jetting, dripping, and recirculation. Phys Rev E 78(036):323
3. Herrada MA, López-Herrera JM, Gañán-Calvo AM, Vega EJ, Montanero JM, Popinet S (2012) Numerical simulation of electrospray in the cone-jet mode. Phys Rev E 86(026):305
4. Anna SL, Bontoux N, Stone HA (2003) Formation of dispersions using flow focusing in microchannels. Appl Phys Lett 82:364–366
5. Castro-Hernández E, van Hoeve W, Lohse D, Gordillo JM (2011) Microbubble generation in a co-flow device operated in a new regime. Lab Chip 11:2023–2029
6. Duffy DC, McDonald JC, Schueller OJA, Whitesides GM (1998) Rapid prototyping of microfluidic systems in poly(dimethylsiloxane). Anal Chem 70:4974–4984
7. Unger MA, Chou HP, Thorsen T, Scherer A, Quake SR (2000) Monolithic microfabricated valves and pumps by multilayer soft lithography. Science 288:113–116
8. Takeuchi S, Garstecki P, Weibel DB, Whitesides GM (2005) An axisymmetric flow-focusing microfluidic device. Adv Mater 17:1067–1072
9. Utada AS, Lorenceau E, Link DR, Kaplan PD, Stone HA, Weitz DA (2005) Monodisperse double emulsions generated from a microcapillary device. Science 308:537–541
10. DePonte DP, Weierstall U, Schmidt K, Warner J, Starodub D, Spence JCH, Doak RB (2008) Gas dynamic virtual nozzle for generation of microscopic droplet streams. J Phys D Appl Phys 41(195):505
11. Muñoz-Sánchez BN, Cabezas MG (2018) Borosilicate nozzles manufactured by reproducible fire shaping. J Mater Proces Tech 261:173–183
12. Muñoz-Sánchez BN, Gañán-Calvo AM, Cabezas MG (2019) A new fire shaping approach to produce highly axisymmetric and reproducible nozzles. J Mater Process Tech 270:241–253
13. Si T, Feng HX, Luo XS, Xu RX (2015) Formation of steady compound cone-jet modes and multilayered droplets in a tri-axial capillary flow focusing device. Microfluid Nanofluid 18:967–977
14. Trebbin M, Kruger K, DePonte D, Roth SV, Chapman HN, Forster S (2014) Microfluidic liquid jet system with compatibility for atmospheric and high-vacuum conditions. Lab Chip 14:1733–1745
15. Beyerlein KR, Adriano L, Heymann M, Kirian R, Knoska J, Wilde F, Chapman HN, Bajt S (2015) Ceramic micro-injection molded nozzles for serial femtosecond crystallography sample delivery. Rev Sci Instrum 86(125):104
16. Knoska J et al (2020) Ultracompact 3D microfluidics for time-resolved structural biology. Nat Commun 11:657
17. Rubio A, Rodríguez S, Cabezas MG (2020) Capabilities and limitations of fire-shaping to produce glass nozzles. Materials 13:5477
18. Nelson G et al (2016) Three-dimensional-printed gas dynamic virtual nozzles for X-ray laser sample delivery. Opt Express 24:11,515–11,530
19. Stone HA, Stroock A, Ajdari A (2004) Engineering flows in small devices: Microfluidics toward a lab-on-a-chip. Annu Rev Fluid Mech 36:381–411
20. Cristini V, Tan YC (2004) Theory and numerical simulation of droplet dynamics in complex flows-a review. Lab Chip 4:257–264
21. Whitesides GM (2006) The origins and the future of microfluidics. Nature 442:368–373
22. Christopher GF, Anna SL (2007) Microfluidic methods for generating continuous droplet streams. J Phys D Appl Phys 40:R319–R336
23. Anna SL (2016) Droplets and bubbles in microfluidic devices. Annu Rev Fluid Mech 48:285–309
24. Barrero A, Loscertales IG (2007) Micro and nanoparticles via capillary flows. Annu Rev Fluid Mech 39:89–106

25. Gu H, Duits MHG, Mugele F (2011) Droplets formation and merging in two-phase flow microfluidics. Int J Mol Sci 12:2572–2597
26. Vladisavljevic GT, Kobayashi I, Nakajima M (2012) Production of uniform droplets using membrane, microchannel and microfluidic emulsification devices. Microfluid Nanofluid 13:151–178
27. Zhuab P, Wang L (2017) Passive and active droplet generation with microfluidics: a review. Lab Chip 17:34–75
28. Montanero JM, Gañán-Calvo AM (2020) Dripping, jetting and tip streaming. Rep Prog Phys 83(097):001
29. Snoeijer JH, Andreotti B (2013) Moving contact lines: scales, regimes, and dynamical transitions. Annu Rev Fluid Mech 45:269–292
30. Sui Y, Ding H, Spelt PDM (2014) Numerical simulations of flows with moving contact lines. Annu Rev Fluid Mech 46:97–119
31. Oberthuer D et al (2017) Double-flow focused liquid injector for efficient serial femtosecond crystallography. Sci Rep 7(44):628
32. Acero AJ, Montanero JM, Ferrera C, Herrada MA, Gañán-Calvo AM (2012) Enhancement of the stability of the flow focusing technique for low-viscosity liquids. J Micromech Microeng 22(115):039
33. Morad M, Rajabi A, Razavi M, Sereshkeh SP (2016) A very stable high throughput Taylor cone-jet in electrohydrodynamics. Sci Rep 6(38):509
34. Sen AK, Darabi J, Knapp DR, Liu J (2006) Modeling and characterization of a carbon fiber emitter for electrospray ionization. J Micromech Microeng 16:620–630
35. Lozano P, Martínez-Sánchez M (2005) On the dynamic response of externally wetted ionic liquid ion sources. J Phys D 38:2371–2377
36. Dole M, Mack LL, Hines RL, Mobley RC, Ferguson LD, Alice MB (1968) Molecular beams of macroions. J Chem Phys 49:2240–2249
37. Fenn JB, Mann M, Meng CK, Wong SF, Whitehouse CM (1989) Electrospray ionization for mass spectrometry of large biomolecules. Science 246:64–71
38. Whitehouse CM, Dreyer RN, Yamashita M, Fenn JB (1985) Electrospray interface for liquid chromatographs and mass spectrometers. Anal Chem 57:675–679
39. Acero AJ, Rebollo-Muñoz N, Montanero JM, Gañán-Calvo AM, Vega EJ (2013) A new flow focusing technique to produce very thin jets. J Micromech Microeng 23(065):009
40. Rebollo-Muñoz N, Acero AJ, Marcos de León JZ, Montanero J, Gañán-Calvo A (2016) A hybrid flow focusing nozzle design to produce micron and sub-micron capillary jets. Int J Mass Spectrom 403:32–38
41. Nazari R, Zaare S, Alvarez RC, Karpos K, Engelman T, Madsen C, Nelson G, Spence JCH, Weierstall U, Adrian RJ, Kirian RA (2020) 3D printing of gas-dynamic virtual nozzles and optical characterization of high-speed microjets. Opt Express 28:21,749–21,765
42. Ponce-Torres A, Rebollo-Muñoz N, Herrada MA, Gañán-Calvo AM, Montanero JM (2018) The steady cone-jet mode of electrospraying close to the minimum volume stability limit. J Fluid Mech 857:142–172
43. Gañán-Calvo AM, Rebollo-Muñoz N, Montanero JM (2013) Physical symmetries and scaling laws for the minimum or natural rate of flow and droplet size ejected by Taylor cone-jets. New J Phys 15(033):035
44. Suryo R, Basaran OA (2006) Tip streaming from a liquid drop forming from a tube in a co-flowing outer fluid. Phys Fluids 18(082):102
45. Guillot P, Colin A, Utada AS, Ajdari A (2007) Stability of a jet in confined pressure-driven biphasic flows at low Reynolds numbers. Phys Rev Lett 99(104):502
46. Guillot P, Colin A, Ajdari A (2008) Stability of a jet in confined pressure-driven biphasic flows at low Reynolds number in various geometries. Phys Rev E 78(016):307
47. Cramer C, Fischer P, Windhab EJ (2004) Drop formation in a co-flowing ambient fluid. Chem Engin Sci 59:3045–3058
48. Utada AS, Fernandez-Nieves A, Gordillo JM, Weitz DA (2008) Absolute instability of a liquid jet in a coflowing stream. Phys Rev Lett 100(014):502

49. Cabezas MG, Herrada MA, Montanero JM (2019) Stability of a jet moving in a rectangular microchannel. Phys Rev E 100(053):104
50. Gañán-Calvo AM (1998) Generation of steady liquid microthreads and micron-sized monodisperse sprays in gas streams. Phys Rev Lett 80:285–288
51. Gañán-Calvo AM, Montanero JM (2009) Revision of capillary cone-jet physics: electrospray and flow focusing. Phys Rev E 79(066):305
52. Vega EJ, Montanero JM, Herrada MA, Gañán-Calvo AM (2010) Global and local instability of flow focusing: the influence of the geometry. Phys Fluids 22(064):105
53. Cruz-Mazo F, Montanero JM, Gañán-Calvo AM (2016) Monosized dripping mode of axisymmetric flow focusing. Phys Rev E 94(053):122
54. Rubio M, Rubio A, Cabezas MG, Herrada MA, Gañán-Calvo AM, Montanero JM (2021) Transonic flow focusing: stability analysis and jet diameter. Int J Multiphase Flow 142(103):720
55. Gañán-Calvo AM, Gordillo JM (2001) Perfectly monodisperse microbubbling by capillary flow focusing. Phys Rev Lett 87(274):501
56. Gañán-Calvo AM (2004) Perfectly monodisperse microbubbling by capillary flow focusing: an alternate physical description and universal scaling. Phys Rev E 69(027):301
57. Jensen MJ, Stone HA, Bruus H (2006) A numerical study of two-phase Stokes flow in an axisymmetric flow-focusing device. Phys Fluids 18(077):103
58. Vega EJ, Acero AJ, Montanero JM, Herrada MA, Gañán-Calvo AM (2014) Production of microbubbles from axisymmetric flow focusing in the jetting regime for moderate Reynolds numbers. Phys Rev E 89(063):012
59. Lister JR (1989) Selective withdrawal from a viscous two-layer system. J Fluid Mech 198:231–254
60. Cohen I, Li H, Hougland JL, Mrksich M, Nagel SR (2001) Using selective withdrawal to coat microparticles. Science 292:265–267
61. Case SC, Nagel SR (2007) Spout states in the selective withdrawal of immiscible fluids through a nozzle suspended above a two-fluid interface. Phys Rev Lett 98(114):501
62. Blanchette F, Zhang WW (2009) Force balance at the transition from selective withdrawal to viscous entrainment. Phys Rev Lett 102(144):501
63. Rubio M, Montanero JM (2023) Influence of a soluble surfactant on the transition to tip streaming. Exp Therm Fluid Sci 141(110):776
64. Oddershede L, Nagel SR (2000) Singularity during the onset of an electrohydrodynamic spout. Phys Rev Lett 85:1234–1237
65. Collins RT, Jones JJ, Harris MT, Basaran OA (2008) Electrohydrodynamic tip streaming and emission of charged drops from liquid cones. Nat Phys 4:149–154
66. Ponce-Torres A, Montanero JM, Herrada MA, Vega EJ, Vega JM (2017) Influence of the surface viscosity on the breakup of a surfactant-laden drop. Phys Rev Lett 118(024):501
67. Ponce-Torres A, Rubio M, Herrada MA, Eggers J, Montanero JM (2020) Influence of the surface viscous stress on the pinch-off of free surfaces loaded with nearly-inviscid surfactants. Sci Rep 10(16):065
68. Rubio A, Galindo F, Vega EJ, Montanero JM, Cabezas MG (2022) Viscoelastic transition in transonic flow focusing. Phys Rev Fluids 7(074):201

Chapter 8
Electrospray

Abstract This chapter focuses on some fundamental results of the classical steady cone-jet mode of electrospray. Specifically, we review the scaling laws that allow one to estimate the droplet diameter, electric current, breakup length, and minimum flow rate stability limit. The numerical solution of the leaky-dielectric model enables us to describe the electrospray phenomenon for flow rates relatively close to the minimum value. We present the results of the global stability analysis, paying attention to charge relaxation. The characteristics of coaxial Taylor cone-jets and AC electrospray are also briefly reviewed. Little attention is paid to the electrospinning configuration, which has been nicely reviewed in several works [1, 2].

Keywords Steady cone-jet mode · Scaling laws · Minimum flow rate · Coaxial electrospray

8.1 Introduction

8.1.1 The Steady Cone-Jet Mode

Electrospray [3, 4] is a microfluidic technique for liquid atomization driven by an intense electric field applied to a droplet. The interaction between electric, capillary, and viscous stresses gives rise to the formation of charged jets and droplets. Electrospray can operate in the standard dripping and jetting modes, producing droplets with sizes similar to or even larger than the needle [5–7]. More interestingly, electrospray can produce tiny droplets via microdripping and microjetting [8, 9], two remarkable examples of tip streaming driven by electrohydrodynamic forces.

The microjetting mode of electrospray, known as the steady cone-jet mode (Fig. 8.1), is the most popular realization of tip streaming. In the typical configuration, a low-conductivity liquid is injected into a passive dielectric medium across a metallic feeding capillary hundreds of microns in diameter. The injection occurs at constant

J. M. Montanero, *Tip Streaming of Simple and Complex Fluids*, Fluid Mechanics and Its Applications 137, https://doi.org/10.1007/978-3-031-52768-5_8

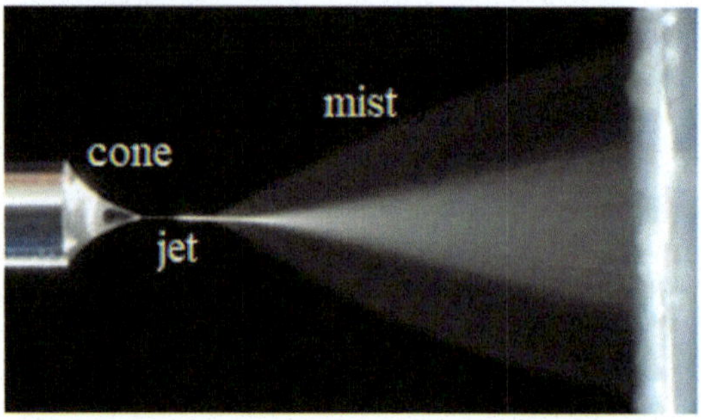

Fig. 8.1 The cone-jet mode of electrospray (reprinted from Ref. [12] with permission of Elsevier)

flow rates in the range 10^{-3}–10 ml. One applies an external electric field by imposing a DC voltage of the order of kilovolts in the feeding capillary. The grounded electrode is located some millimeters downstream from the capillary. When the liquid properties, flow rate, and voltage are selected adequately, the cone-jet mode of electrospraying is obtained. In this mode, the liquid meniscus attached to the needle adopts a stable quasi-conical shape (the Taylor cone) whose tip steadily ejects a very thin jet. This electrified jet eventually breaks into droplets due to capillary forces, giving rise to a charged mist. This tip streaming flow can be seen as a way to escape the singularity arising in the vertex of the Taylor cone, which results from the exact static balance between surface tension and electric forces [10, 11].

In the cone-jet mode of electrospray, the net free charge migrates toward the free surface and vanishes in the bulk. In most realizations, the charge accumulated on the meniscus surface practically screens the outer electric field (the inner electric field vanishes). The liquid meniscus is essentially equipotential and remains practically at rest. The most interesting electrohydrodynamic phenomena occur in the critical cone-jet transition region, whose size L is typically much smaller than the meniscus (Fig. 8.2).

In the cone-jet transition region, the liquid sharply accelerates, and the surface charge significantly increases, but it no longer screens the electric field. Consequently, a significant tangential electric field arises, leading to a Maxwell shear stress (see Chap. 2) This stress collaborates with the drop of hydrostatic pressure (the electrostatic suction) to drive the liquid ejection [13, 14]. The charge is conducted across the meniscus bulk and convected by the jet's free surface. Both mechanisms are commensurate in the cone-jet transition region [15].

The steady flow analyzed in this chapter is fundamentally different from the transient tip streaming arising when a droplet bursts due to the action of an intense electric field (see Chap. 5). The injection of liquid at a constant flow rate allows the system to reach a steady flow characterized by conservation equations that do not apply to the droplet bursting. Therefore, the analysis presented in this chapter cannot be extended to the electrohydrodynamic tip streaming occurring during a droplet breakup.

As mentioned above, the net free charge vanishes in the hydrodynamic bulk and accumulates in the liquid surface. The ohmic conduction model accurately predicts

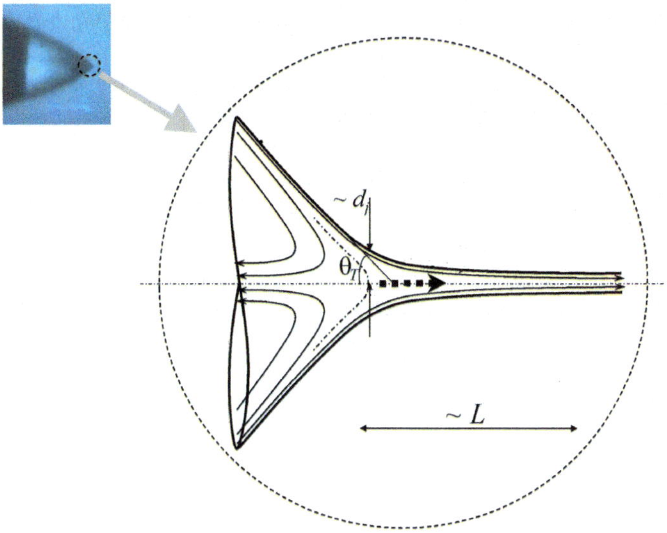

Fig. 8.2 Sketch of the critical cone-jet transition region of the cone-jet mode of electrospray. Its radial d_j and axial L length scales are very small compared to the feeding capillary diameter. θ_T is the Taylor cone angle

the charge transferred to the surface. These two conditions are at the root of the leaky-dielectric model. For this reason, this approximation provides reliable results for the steady cone-jet mode of electrospray [16]. In addition, the critical cone-jet transition region adopts a slender shape, which elicits the use of the 1D model described in Chap. 2 [17].

8.1.2 Applications

The steady cone-jet mode is observed over a considerable range of length scales. As explained below, the increase of the conductivity allows reducing the jet diameter from tens of microns to a few nanometers [18, 19]. The richness and complexity of electrospray are not found in other fields of mesoscale physics and have long ago attracted the attention of many investigators [3, 4, 20–22].

The cone-jet mode applications are countless. For instance, it has been applied to produce charged droplets in mass spectrometry [8] and to generate electric propulsion for spacecraft [23]. Electrospray is also used for high-resolution electrohydrodynamic jet printing [24], improving the quality of Electrostatic Inkjet Printing [25] with nanoparticle inks [26]. New applications are continuously emerging. For example, electrospray can constitute an advantageous alternative to flow focusing for sample injection for single-particle imaging with X-ray lasers [27].

In the coaxial version of the cone-jet mode of electrospray, two immiscible liquids are injected into a compound pendant droplet from whose tip a jet carrying the two phases is ejected. Different capillary shapes and structures, such as emulsions and microparticles, are fabricated in the food industry, drug encapsulation [28, 29] and tissue engineering [30] with coaxial electrohydrodynamic atomization [31–34]. This technique has also been used for producing polymer coated starch-protein microspheres [35], suspensions containing microbubbles [36], as well as organic and inorganic macro and nano-sized emulsions [33]. Recent advances in coaxial electrohydrodynamic atomization have enabled the fabrication of micro- and nano-sized biodegradable polymeric particles for controlled drug release and biochemical applications [29, 37].

8.1.3 Electrospinning

In electrospinning, the electric field is applied to a pendant droplet of a viscoelastic liquid containing polymer macromolecules dissolved in a Newtonian solvent. The solvent evaporation during the jet flight transforms the liquid thread into a fiber. Electrospinning of polymers has become a highly recognized method for the preparation of simple, compound, and hollow fibers [33] with diameters ranging from tens of microns to a few nanometers [1, 34, 38–47].

Electrospinning is frequently applied in medical areas such as tissue engineering and drug delivery [48], as well as in technical fields demanding nanofibers with specific electronic, photonic, photocatalytic, and magnetic properties. A wide range of complex architectures and morphologies of nanofibers and nonwovens can be produced with this technique. Nanofiber-based architectures have facilitated the development of energy storage systems, fuel and solar cells, lithium-ion batteries, and electronic sensors, among many others. Textile and filter applications have also benefited from electrospinning [49]. Sophisticated versions of this technique are continuously emerging. For instance, Liashenko et al. [50] have shown that ultrafast 3D printing with submicrometer resolution can be achieved by electrostatically deflecting the emitted jet with electrodes around it.

The polymer molecular weight and concentration are two key parameters controlling the quality of the electrospinning outcome. Lower molecular weights and concentrations lead to thinner fibers, which is desirable because it enhances fiber functionality. However, reducing the values of these two parameters decreases the solution viscosity and elasticity (stress relaxation time), two inhibitors of the capillary instability responsible for the appearance of the undesired droplets or beads. Therefore, uniform (bead-free) fibers are produced at the expense of increasing their diameters, and *vice versa*.

In electrospinning, fibers are produced chaotically owing to the bending (whipping) instability suffered by the jet during its flight. This instability constitutes a serious obstacle in the fabrication of devices that demand arranged or patterned micro/nanoscale fibrous structures. In near-field electrospinning, the substrate collecting the fibers is brought much closer to the droplet. This enables the deposition of ultrafine fibers in a direct, continuous, and controllable manner [51]. Fibers with unique physical and chemical properties are fabricated with near-field electrospinning and can be used in electronic components, flexible sensors, energy harvesting, and tissue engineering [52]. The near-field electrospinning configuration has also been used to produce ceramic fibers with potential applications in 3D printing scaffolds [53].

8.2 Governing Parameters

The cone-jet mode of electrospray is affected by the liquid density ρ, viscosity μ, surface tension σ, electrical conductivity K, and electrical permittivity ε_i. If the outer environment is either a vacuum or gas, its dynamic effect on the system can be neglected. The only additional parameter involved in the problem is the outer electrical permittivity ε_o.

Based on the above quantities, one defines the characteristic radial length $d_o = [\sigma\varepsilon_o^2/(\rho K^2)]^{1/3}$ and axial velocity $v_o = [\sigma K/(\rho\varepsilon_o)]^{1/3}$. Two intrinsic dimensionless numbers can be formed with the outer permittivity, the liquid properties, and these characteristic scales: the relative permittivity $\varepsilon = \varepsilon_i/\varepsilon_o$ and the electrohydrodynamic Reynolds number $\delta_\mu = \rho v_o d_o/\mu = [\sigma^2\rho\varepsilon_o/(\mu^3 K)]^{1/3}$ [54]. The dimensionless numbers ε and δ_μ are the same as those characterizing the transient bursting of an electrified droplet (see Chap. 5).

The liquid properties and the outer medium permittivity also allow us to define the characteristic electric field $E_o = (\sigma^2\rho K^2/\varepsilon_o^5)^{1/6}$ and electric current $I_o = \sigma\rho^{-1/2}\varepsilon_o^{1/2}$.

In the steady cone-jet mode of electrospray, the injected flow rate Q_i becomes the major control parameter. Its dimensionless counterpart is $Q = Q_i/Q_c$, where $Q_c = v_o d_o^2 = \sigma\varepsilon_o/(\rho K)$ is the characteristic flow rate. There is a narrow interval of applied voltage V within which the steady cone-jet mode can be established. One typically selects that quantity within that interval so that the apparent semiangle θ_T of the conical meniscus (Fig. 8.2) is approximately that of Taylor's solution, $\theta_T = 49°$ [55]. Therefore, the applied voltage V is not regarded as an independent variable in most cases. Some studies have examined the influence of V on the electrospray cone-jet mode. This quantity enters the problem through the electric Bond (Taylor) number $\chi = \varepsilon_o V^2/(R_i\sigma)$, where R_i is the radius of the triple contact line anchored at the feeding capillary end.

In the steady cone-jet mode, conduction gives way to dominant charge convection over the liquid surface within the cone-jet transition region [56]. This critical region is typically much smaller than the electrospray device, which confers local character on the phenomenon (Fig. 8.2). For this reason, the device's geometrical features and associated lengths generally play a secondary role in electrospray.

When the radial or axial dimension of the cone-jet transition region is commensurate with the Taylor cone size, the latter may affect some features of the cone-jet mode. In this case, one defines the diameter ratio $\Lambda = d_o/(2R_i) = [\sigma\varepsilon_o^2/(8\rho K^2 R_i^3)]^{1/3}$ to account for the feeding capillary size. For $\Lambda \gtrsim 10^{-2}$, the cone-jet mode stability can be influenced by the radius R_i [57, 58]. For instance, it is well known that the feeding capillary stabilizes the cone-jet mode of electrospray in nanoelectrospray [59, 60]. Finally, when the jet tapers from a millimeter meniscus, gravity may play a significant role, which is taken into account through the gravitational Bond number $B = \rho g R_i^2/\sigma$.

8.3 Scaling Laws

As explained below, there is a minimum flow rate below which the cone-jet mode of electrospray becomes unstable. Some useful scaling laws can be straightforwardly obtained when the cone-jet mode safely operates far from that stability limit. We devote this section to deriving those scaling laws.

8.3.1 Droplet Diameter

The tangential component E_t of the electric field in the cone-jet transition region is essentially determined by the Taylor cone [11, 56, 61] (Fig. 8.3), i.e.,

$$E_t \sim \left(\frac{\sigma}{\varepsilon_o L}\right)^{1/2}, \tag{8.1}$$

where L is the length of that region. This means that E_t in the cone-jet transition region is commensurate with the normal outer component in the Taylor cone (Fig. 8.3) [54].

The energy consumed by surface tension, viscosity, polarization force, and Joule effect in the cone-jet transition region is at most comparable with the increase of kinetic energy there [62]. This implies that the power supplied by the electric field to the liquid in that region is essentially converted into kinetic energy. The electric power can be calculated as $I\Delta V$, where $I \sim K E_t d_j^2$ is the electric current, d_j is the jet's diameter, and $\Delta V \sim E_t L$ is the voltage drop in the cone-jet transition region. Therefore,

$$K E_t d_j^2 (E_t L) \sim \rho v_j^2 Q_i, \tag{8.2}$$

where $v_j \sim Q_i/d_j^2$ is the jet's velocity. Taking into account Eqs. (8.1) and (8.2), the jet's diameter and velocity scale as

$$d_j \sim d_o Q^{1/2} \quad \text{and} \quad v_j \sim v_o\,. \tag{8.3}$$

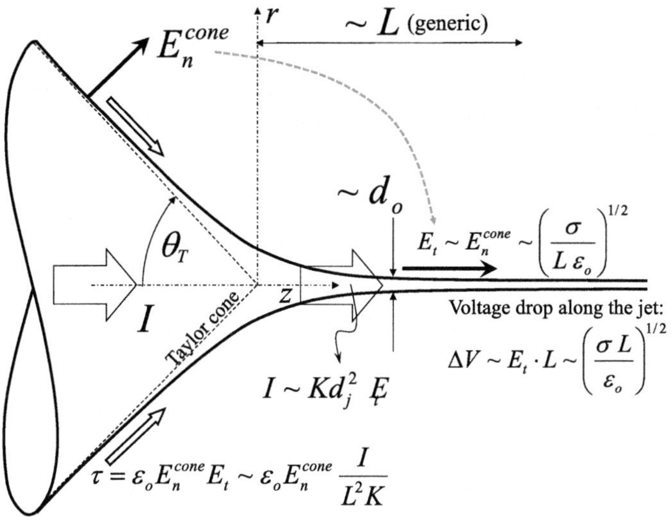

Fig. 8.3 Characteristic scales of the cone-jet mode of electrospray

The diameter of the droplets resulting from the capillary breakup of the emitted jet is commensurate with d_j. Therefore, the droplet diameter d_d scales as [10, 56]

$$d_d \sim \left(\frac{\rho\varepsilon_o Q_i^3}{\sigma K}\right)^{1/6}. \tag{8.4}$$

It is worth noting that this result is derived from the Taylor cone solution (8.1) and the energy balance (8.2) in the cone-jet transition region, without resorting to any additional hypothesis about the flow or the electric field in that region.

The scaling law (8.4) exhibits remarkable agreement with experimental data gathered from many authors (Fig. 8.4) [63, 64]. Significant deviations can be expected in the viscous and polarization regimes, corresponding to small and large values of δ_μ and ε, respectively, where viscosity and polarization forces can translate into considerable sinks of energy.

Equation (8.4) differs from that derived by Fernandez de la Mora and Loscertales [15], who assumed that (i) the flow in front of the jet corresponds to a spherically symmetric sink, and (ii) the electric relaxation time is of the same order of magnitude as that of the fluid particle residence time in that region.

A simple way of describing most electrospray realizations is to assume that an effective pressure drop fuels the liquid inertia. Considering the above results, this pressure drop can be estimated as [63]

$$\Delta p_e \sim \rho v_o^2 \sim \left(\frac{\sigma^2 K^2 \rho}{\varepsilon_o^2}\right)^{1/3}. \tag{8.5}$$

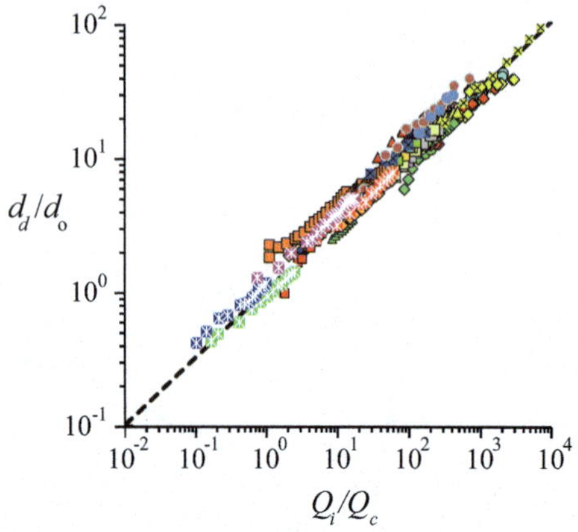

Fig. 8.4 Diameter d_d of the droplet emitted in the cone-jet mode of electrospray and measured by different authors. The dashed line corresponds to the scaling law (8.4) (reprinted from Ref. [64] with permission of Elsevier)

Interestingly, Δp_e is a function of the liquid properties (and ε_o) exclusively, i.e., it does not depend on the injected flow rate Q_i and the applied voltage V. It must be pointed out that the surface tension σ affects the fluid dynamics only through its influence on the Taylor cone electric field [Eq. (8.1)], not because it constitutes a significant sink of energy.

Equation (8.5) constitutes the link between electrospray and gaseous flow focusing [63], where the focused liquid current is driven by the pressure drop applied to the focusing gas stream (see Chap. 10). The liquid density and surface tension take similar values in many electrospray realizations. This means that one can consider the electrical conductivity K as the *effective* operational parameter in electrospray. Equation (8.5) allows one to determine the corresponding driving pressure drop. As occurs in flow focusing, the jet diameter d_j can be estimated from that value as

$$d_j \sim \left(\frac{\Delta p_e}{\rho Q_i^2}\right)^{1/4}. \tag{8.6}$$

Equations (8.5) and (8.6) have frequently been used by experimentalists to estimate the electrosprayed jet diameter.

8.3.2 Electric Current

Charge conduction in the cone-jet transition region scales as $K E_t d_j^2$. In addition, the surface charge density scales as $\sigma_e \sim \varepsilon_o E_n^o$ (see Chap. 2), where we have considered that $E_n^o - \varepsilon E_n^i \sim E_n^o$. Here, E_n^o and E_n^i are the outer and inner normal components of the electric field at the free surface, respectively. Therefore, charge surface convection in the cone-jet transition region, $\sigma_e v_j d_j$, scales as $\varepsilon_o E_n^o v_j d_j$.

In the cone-jet transition region, conduction gives way to surface convection, i.e.,

$$K E_t d_j^2 \sim \varepsilon_o E_n^o v_j d_j . \tag{8.7}$$

The above relationship and the scaling laws derived in the previous section lead to

$$E_t \sim E_n^o Q^{-1/2} . \tag{8.8}$$

Since $Q \gg 1$ ($Q_i \gg Q_c$) in the considered electrospray realizations, $E_t \ll E_n^o$.

Interestingly, Eqs. (8.1) and (8.3) allow one to re-write (8.7) as

$$\rho Q_i^2 d_j^{-4} \sim L \varepsilon_o E_n^o E_t / d_j . \tag{8.9}$$

The work per unit volume done by the electric shear stress $\varepsilon_o E_n^o E_t$ on a liquid thread slice of thickness Δz approximately is

$$\varepsilon_o E_n^o E_t \, (d_j \Delta z) \, L \, (d_j^2 \Delta z)^{-1} \sim L \varepsilon_o E_n^o E_t / d_j . \tag{8.10}$$

Therefore, Eq. (8.9) indicates that the jet's kinetic energy is commensurate with the work done by the electric shear stress [56].

Assuming that the kinetic energy ρv_j^2 also scales as the electrostatic suction $\varepsilon_o (E_n^o)^2$, one gets [Eq. (8.5)]

$$E_n^o \sim \left(\frac{\sigma^2 \rho K^2}{\varepsilon_o^5} \right)^{1/6} \equiv E_o . \tag{8.11}$$

This means that the scale of the surface charge density $\sigma_e \sim \varepsilon_o E_o$ is a universal value independent of the operating conditions (flow rate, applied voltage, and geometry) [56].

The scale of the electric current transported by the jet is [56]

$$I \sim \sigma_e v_j d_j \sim \varepsilon_o E_o \left(Q_i d_j^{-2} \right) d_j \Longrightarrow I \sim (\sigma K Q_i)^{1/2} \sim I_o Q^{1/2} . \tag{8.12}$$

Experiments show that the above scaling law verifies for small and moderate polarity, even for flow rates relatively close to the stability limit (Fig. 8.5).

Equation (8.9) shows that the axial characteristic length L scales as

$$L \sim d_o Q \sim d_j Q^{1/2} . \tag{8.13}$$

The ratio of the electric relaxation time $t_e = \varepsilon_i / K$ to the residence time $t_r \sim L / v_j$ scales as

$$t_e / t_r \sim \varepsilon Q^{-1} . \tag{8.14}$$

We have assumed that, for $Q \gg 1$, the electrostatic suction $\varepsilon_o (E_n^o)^2$ collaborates with the electric shear stress $\varepsilon_o E_n^o E_t$ to power the liquid jet even though $E_t \ll E_n^o$ [Eq. (8.8)]. This is licit due to the disparity between the areas of the surfaces where

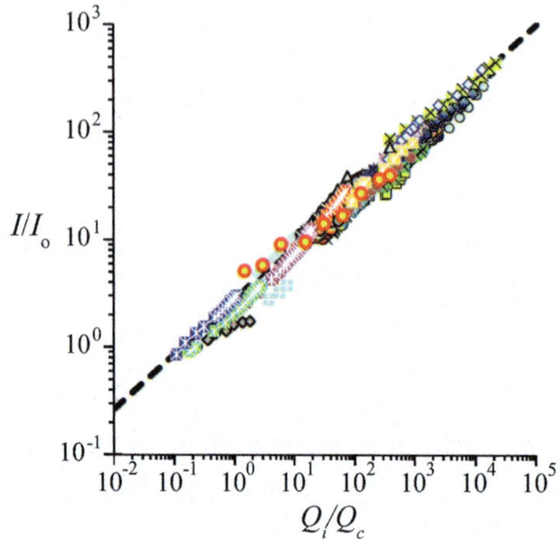

Fig. 8.5 Electric current I emitted in the cone-jet mode of electrospray and measured by different authors. The dashed line is the scaling law (8.12) (reprinted from Ref. [64] with permission of Elsevier)

those stresses are applied. While the shear stress acts over the lateral surface of the slender cone-jet transition region, the electrostatic suction pushes the liquid across the jet cross-section. Equations (8.8) and (8.13) show that the forces exerted by these two stresses are of the same order of magnitude, which constitutes a consistency test of the scaling analysis.

According to the present scaling analysis, v_j and E_n^o are intrinsic properties of the liquid, while d_j, E_t, I, and L depend on the injected flow rate as well. The results hold for flow rates much larger than those leading to instability. Nevertheless, the simulations show that they are qualitatively valid close to the minimum flow rate stability limit, even when viscous effects are important [58]. However, there are significant deviations in the current intensity in the case of very polar liquids.

It is worth noting that the above scaling laws cannot be applied to electrospinning, where viscous forces cannot be neglected even for large flow rates. Besides, the locality of the phenomenon analyzed here demands the characteristic electrohydrodynamic length L to be at most comparable with the characteristic size R_i of the feeding capillary. Thus, the cases where $R_i < L$ (e.g., "nanoelectrospray") are also excluded from this analysis.

8.3.3 *Breakup Length*

The breakup of a jet issuing from a nozzle or a tapering meniscus (as occurs in electrospray) is frequently explained in terms of the growth of the dominant temporal mode (i.e., that with the largest growth rate) (see Chap. 3). This mode is supposed to be triggered by a perturbation next to the jet inception region and convected by the jet. The mode grows in amplitude while traveling along the jet until the interface pinches, determining the breakup length.

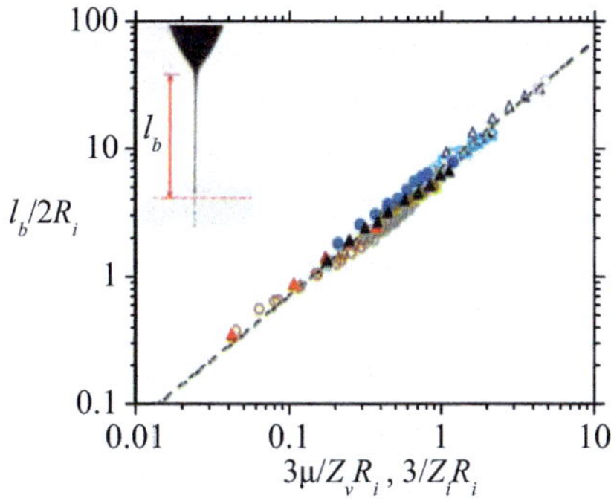

Fig. 8.6 Experimental values of the breakup length l_b of a jet emitted in the cone-jet mode of electrospray for low (open symbols) and high (solid symbols) Reynolds numbers. The dashed line is the law (8.15) (reprinted with permission from Ref. [26]. Copyright (2018) by the American Physical Society)

Based on these ideas, Ismail et al. [26] have estimated the breakup length l_b of a jet emitted in the cone-jet mode of electrospray. This length was calculated by multiplying the jet velocity and the maximum growth rate calculated by Saville [65] for a leaky-dielectric jet moving in an axial electric field [26]. The result for low and high Reynolds numbers is

$$l_b \simeq 42\ Z_v^{-1} \quad \text{and} \quad l_b \simeq 42\ Z_i^{-1}, \tag{8.15}$$

respectively, where

$$Z_v \equiv \mu^{-1}\left[\left(\frac{\rho\sigma^5\varepsilon_o}{Q_i^3\,K}\right)^{1/6} - \frac{1}{4\pi}\left(\frac{\rho^2\sigma K}{\varepsilon_o}\right)^{1/3}\right] \tag{8.16}$$

and

$$Z_i \equiv \left[\left(\frac{\sigma^5\varepsilon_o}{\rho^5 Q_i^9\,K}\right)^{1/6} - \frac{1}{4\pi}\left(\frac{\sigma K}{\rho Q_i^3\varepsilon_o}\right)^{1/3}\right]^{1/2}. \tag{8.17}$$

The above theoretical predictions agree with experimental data of liquids with low polarity (Fig. 8.6).

The physical mechanism underlying the above results does not explain the apparent insensitivity of the breakup length to the perturbations introduced in the liquid nozzle. In fact, l_b should scale as the logarithm of the initial perturbation magnitude, which may differ in several orders of magnitude between different experimental conditions.

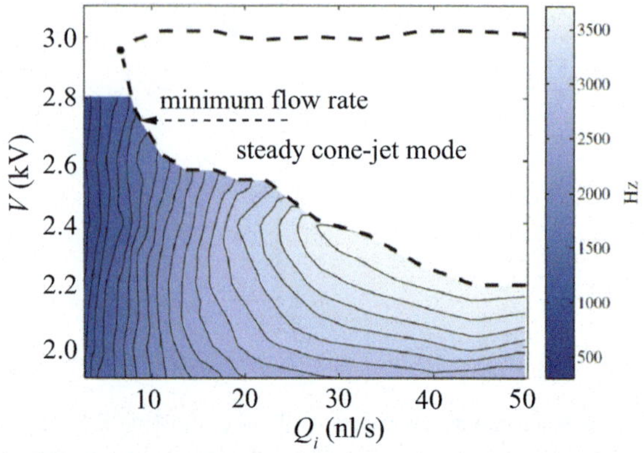

Fig. 8.7 Parameter window for the steady cone-jet mode. The contours show the frequency of the dripping mode (intermittent cone-jet) for flow rates below the minimum flow rate stability limit. The circle indicates the absolute minimum flow rate for the configuration analyzed (reproduced with permission, Ref. [67])

8.4 Minimum Flow Rate Stability Limit

In microjetting, the jet diameter could be indefinitely reduced by lowering the injected flow rate. However, experiments and numerical simulations have repeatedly shown that the flow inevitably becomes unstable at a critical value of Q_i, which sets a minimum value of the jet diameter [66]. The steady cone-jet mode of electrospray is a remarkable example of this behavior. It is adopted within a narrow parameter window [6, 67] (Fig. 8.7). For a fixed applied voltage V, the flow becomes unstable if the injected flow rate Q_i falls below its minimum value $Q_{i\min}$.

The disparity between the spatial and temporal scales of the tapering meniscus and the cone-jet region allows one to decouple the flow in these two regions. The instability at $Q_i = Q_{i\min}$ originates in the critical cone-jet region. Specifically, the balance between the electric tangential stress driving the liquid ejection and the polarity/viscosity ones opposing it breaks down in this region, and the cone-jet mode destabilizes. This instability mechanism is entirely different from that of most hydrodynamic-focusing realizations, where instability has been linked to the growth of recirculation cells in the tapering meniscus (see Chaps. 9 and 10).

Dimensional analysis dictates that the dimensionless minimum flow rate $Q_{\min} = Q_{i\min}/Q_c$ obeys the formal relationship $Q_{\min} = Q_{\min}(\varepsilon, \delta_\mu)$ as long as the stability of the cone-jet mode essentially relies on the local flow in the cone-jet transition region. It is natural to hypothesize that $Q_{\min}$ depends essentially on ε (δ_μ) for $\varepsilon \gg \delta_\mu^{-1}$ ($\varepsilon \ll \delta_\mu^{-1}$). Experimental results are consistent with the simple scaling laws $Q_{\min} \sim \varepsilon$ and $Q_{\min} \sim \delta_\mu^{-1}$ for the polar and viscous limits, respectively (Fig. 8.8). This agreement can be explained in terms of the forces opposing the liquid ejection in those limits [68]. In the region $\varepsilon \sim \delta_\mu^{-1}$ of the parameter plane (ε, δ_μ^{-1}), the polarization and viscosity forces are comparable to each other, and the minimum flow rate $Q_{\min} = f(\varepsilon, \delta_\mu)$ is not expected to verify any of those scaling laws.

It must be noted that these results rely on the local character of the jet emission. For very viscous liquids, the cone-jet transitional region spans a length scale comparable

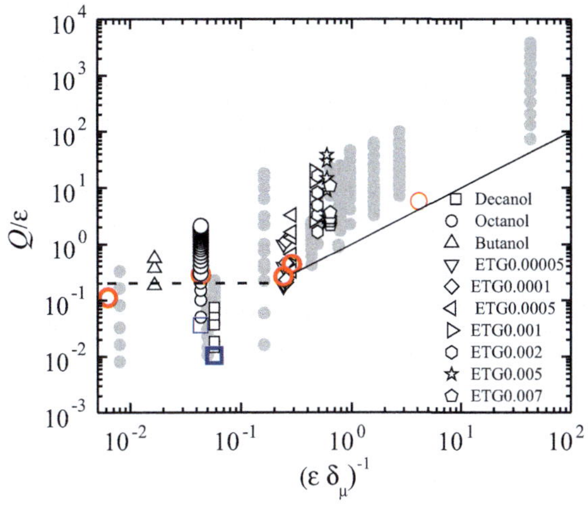

Fig. 8.8 Steady cone-jet mode realizations. The grey symbols are the data gathered from different authors. The white symbols correspond to the experimental results obtained by Ponce-Torres et al. [58] with a capillary $R_i = 105\,\mu\text{m}$ in radius. The big white circles show the experiments with 1-octanol and $R_i = 550\,\mu\text{m}$ [58]. The big red circles and big blue squares are marginally stable basic flows corresponding to $R_i = 105\,\mu\text{m}$ and $R_i = 550\,\mu\text{m}$, respectively. The dashed and solid lines are the scaling laws $Q = 0.2\varepsilon$ and $Q = \delta_\mu^{-1}$ for the polar and viscous limits, respectively (reproduced with permission, Ref. [58])

to the nozzle diameter, and the minimum flow rate can be affected by that diameter [57]. Figure 8.8 shows the stabilizing effect of the feeding capillary ($Q_{\min}$ decreases as d_o/R_i increases). In addition, small conductivities may lead to jet diameters only one order of magnitude smaller than those of the nozzle, which does not ensure the locality of the jet emission.

The scaling law $Q_{\min} \sim \varepsilon$ coincides with that proposed by Fernandez de la Mora and Loscertales [15] and Higuera [14]. However, the scaling laws derived by Higuera [14] for large δ_μ^{-1} differ from $Q_{\min} \sim \delta_\mu^{-1}$.

Gamero-Castaño and Magnani [69] examined both experimentally and numerically the limit of the minimum flow rate of the cone-jet mode of electrospray. Their major finding is that the regime dominated by polarity was not reached even for realizations with $\varepsilon \gg \delta_\mu^{-1}$ because of the significant influence of viscosity in those cases. The numerical solution of the leaky-dielectric model allowed them to hypothesize that instability occurs because viscous dissipation takes place before electric shear stresses inject energy into the system.

8.4.1 Governing Equations

The global stability analysis of the steady cone-jet mode considers not only the cone-jet region but also the Taylor cone and the emitted jet. In this section, we formulate the leaky-dielectric model (see Chap. 2) as applied to this problem. All the quantities have been made dimensionless with the triple contact line radius R_i, the liquid density ρ, the surface tension σ, and the applied voltage V, which yields the characteristic time, velocity, pressure, and electric field scales $t_{\rho\sigma} = (\rho R_i^3/\sigma)^{1/2}$, $U_{\rho\sigma} = R_i/t_{\rho\sigma}$, $p_c = \sigma/R_i$ and $E_c = V/R_i$, respectively. As can be observed, these scales differ from those used to describe the local flow in the cone-jet transition region (see Sect. 8.3).

Using cylindrical coordinates, the axisymmetric and incompressible Navier-Stokes equations are

$$(ru)_r + rw_z = 0, \tag{8.18}$$

$$\frac{\partial u}{\partial t} + uu_r + wu_z = -p_r + \mathrm{Oh}[u_{rr} + (u/r)_r + u_{zz}], \tag{8.19}$$

$$\frac{\partial w}{\partial t} + uw_r + ww_z = -p_z + \mathrm{Oh}[w_{rr} + w_r/r + w_{zz}], \tag{8.20}$$

where t is the time, r/z is the radial/axial coordinate, u/w is the radial/axial velocity component in the liquid phase, p is the pressure field, and $\mathrm{Oh} = \mu(\rho\sigma R_i)^{-1/2}$ is the Ohnesorge number. The subscripts r and z here and henceforth denote the partial derivatives with respect to the corresponding spatial variables.

The electric potentials ϕ^i and ϕ^o in both the inner (liquid) and outer (gas) domains obey the Laplace equation

$$\phi_{zz}^{i,o} + \phi_{rr}^{i,o} + \phi_r^{i,o}/r = 0. \tag{8.21}$$

The boundary conditions at the free surface $r = F(z,t)$ are:

$$\frac{\partial F}{\partial t} + F_z w - u = 0, \tag{8.22}$$

$$p + \frac{FF_{zz} - 1 - F_z^2}{F(1+F_z^2)^{3/2}} - \frac{2\mathrm{Oh}[u_r - F_z(w_r + u_z) + F_z^2 w_z]}{1+F_z^2} = -\frac{\chi}{2}\left[(E_n^o)^2 - \varepsilon(E_n^i)^2\right] - \chi\frac{\varepsilon - 1}{2}(E_t)^2, \tag{8.23}$$

$$\mathrm{Oh}\frac{(1-F_z^2)(w_r + u_z) + 2F_z(u_r - w_z)}{(1+F_z^2)^{1/2}} = \sigma_e E_t, \tag{8.24}$$

where $\chi = \varepsilon_o V^2/(R_i\sigma)$ is the electric Bond number. Equation (8.22) is the kinematic compatibility condition, while Eqs. (8.23) and (8.24) express the balance of normal and tangential stresses on the two sides of the free surface, respectively. The right-hand sides of these equations are the Maxwell stresses resulting from the accumulation of free charge at the interface and the jump of permittivity across this surface.

The electrical field at the free surface and the surface charge density are calculated as

$$E_n^i = \frac{F_z\phi_z^i - \phi_r^i}{\sqrt{1+F_z^2}}, \quad E_n^o = \frac{F_z\phi_z^o - \phi_r^o}{\sqrt{1+F_z^2}}, \tag{8.25}$$

$$E_t = \frac{-F_z\phi_r^o + \phi_z^o}{\sqrt{1+F_z^2}} = \frac{-F_z\phi_r^i + \phi_z^i}{\sqrt{1+F_z^2}}, \tag{8.26}$$

$$\sigma_e = \chi(E_n^o - \beta E_n^i). \tag{8.27}$$

It must be noted that the continuity of the electric potential across the free surface, $\phi^i = \phi^o$, has been considered in Eq. (8.26).

The free surface equations are completed by imposing the surface charge conservation at $r = F(z)$,

$$\frac{\partial \sigma_e}{\partial t} + \mathbf{\nabla}_s \cdot (\sigma_e \mathbf{v}) = \alpha\chi E_n^i, \tag{8.28}$$

where $\mathbf{\nabla}_s$ is the tangential intrinsic gradient along the free surface, $\mathbf{v}$ is the 3D fluid velocity on the free surface, and $\alpha = K\left[\rho R_i^3/(\sigma\varepsilon_o^2)\right]^{1/2}$ is the dimensionless electrical conductivity.

As mentioned above, the Hagen-Poiseuille velocity profile is prescribed at the entrance of the liquid domain $z = 0$ (Fig. 8.9):

$$u = 0, \quad w = 2\widetilde{Q}(1 - r^2), \tag{8.29}$$

where $\widetilde{Q} = Q_i/(\pi R_i^2 U_{\rho\sigma})$. At the capillary wall, we fix the electric potential and impose no-slip boundary conditions, i.e.,

$$\phi^i = \phi^e = 1 \quad \text{and} \quad u = w = 0. \tag{8.30}$$

The triple contact line is anchored at the end of the capillary:

$$F = 1 \quad \text{at} \quad z = L_n. \tag{8.31}$$

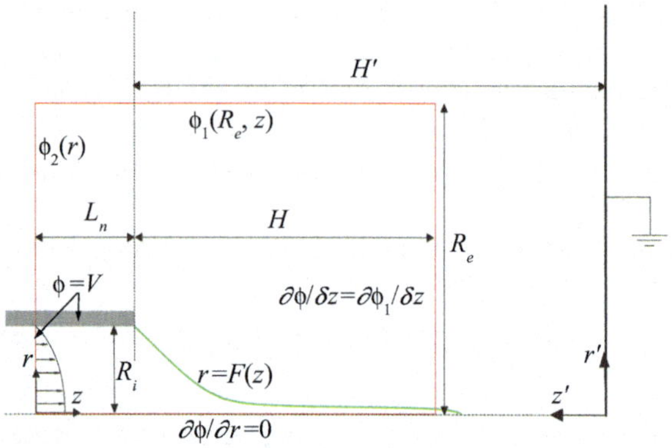

Fig. 8.9 Sketch of the problem's formulation. The rectangle (red online) denotes the limits of the computational domain

The standard regularity conditions

$$\phi_r^i = u = w_r = 0 \tag{8.32}$$

are prescribed on the symmetry axis, and the outflow conditions

$$u_z = w_z = F_z = \sigma_z = 0 \tag{8.33}$$

are considered at the right-hand end $z_e = H + L_n$ of the computational domain. The problem is closed by imposing the electric boundary conditions at the outer boundaries $z = 0$ and $r = R_e$. Analytical solutions for the far-field electric potential can be used for this purpose [58].

The governing equations are formulated in terms of the dimensionless numbers $\{\text{Oh}, \chi, \varepsilon, \alpha, \widetilde{Q}\}$ and those characterizing the rest of boundary conditions. The dimensionless numbers $\{\text{Oh}, \alpha, \widetilde{Q}\}$ can be combined to obtain the parameters defined in Sect. 8.2; specifically, $\delta_\mu = \alpha^{-1/3}\text{Oh}^{1/2}$ and $Q = \pi\,\alpha\,\widetilde{Q}$.

8.4.2 The Critical Eigenmode and Stabilization Methods

In the global stability analysis, the base flow steady solution to the leaky-dielectric model is perturbed by adding a linear eigenmode proportional to $e^{-i\omega t}$ (see Chap. 3). The eigenfrequency $\omega = \omega_r + i\omega_i$ is calculated to fulfill the hydrodynamic equations. When the growth rate ω_i becomes positive, the steady flow destabilizes and evolves towards a different state. As applied to electrospray, one calculates $\omega_i(\widetilde{Q})$ and determines the critical flow rate at which $\omega_i = 0$. This critical flow rate is the minimum flow rate of the cone-jet mode.

Dharmansh and Chokshi [70] conducted the global linear stability analysis of a leaky-dielectric liquid in the framework of the 1D model. They concluded that the jet thinning produces a stabilizing effect due to both the variation of the surface

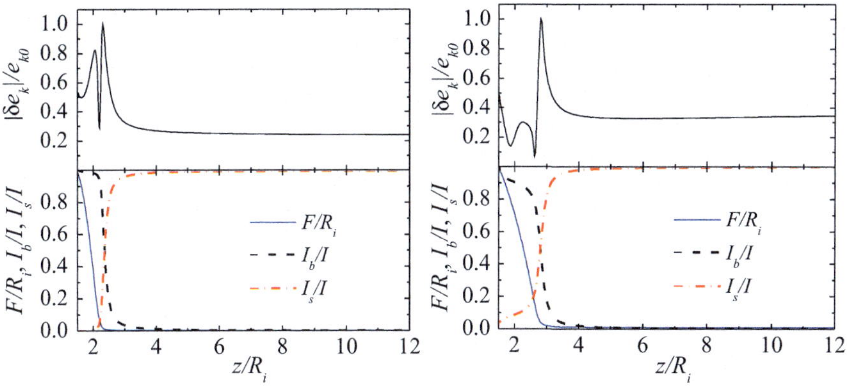

Fig. 8.10 (Upper graphs) Perturbation amplitude of the kinetic energy $|\delta e_k|$ along the free surface relative to its value e_{k0} in the base flow. The perturbations are normalized so that the maximum value of $|\delta e_k|/e_{k0}$ equals one. (Lower graphs) Free surface position F, bulk current intensity I_b, and surface current intensity I_s as a function of the axial position z. The results were calculated for $\varepsilon = 23.7$, $\delta_\mu = 0.147$ and $Q_r = 7.74$ (left), and $\varepsilon = 10$, $\delta_\mu = 2.29$ and $Q_r = 2.98$ (right) (reproduced with permission, Ref. [58])

charge density and the extensional viscous stresses in the base flow. Ponce-Torres et al. [58] showed that the minimum flow rate stability limit could be predicted from the linear stability analysis of the solution to the full leaky-dielectric model (Fig. 8.8). The theory also predicts the stabilization caused by the small size of the feeding capillary.

Ponce-Torres et al. [58] calculated the kinetic energy magnitude of the critical perturbation, $|\delta e_k| = |u_0 \delta u + w_0 \delta w|$, relative to its value $e_{k0} = 1/2\ (u_0^2 + w_0^2)$ in the base flow. In these expressions, u_0 (w_0) and δu (δw) represent the base flow value and perturbation amplitude of the radial (axial) component of the velocity field, respectively, calculated along the free surface. The quantity $|\delta e_k(z)|/e_{k0}(z)$ increases with the free surface deformation, and, therefore, it indicates where the instability initiates.

In the two cases analyzed in Fig. 8.10, $|\delta e_k|/e_{k0}$ reaches its maximum value in the cone-jet transition region, close to the jet's section where the conduction intensity equals the surface convection of charges. This result confirms the critical role played by this singular fluidic structure in the onset of the instability.

Different methods have been proposed to stabilize the cone-jet mode of electrospray. These can be categorized into those modifying the applied electric field [71, 72] and the emitter geometry. In this latter case, use has been made of carbon fiber emitters [73], ballpoint pen nozzles [74], or hemispherical caps [75]. Cruz-Mazo et al. [76] have proposed a way to stabilize steady micro/nanojets issuing from Taylor cones by applying a coflowing gas stream, significantly reducing the minimum flow rate for a given set of geometrical parameters.

The maximum flow rate limit is very relevant in those applications where the droplet production rate must be maximized. Its analysis is complicated because it involves a wide variety of complex issues. The excess of charge carried by the liquid jet limits the electric field on the cone surface [6, 77] and can cause the surrounding gas ionization. There can also be polarity effects related to the different mobility and nature of ions that transport the charge [78].

The minimum flow rate stability limit is the most interesting parameter region at fundamental and practical levels. For this reason, we devote the following two sections to present some numerical results for electrospray realizations relatively close to that limit.

8.4.3 *Charge Relaxation*

The surface charge conservation equation in a Lagrangian formulation can be derived as follows. Consider the electric current $I_s = 2\pi F v_s \sigma_e$ convected by an annular free surface element of radius $F(s)$ (Fig. 8.11). Here, s and v_s are the intrinsic (surface) coordinate and velocity along the free surface, respectively, while $\sigma_e = \varepsilon_o(E_n^o - \varepsilon E_n^i)$ is the surface charge density. We define the electrostatic surface current $I_{se}(s) = 2\pi F v_s \varepsilon_o E_n^o$ as the value for zero inner electric field.

Suppose we neglect conduction and diffusion of electrical charges *over the free surface* in the cone-jet transition region. In that case, the surface charge conservation equation reads $dI_s = 2\pi F K E_n^i ds$, where dI_s stands for the increment along ds of the current convected by the surface element. The term $2\pi F K E_n^i ds$ represents the charge transfer from the bulk. Taking into account that $ds = v_s dt$, the temporal integration of that conservation equation leads to the linear relaxation relationship [15]

$$t_e \frac{dI_s}{dt} = -(I_s(t) - I_{se}(s(t))). \tag{8.34}$$

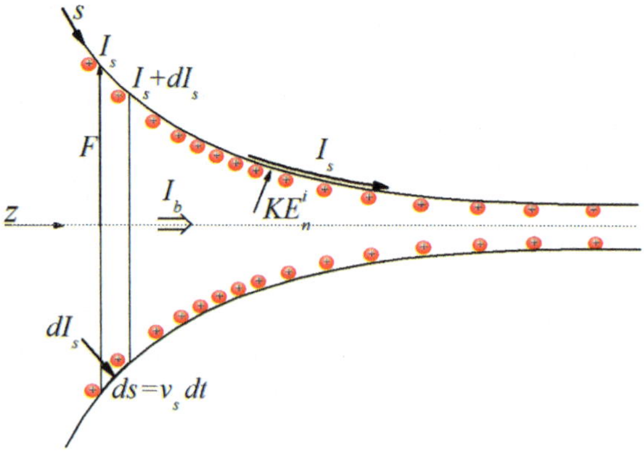

Fig. 8.11 Sketch of the transfer of charge from the bulk to the interface in the cone-jet transition region

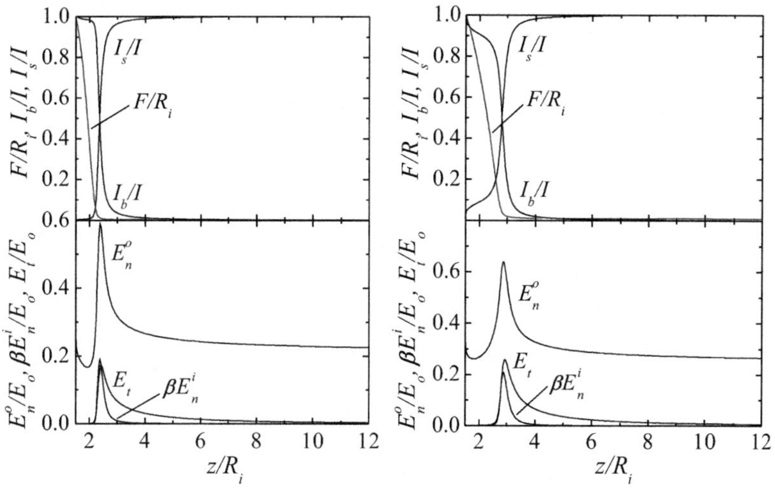

Fig. 8.12 (Upper graphs) Free surface position F, bulk current intensity $I_b = I - I_s$ and surface current intensity I_s as a function of the axial position z. (Lower graphs) Tangential and normal components of the inner and outer electric fields at the free surface. The results were calculated for $\varepsilon = 23.7$, $\delta_\mu = 0.147$ and $Q = 7.74$ (left), and for $\varepsilon = 10$, $\delta_\mu = 2.29$ and $Q = 2.98$ (right) (reproduced with permission, Ref. [58])

The above equation shows how the convected current I_s tries to adapt to the local electrostatic value I_{se} while traveling along the free surface. The speed at which this adaptation takes place is inversely proportional to the electric relaxation time t_e. The latter must be compared with the residence time t_r of the surface element in the cone-jet transition region, i.e.,

$$t_r = \int_{s_i}^{s_f} \frac{ds}{v_s}, \tag{8.35}$$

where s_i and s_f are the intrinsic coordinates of the inlet and outlet sections of the cone-jet transition region. For $t_e \gg t_r$, I_s would remain constant (the surface current would "freeze") [11] throughout the surface element motion. This situation does not occur in any stable electrospraying realization. For $t_e \ll t_r$, the current transported by the surface element relaxes to its local electrostatic value almost instantaneously and, therefore, $I_s(s) \simeq I_{se}(s)$ ($\varepsilon E_n^i \ll E_n^o$).

The simulations at the minimum flow rate stability limit [58] show that t_e is considerably smaller than t_r except for the high-polarity case, where $t_e \simeq t_r$ [68]. However, εE_n^i is not always negligible compared to E_n^o (Fig. 8.12). Therefore, one can safely assume that εE_n^i is negligible in the cone-jet transition region only if there is true disparity (two or more orders of magnitude) between the two characteristic times t_e and t_r.

8.4.4 Electric Forces

In the 1D (slenderness) approximation (see Chap. 2), the momentum equation in the z-direction becomes [56]:

$$\underbrace{\frac{\chi}{2}\left[(E_n^o)^2-\varepsilon(E_n^i)^2\right]_z}_{\text{I}}+\underbrace{\chi\frac{\varepsilon-1}{2}\left(E_t^2\right)_z}_{\text{II}}+\underbrace{\frac{2\sigma_e E_t}{F}}_{\text{III}}=\underbrace{\left(\frac{1}{F}\right)_z}_{\text{IV}}+\underbrace{\left(\frac{\widetilde{Q}^2}{2F^4}\right)_z}_{\text{V}}+\underbrace{\frac{6\text{Oh}\widetilde{Q}}{F^2}\left(\frac{F_z}{F}\right)_z}_{\text{VI}}, \tag{8.36}$$

where the quantities have been made dimensionless with the characteristic scales introduced in Sect. 8.4.1.

The terms of Eq. (8.36) have been grouped into electric (left-hand side) and hydrodynamic (right-hand side) forces. The first, second, and last addends on the left-hand side of Eq. (8.36) are generally referred to as the electrostatic, polarization, and electric tangential forces per unit volume, respectively [56]. The terms on the right-hand side of Eq. (8.36) correspond to surface tension, inertia, and viscosity, respectively. Figure 8.13 shows the values taken by the six terms of Eq. (8.36) calculated from the 2D solution of the leaky-dielectric model [58]. The area enclosed by the curves equals the work per unit volume associated with the corresponding force.

The electric tangential force III is the primary energy source in the cone-jet transition region and throughout the jet. Using numerical simulations in the Stokes limit, [13] pointed out the dominant role played by that force. Both the electrostatic and polarization forces are subdominant in these electrospray realizations. The fact that the electrostatic suction transfers much less energy than the tangential force differs from the scaling analysis prediction for larger flow rates (Sect. 8.3), which concludes that these effects are commensurate with each other.

Most of the work done by the electric forces is converted into kinetic energy. The viscous force arises against the liquid ejection for $\delta_\mu = 0.147$, while surface tension becomes the largest resistant force for $\delta_\mu = 2.29$. The electrostatic and polarization forces push the liquid in the cone-jet transition region. They stand against the liquid ejection behind that region. Graph (c) shows the accuracy of the 1D approximation throughout the whole liquid domain, even for the least viscous case, which exhibits recirculation in the electrified meniscus [58].

8.5 Coaxial Taylor Cone-Jets and AC Electrospray

Micrometer coaxial jets were first produced with gaseous flow focusing in 1998 [79]. Subsequently, the same idea was applied to produce electrified coaxial jets [31] based on the analogies between flow-focused (mechanical) and Taylor (electrical) cone-jets. This result has produced a remarkable impact on microfluidics and has been followed by the development of numerous applications since then. Fibers [80,

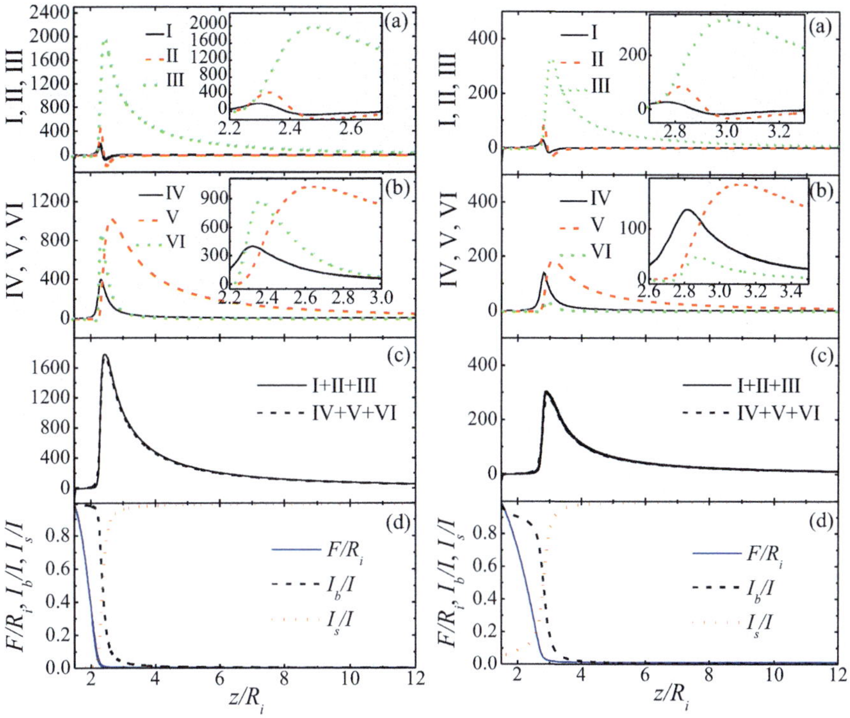

Fig. 8.13 [Graphs (**a**), (**b**) and (**c**)] Terms I-VI of the slenderness model (8.36) as a function of the axial position z. [Graph (**d**)] Free surface position F, bulk current intensity I_b, and surface current intensity I_s as a function of the axial position z. The results were calculated for $\varepsilon = 23.7$, $\delta_\mu = 0.147$, and $Q = 7.74$ (left), and for $\varepsilon = 10$, $\delta_\mu = 2.29$ and $Q = 2.98$ (right) (reproduced with permission, Ref. [58])

81] and particles [32] with complex morphology have been fabricated with this technique. Large area and highly aligned polymer semiconductor sub-microwires have recently been fabricated using the coaxial electrospray [82].

The working modes of coaxial electrospray have been described by [83], who related them to those in the single-phase configuration. The steady regime and the initial transient phase were studied by López-Herrera et al. [84] and Reznik et al. [85], respectively. The conclusions can be summarized as follows.

The liquid with the largest electrical conductivity drives the flow. In most electrospray realizations, that role is played by the external liquid phase, which is in contact with a vacuum, gas, or dielectric liquid [32, 83] (Fig. 8.14a). If the flow rate of the driver liquid is much larger than that of the inner one, the differences with respect to the corresponding single-phase configuration are minimal.

The outer liquid shields the inner one from the electric field, and the two-phase bulk becomes almost equipotential. When the outer viscosity is not too large or small compared to the inner one, the compound jet diameter is similar to that of the

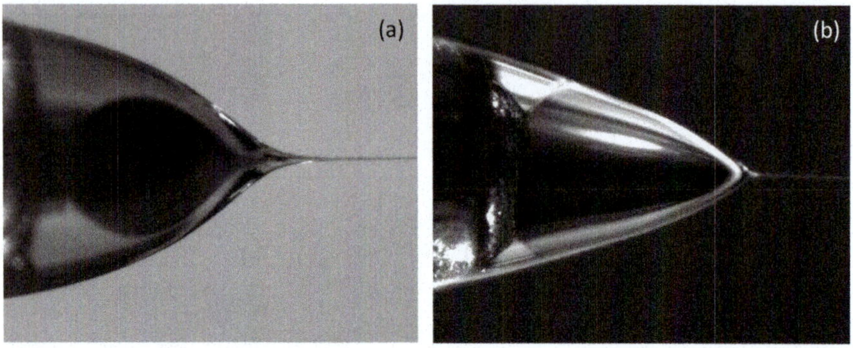

Fig. 8.14 **a** Coaxial Taylor cone-jet driven by the outer liquid (Cortijo and Loscertales 2000, private communication). The outer and inner liquids were Somos additive ($K = 6.25 \times 10^{-5}$ S/m, $\mu = 7.5 \times 10^{-1}$ Pa s) and colored ethylene glycol ($K \simeq 10^{-5}$ S/m, $\mu = 1.6 \times 10^{-2}$ Pa s), respectively. **b** Coaxial Taylor cone-jet with an inner driver. The outer and inner liquids are olive oil ($K < 10^{-8}$ S/m, $\mu \simeq 7 \times 10^{-2}$ Pa s) and ethylene glycol ($K \simeq 1.1 \times 10^{-5}$ S/m, $\mu = 1.6 \times 10^{-2}$ Pa s), respectively (reprinted from Ref. [64] with permission of Elsevier)

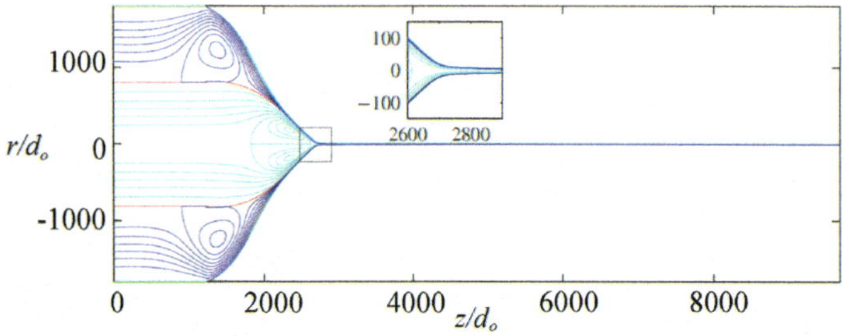

Fig. 8.15 Shape and streamlines of a coaxial electrospray using deionized water and tributyl phosphate as the inner and outer fluids, respectively. The inner and outer flow rates are 234 μl/h and 60 μl/h, respectively, while the applied voltage is 4570 V (reproduced with permission, Ref. [86])

single-phase configuration for the outer liquid flow rate. In this case, the inner liquid practically acts as a hollow core since the driver fixes the jet velocity.

In the original configuration of coaxial electrospray [31], the flow was driven by the inner liquid, which tends to reach the meniscus free surface, expanding the interface exposed to the applied electric field (Fig. 8.14b). In this case, one expects the outer flow rate to be small compared to the driver's. This occurs only if the outer viscosity is much larger than the driver's. Otherwise, momentum is efficiently transferred to the external liquid bulk, and the outer flow rate can become larger than that of the driver [84].

[86] have integrated the 2D leaky-dielectric model for the cone-jet mode of coaxial electrospray (Fig. 8.15). A segment of the opposite charge (negative if the polarity of the applied potential is positive, and *vice versa*) can appear on the inner interface at the transition region.

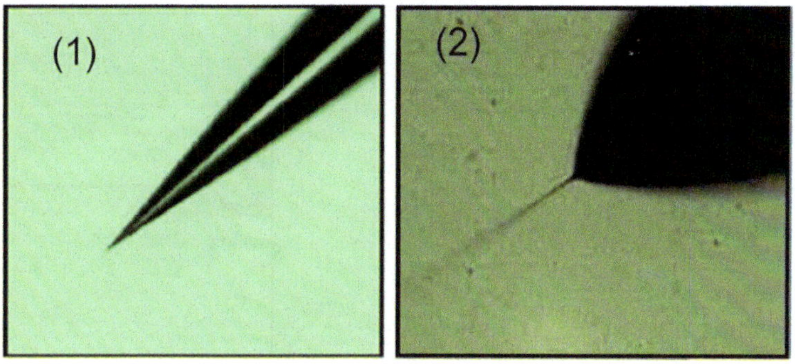

Fig. 8.16 (**1**) AC and (**2**) DC cone-jets of ethanol (reprinted with permission from Ref. [88]. Copyright (2010) American Physical Society)

The cone-jet mode of electrospray can also be produced with AC electric fields at frequencies greater than 50 kHz (Fig. 8.16). In this case, the cone is more slender and is sustained by the entrainment of net space free charge within the cone. This physical mechanism is entirely different from that of DC electrospray and cannot be described from the leaky-dielectric model [87]. AC electrospray can be used as an ionization source for mass spectrometry of biomolecules [88].

8.6 Electrospinning

When the electric field is applied to a polymeric solution, the solvent evaporates during the jet flight, and a fiber is extruded from the tapering meniscus. This technique, called electrospinning, has been profusely studied over the last decades due to its enormous relevance in diverse fields. Lauricella et al. [2] have reviewed the theoretical description and numerical simulation of electrospun jets from polymeric solutions, paying particular attention to the critical role played by instability phenomena of electrical and hydrodynamic origin. Interested readers are referred to that excellent work to gain insight into the physical mechanisms governing the flow. Here, we only describe some results obtained from the global stability analysis.

To analyze the linear stability of the cone-jet mode of electrospinning [34, 41, 44, 46, 89] modeled the rheological effects using the Oldroyd-B and XPP approximations. As in the Newtonian case, the results showed the stabilizing effect of the jet thinning.

[90] solved the 2D hydrodynamic equations with the Oldroyd-B constitutive model. They considered as a reference the physical properties of 1-octanol and added polymeric stresses characterized by relatively short relaxation times in terms of the capillary time (small Deborah numbers). The relaxation time was commensurate with the inverse of the strain rate in the cone-jet transition region, making the dissolved polymers stretch. The polymer coil-stretch transition produced axial

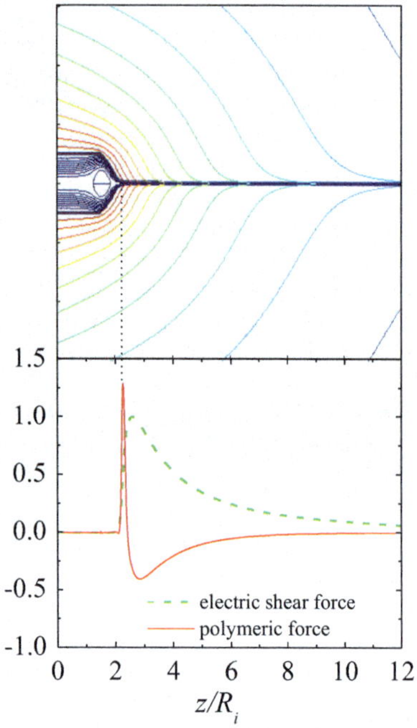

Fig. 8.17 (Upper graph) Streamlines (inner domain) and equipotential lines (outer domain) calculated from the 2D leaky-dielectric Oldroyd-B model for $\varepsilon = 10$, $\delta_\mu = 2.29$, $Q = 7.72$, $\lambda_s^{(i)} = 2.656\, t_e$ and $\lambda_r^{(i)} = 0.0926\lambda_s^{(i)}$. (Lower graph) Electric shear force and polymeric force both per unit volume and normalized with the maximum value of the former (reprinted with permission from Ref. [90]. Copyright (2019) by The American Physical Society)

polymeric stresses much larger than those caused by the solvent viscosity. These stresses pull from the liquid in front of the meniscus tip (Fig. 8.17), accelerating the polymeric solution and making the electrified liquid meniscus shrink. The polymers relax to their coiling state in the jet, and the kinetic energy gained in the meniscus is lost in the jet region. One may say that the liquid borrows energy from the polymers in the most unstable region and returns it when safely moving downstream. The viscoelastic stress in the meniscus tip stabilizes the flow, which explains the reduction of the minimum flow rate obtained in global stability analysis [90].

The meniscus stabilization caused by the growth of the polymeric stress in the meniscus tip has been observed experimentally in the gaseous flow focusing configuration (see Chap. 11). The phenomenon described above is essentially the same as that occurring when a tiny amount of a low-molecular-weight polymer is dissolved in water at the appropriate concentration, and the tapering meniscus is focused with a transonic gaseous current [91, 92].

References

1. Yarin AL, Pourdeyhimi B, Ramakrishna S (2014) Fundamentals and applications of micro- and nanofibers. Cambridge University Press, Cambridge
2. Lauricella M, Succi S, Zussman E, Pisignano D, Yarin AL (2020) Models of polymer solutions in electrified jets and solution blowing. Rev Mod Phys 92(035):004

3. Zeleny J (1914) The electrical discharge from liquid points, and a hydrostatic method of measuring the electric intensity at their surfaces. Phys Rev 3:69–91
4. Zeleny J (1917) Instability of electrified liquid surfaces. Phys Rev 10:1–6
5. Jaworek A, Krupa A (1999) Classification of the modes of EHD spraying. J Aerosol Sci 30:873–893
6. Cloupeau M, Prunet-Foch B (1989) Electrostatic spraying of liquids in cone-jet mode. J Electrostatics 22:135–159
7. Rosell-Llompart J, Grifoll J, Loscertales IG (2018) Electrosprays in the cone-jet mode: from Taylor cone formation to spray development. J Aerosol Sci 125:2–31
8. Fenn JB, Mann M, Meng CK, Wong SF, Whitehouse CM (1989) Electrospray ionization for mass spectrometry of large biomolecules. Science 246:64–71
9. Whitehouse CM, Dreyer RN, Yamashita M, Fenn JB (1985) Electrospray interface for liquid chromatographs and mass spectrometers. Anal Chem 57:675–679
10. Gañán-Calvo AM (1997) Cone-jet analytical extension of Taylor's electrostatic solution and the asymptotic universal scaling laws in electrospraying. Phys Rev Lett 79:217–220
11. Fernández de la Mora J (2007) The fluid dynamics of Taylor cones. Annu Rev Fluid Mech 39:217–243
12. Borra JP (2018) Review on water electro-sprays and applications of charged dropswith focus on the corona-assisted cone-jet mode for high efficiency air filtration by wet electro-scrubbing of aerosols. J Aerosol Sci 125:208–236
13. Higuera FJ (2010) Numerical computation of the domain of operation of an electrospray of a very viscous liquid. J Fluid Mech 648:35–52
14. Higuera FJ (2017) Qualitative analysis of the minimum flow rate of a cone-jet of a very polar liquid. J Fluid Mech 816:428–441
15. Fernandez de la Mora J, Loscertales IG (1994) The current transmitted through an electrified conical meniscus. J Fluid Mech 260:155–184
16. Herrada MA, López-Herrera JM, Gañán-Calvo AM, Vega EJ, Montanero JM, Popinet S (2012) Numerical simulation of electrospray in the cone-jet mode. Phys Rev E 86(026):305
17. López-Herrera JM, Herrada MA, Montanero JM, Rebollo-Muñoz N, Gañán-Calvo AM (2013) On the validity and applicability of the one-dimensional approximation in cone-jet electrospray. J Aerosol Sci 61:60–69
18. Benassayag G, Sudraud P, BJouffrey, (1985) In situ high voltage TEM observation of an electrohydrodynamic (EHD) ion source. Ultramicroscopy 16:1–8
19. Driesel W, Dietzsch C, Niedrig H, Praprotnik B (1995) HV TEM in situ investigations of the tip shape of a gallium liquid-metal ion/electron emitter. Ultramicroscopy 57:45–58
20. Carswell DJ, Milsted J (1957) A new method for the preparation of thin films of radioactive material. J Nucl Energy 4:51–54
21. Michelson D, Richardson HOW (1963) A note on making thin β-ray sources by electrostatic spraying. Nucl Instrum Methods 21:355
22. Shorey JD, Michelson D (1969) Electrospray ionization for mass spectrometry of large biomolecules. Science 246:64–71
23. Gamero-Castaño M, Hruby V (2001) Electrospray as a source of nanoparticles for efficient colloid thrusters. J Propul Power 17:977–987
24. Park JU et al (2007) High-resolution electrohydrodynamic jet printing. Nat Mater 6:782–789
25. Onses MS, Sutanto E, Ferreira PM, Alleyne AG, Rogers JA (2015) Mechanisms, capabilities, and applications of high-resolution electrohydrodynamic jet printing. Small 11:4237–4266
26. Ismail AS, Yao J, Xia HH, Stark JPW (2018) Breakup length of electrified liquid jets: scaling laws and applications. Phys Rev Appl 10(064):010
27. Bielecki J et al (2019) Electrospray sample injection for single-particle imaging with X-ray lasers. Sci Adv 5:1–9
28. Mei F, Chen DR (2007) Investigation of compound jet electrospray: particle encapsulation. Phys Fluids 19(103):303
29. Chakraborty S, Liao IC, Adler A, Leong KW (2009) Electrohydrodynamics: a facile technique to fabricate drug delivery systems. Adv Drug Deliv Rev 61:1043–1054

30. Sill TA, von Recum HA (2008) Electrospinning: applications in drug delivery and tissue engineering. Biomaterials 29:1989–2006
31. Loscertales IG, Barrero A, Guerrero I, Cortijo R, Marquez M, Gañán-Calvo AM (2002) Micro/nano encapsulation via electrified coaxial liquid jets. Science 295:1695–1698
32. Marín AG, Loscertales IG, Márquez M, Barrero A (2007) Simple and double emulsions via coaxial jet electrosprays. Phys Rev Lett 98(014):502
33. Barrero A, Loscertales IG (2007) Micro and nanoparticles via capillary flows. Annu Rev Fluid Mech 39:89–106
34. Yarin AL (2011) Coaxial electrospinning and emulsion electrospinning of core-shell fibers. Polym Adv Technol 22:310–317
35. Pareta R, Edirisinghe M (2006) A novel method for the preparation of biodegradable microspheres for protein drug delivery. J R Soc Interface 3:573–582
36. Farook U, Zhang HB, Edirisinghe M, Stride E, Saffari N (2007) Preparation of microbubble suspensions by co-axial electrohydrodynamic atomization. Med Eng Phys 29:749–754
37. Enayati M, Ahmad Z, Stride E, Edirisinghe M (2010) Size mapping of electric field-assisted production of polycaprolactone particles. J R Soc Interface 7:S393–S402
38. Reneker DH, Yarin A, Fong H, Koombhongse S (2000) Bending instability of electrically charged liquid jets of polymer solutions in electrospinning. J Appl Phys 87:4531–4547
39. Yarin AL, Koombhongse S, Reneker DH (2001) Bending instability in electrospinning of nanofibers. J Appl Phys 89:3018–3026
40. Yarin AL, Koombhongse S, Reneker DH (2001) Taylor cone and jetting from liquid droplets in electrospinning of nanofibers. J Appl Phys 90:4836–4846
41. Theron SA, Zussman E, Yarin AL (2004) Experimental investigation of the governing parameters in the electrospinning of polymer solutions. Polymer 45:2017–2030
42. Yarin AL, Kataphinan W, Reneker DH (2005) Branching in electrospinning of nanofibers. J Appl Phys 98(064):501
43. Han T, Reneker D, Yarin A (2007) Buckling of jets in electrospinning. Polymer 48:6064–6076
44. Thompson CJ, Chase GG, Yarin AL, Reneker DH (2007) Effects of parameters on nanofiber diameter determined from electrospinning model. Polymer 48:6913–6922
45. Han T, Yarin AL, Reneker DH (2008) Viscoelastic electrospun jets: initial stresses and elongational rheometry. Polymer 49:1651–1658
46. Reneker DH, Yarin AL (2008) Electrospinning jets and polymer nanofibers. Polymer 49:2387–2425
47. Miloh T, Spivak B, Yarin AL (2009) Needleless electrospinning: electrically driven instability and multiple jetting from the free surface of a spherical liquid layer. J Appl Phys 106(114):910
48. Kanani AG, Bahrami SH (2010) Review on electrospun nanofibers scaffold and biomedical applications. Trends Biomater Artif Organs 24:93–115
49. Agarwala S, Greinera A, Wendorff JH (2013) Functional materials by electrospinning of polymers. Prog Polym Sci 38:963–991
50. Liashenko I, Rosell-Llompart J, Cabot A (2020) Ultrafast 3D printing with submicrometer features using electrostatic jet deflection. Nat Commun 11:753
51. Sun D, Chang C, Li S, Lin L (2006) Near-field electrospinning. Nano Lett 6:839–842
52. He XX, Zheng J, Yu GF, You MH, Yu M, Ning X, Long YZ (2017) Near-field electrospinning: progress and applications. J Phys Chem C 121:8663–8678
53. García-Galán M, Martínez-Vázquez FJ, Rebollo-Muñoz N, Montanero J, Miranda P (2022) Applicability of near-field electrospinning for the development of TCP-based thin fibres and scaffold 3D printing. Bol Soc Esp Ceram Vidr (in Press)
54. Gañán-Calvo AM (2004) On the general scaling theory for electrospraying. J Fluid Mech 507:203–212
55. Taylor G (1964) Disintegration of water drops in electric field. Proc R Soc Lond A 280:383–397
56. Gañán-Calvo AM (1999) The surface charge in electrospraying: its nature and its universal scaling laws. J Aerosol Sci 30:863–872
57. Scheideler WJ, Chena CH (2014) The minimum flow rate scaling of Taylor cone-jets issued from a nozzle. Appl Phys Lett 104(024):103

58. Ponce-Torres A, Rebollo-Muñoz N, Herrada MA, Gañán-Calvo AM, Montanero JM (2018) The steady cone-jet mode of electrospraying close to the minimum volume stability limit. J Fluid Mech 857:142–172
59. Carlier J, Arscott S, Camart JC, Cren-Olivé C, Gac SL (2005) Integrated microfabricated systems including a purification module and an on-chip nano electrospray ionization interface for biological analysis. J Chromatogr A 1071:213–22
60. Yuill EM, Ray NSSJ, Hieftje GM, Baker LA (2013) Electrospray ionization from nanopipette emitters with tip diameters of less than 100 nm. Anal Chem 85:8498–8502
61. Gañán-Calvo AM, Barrero A, Pantano C (1993) The electrohydrodynamics of electrified conical menisci. J Aerosol Sci 24:S19-20
62. Gamero-Castaño M (2010) Energy dissipation in electrosprays and the geometric scaling of the transition region of cone-jets. J Fluid Mech 662:493–513
63. Gañán-Calvo AM, Montanero JM (2009) Revision of capillary cone-jet physics: electrospray and flow focusing. Phys Rev E 79(066):305
64. Gañán-Calvo AM, López-Herrera JM, Herrada MA, Ramos A, Montanero JM (2018) Review on the physics of electrospray: from electrokinetics to the operating conditions of single and coaxial Taylor cone-jets, and AC electrospray. J Aerosol Sci 125:32–56
65. Saville DA (1971) Stability of electrically charged viscous cylinders. Phys Fluids 14:1095–1099
66. Montanero JM, Gañán-Calvo AM (2020) Dripping, jetting and tip streaming. Rep Prog Phys 83(097):001
67. Bober DB, Chen CH (2011) Pulsating electrohydrodynamic cone-jets: from choked jet to oscillating cone. J Fluid Mech 689:552–563
68. Gañán-Calvo AM, Rebollo-Muñoz N, Montanero JM (2013) Physical symmetries and scaling laws for the minimum or natural rate of flow and droplet size ejected by Taylor cone-jets. New J Phys 15(033):035
69. Gamero-Castaño M, Magnani M (2019) Numerical simulation of electrospraying in the cone-jet mode. J Fluid Mech 859:247–267
70. Dharmansh D, Chokshi P (2016) Axisymmetric instability in a thinning electrified jet. Phys Rev E 93(043):124
71. Park H, Kim K, Kim S (2004) Effects of a guard plate on the characteristics of an electrospray in the cone-jet mode. J Aerosol Sci 35:1295–1312
72. Martin S, Perea A, Garcia-Ybarra PL, Castillo JL (2012) Effect of the collector voltage on the stability of the cone-jet mode in electrohydrodynamic spraying. J Aerosol Sci 46:53–63
73. Sen AK, Darabi J, Knapp DR, Liu J (2006) Modeling and characterization of a carbon fiber emitter for electrospray ionization. J Micromech Microeng 16:620–630
74. Li C, Chang M, Yang W, Madden A, Deng W (2014) Ballpoint pen tips as robust cone-jet electrospray emitters. J Aerosol Sci 77:10–15
75. Morad M, Rajabi A, Razavi M, Sereshkeh SP (2016) A very stable high throughput Taylor cone-jet in electrohydrodynamics. Sci Rep 6(38):509
76. Cruz-Mazo F, Wiedorn MO, Herrada MA, Bajt S, Chapman HN, Gañán-Calvo AM (2019) Aerodynamically stabilized Taylor cone jets. Phys Rev E 100(031):101(R)
77. Cloupeau M, Prunet-Foch B (1994) Electrohydrodynamic spraying functioning modes: a critical review. J Aerosol Sci 25:1021–1036
78. Pongrac B, Krcma F, Dostal L, Kim HH, Homola T, Machala Z (2016) Effects of corona space charge polarity and liquid phase ion mobility on the shape and velocities of water jets in the spindle jet and precession modes of water electro-spray. J Aersol Sci 101:196–206
79. Gañán-Calvo AM (1998) Generation of steady liquid microthreads and micron-sized monodisperse sprays in gas streams. Phys Rev Lett 80:285–288
80. Li D, Xia Y (2004) Direct fabrication of composite and ceramic hollow nanofibers by electrospinning. Nano Lett 4:933–38
81. Greiner A, Wendorff JH (2007) Electrospinning: a fascinating method for the preparation of ultrathin fibers. Angew Chem Int Edn 46:5670–5703

82. Wang D et al (2022) Large area polymer semiconductor sub-microwire arrays by coaxial focused electrohydrodynamic jet printing for high-performance OFETs. Nat Commun 13:6214
83. Chen X, Jia L, Yin X, Cheng J, Lu J (2005) Spraying modes in coaxial jet electrospray with outer driving liquid. Phys Fluids 17(032):101
84. López-Herrera JM, Barrero A, Lopez A, Loscertales IG, Márquez M (2003) Coaxial jets generated from electrified Taylor cones. Scaling laws. J Aerosol Sci 34:535–552
85. Reznik SN, Yarin AL, Zussman E, Bercovici L (2006) Evolution of a compound droplet attached to a core-shell nozzle under the action of a strong electric field. Phys Fluids 18(062):101
86. López-Herrera J, Herrada M, Gamero-Castaño M, Gañán-Calvo AM (2020) A numerical simulation of coaxial electrosprays. J Fluid Mech 885:A15
87. Chetwani N, Maheshwari S, Chang HC (2008) Universal cone angle of AC electrosprays due to net charge entrainment. Phys Rev Lett 101(204):501
88. Chetwani N, Cassou C, Go D, Chang HC (2010) High-frequency AC electrospray ionization source for mass spectrometry of biomolecules. J Am Soc Mass Spectrom 21:1852–1856
89. Dharmansh D, Chokshi P (2017) Stability analysis of an electrospinning jet of a polymeric fluid. Polymer 131:34–49
90. Blanco-Trejo S, Herrada MA, Gañán-Calvo AM, Montanero JM (2019) Electrospray cone-jet mode for weakly viscoelastic liquids. Phys Rev E 100(043):114
91. Rubio A, Galindo F, Vega EJ, Montanero JM, Cabezas MG (2022) Viscoelastic transition in transonic flow focusing. Phys Rev Fluids 7(074):201
92. Rubio A, Vega EJ, Gañán-Calvo AM, Montanero JM (2022) Unexpected stability of micrometer weakly viscoelastic jets. Phys Fluids 34(062):014

Chapter 9
Coflowing and Hydrodynamic Focusing

Abstract One of the most salient applications of the tip streaming phenomenon is the production of monodisperse emulsions of micrometer droplets. The microdripping and microjetting modes can be realized in different microfluidic devices for this purpose. These devices can be grouped into three main classes [1]: crossflowing systems, coflowing streams, and hydrodynamic focusing. Only the last two classes correspond to an axisymmetric configuration.

In this chapter, we review some relevant results obtained in this field. Specifically, we describe the coflowing, flow focusing, confined selective withdrawal, and selective withdrawal techniques. We present results for liquid-liquid and liquid-gas systems. We do not consider here the gaseous flow-focusing configuration (a liquid stream focused by a gaseous current), to which we devote two specific chapters.

Two significant aspects of each configuration mentioned above are examined: the stability of the tip streaming flow and the scaling laws predicting the size of the jets and droplets. We present some preliminary numerical results on the effect of a soluble surfactant in liquid-liquid flow focusing. The chapter closes with some results about viscous entrainment and microjetting in a uniaxial extensional flow. This microjetting mode may lead to the stable production of arbitrarily thin fluid jets.

Keywords Coflowing · Flow focusing · Selective withdrawal

9.1 Coflowing

9.1.1 Liquid-Liquid Systems

In an axisymmetric coflowing device, the inner dispersed phase is injected at a constant flow rate across a cylindrical feeding capillary coaxially located inside a cylindrical tube. The continuous outer fluid is injected at a constant flow rate to drag the inner liquid (see Chap. 7).

J. M. Montanero, *Tip Streaming of Simple and Complex Fluids*, Fluid Mechanics and Its Applications 137, https://doi.org/10.1007/978-3-031-52768-5_9

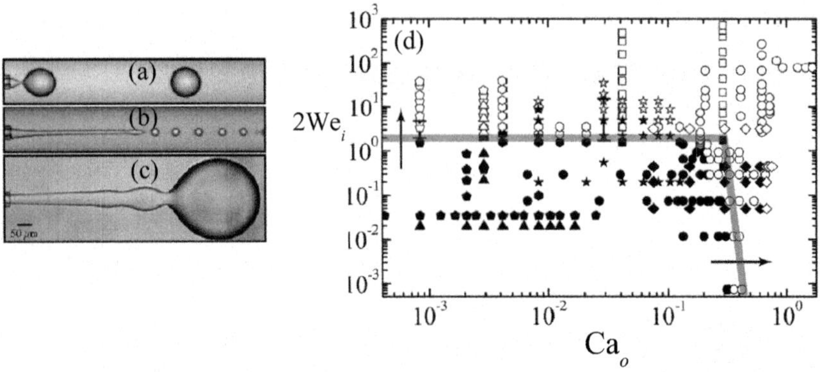

Fig. 9.1 Dripping (**a**), narrowing (**b**), and widening (**c**) regimes in the coflowing configuration. State diagram showing the dripping (solid symbols) and narrowing/widening (open symbols) experimental realizations for varied liquid-liquid configurations (**d**) (reprinted with permission from Ref. [3]. Copyright (2007) by the American Physical Society)

Liquid-liquid coflowing devices usually operate in the standard dripping and jetting modes, in which the droplets and jets' size is commensurate with that of the inner capillary. Cramer et al. [2] have shown that the so-called limiting length of the produced drops increases with the outer fluid flow rate. When this parameter exceeds a certain threshold, dripping gives way to jetting. The jets can remain quasi-cylindrical, narrow, or widen downstream (Fig. 9.1). Utada et al. [3] have found that the dripping-to-jetting transition in coflowing liquid streams with moderate Reynolds numbers can be characterized by a state diagram that depends on the Weber number We_i and the capillary number Ca_o (see Chap. 7). Specifically, jetting takes place only if the sum $\mathrm{Ca}_o + \mathrm{We}_i$ is at least of order unity (Fig. 9.1).

In many microfluidic realizations, the produced jets are quasi-cylindrical, and the velocity profiles are practically parabolic in both the continuous and dispersed phases. However, the wavelength of the perturbation responsible for the jet breakup can still be estimated from Tomotika's dispersion relationship for a uniform velocity distribution [4]. This allows one to calculate the droplet diameter as a function of the flow rate ratio $Q_r = Q_o/Q_i$ and viscosity ratio μ [5]. For Weber numbers at least of order unity, jets in the widening mode break up due to an absolute instability (see Chap. 3) occurring for sufficiently small values of Ca_o [6]. This instability produces droplets with diameters much larger than the precursor jet.

Suryo and Basaran [7] analyzed numerically the transition from the dripping to the microjetting mode of tip streaming in a coflowing configuration. For sufficiently large values of the flow rate ratio Q_r, the meniscus attached to the inner capillary adopts a conical shape within which a recirculation pattern appears. For a narrow region of the parameter space, the meniscus tip steadily emits a jet with a radius several orders of magnitude smaller than that of the inner capillary [8, 9].

$$Q_r \gtrsim (\mu\, \mathrm{Ca}_i)^{-1}. \tag{9.1}$$

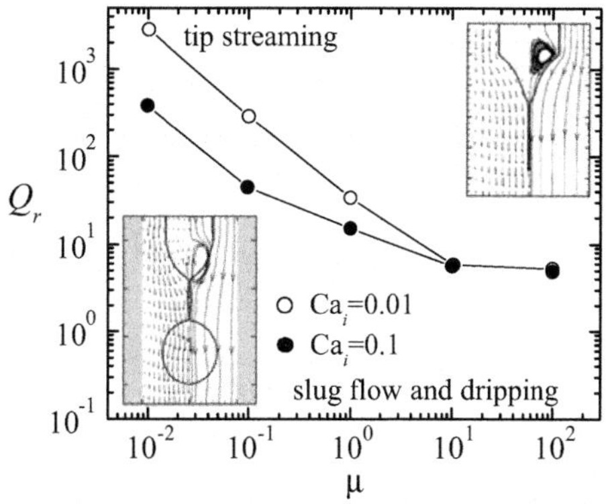

Fig. 9.2 Critical value of the flow rate ratio Q_r for the onset of tip streaming as a function of the viscosity ratio μ calculated numerically (reprinted from Ref. [7] with permission of AIP Publishing)

Numerical simulations agree with this prediction for small μ (Fig. 9.2). For moderate and large values of this parameter, the critical value of the flow rate ratio Q_r becomes independent of the viscosity ratio μ.

Gordillo et al. [9] solved the slenderness model and calculated the linear eigenmodes of the steady base flow. This global stability analysis satisfactorily predicted the parameter conditions to obtain monodisperse micro-emulsions via tip streaming.

It is worth mentioning that, as frequently occurs in the standard dripping-to-jetting transition, it is not easy to distinguish narrowing jetting from microjetting in some experiments. One may find a gradual transition from one mode to another rather than a sharp flow bifurcation.

9.1.2 Liquid-Gas Systems

The production of liquid jets assisted by an outer gaseous stream in a coflowing geometry is much less popular than its liquid-liquid counterpart due to the limited force exerted by the gas in this geometry. Nevertheless, the viscoelastic version of the coflowing technique constitutes an interesting exception.

Submicrometer fibers can be spun from polymeric solutions using the Solution Blow Spinning (SBS) method [10, 11]. SBS has become a popular technique to produce fibers at the micro- and nanometer scales. A polymer solution is extruded in SBS at a constant flow rate across an inner capillary. The liquid phase coflows with a high-speed gas current released from an outer nozzle and driven by a large pressure drop. This current expands around the inner capillary and creates a liquid meniscus similar to Taylor's cone in electrospinning (Fig. 9.3). For sufficiently large pressure drops, the meniscus apex tapers a thin liquid jet that flies toward the collection target. The solvent evaporation leaves behind polymer fibers. SBS can be classified as a microjetting realization. The physics governing this flow has been nicely reviewed by Lauricella et al. [12].

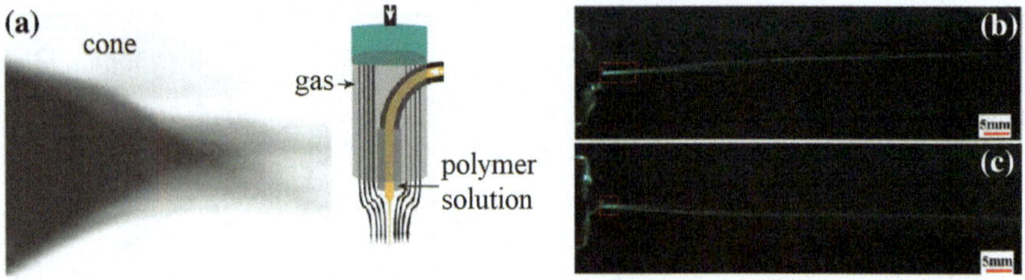

Fig. 9.3 **a** Image of a liquid cone at the tip of an SBS device (reprinted with permission from Ref. [13]). The sketch shows the SBS configuration (reprinted with permission from Ref. [11]. Copyright (2016) American Chemical Society). **b** and **c** Images of the polymer jet produced by SBS at gas pressures of 1.121 bar (**b**) and 1.363 bar (**c**) (reprinted with permission from Ref. [14]. Copyright (2014) American Chemical Society)

It is worth mentioning that several techniques have used high-speed gaseous streams for spinning fibers. For instance, the airbrushing method [10] relies on a mechanism similar to that described above for SBS. In the so-called Gas Jet Process [15], a drop of polymer solution is exposed to a high-speed gas stream ejected independently. The gaseous current pulls a thin jet out of the droplet. The jet transforms into a fiber after solvent evaporation. Nanofibers can also be produced from gas-assisted polymer melt electrospinning [16], where the drag force exerted by the coflowing gas can significantly reduce the fiber diameter.

9.2 Flow Focusing

9.2.1 Liquid Stream Focused by a Liquid Current

In the liquid-liquid version of axisymmetric flow focusing, both the continuous (outer) and dispersed (inner) phases are forced to cross a discharge orifice located in front of the capillary that feeds the inner stream (see Chap. 7). The resulting stresses accelerate that stream until a thin fluid thread is emitted from the tip of the tapering meniscus attached to the capillary. In the microdripping mode (Fig. 9.4-left), the thread breaks up right behind the discharge orifice, and the size of the droplets formed from the breakup is commensurate with that of the orifice. The microjetting regime (Fig. 9.4-right) produces longer and thinner jets giving rise to smaller droplets with an acceptable degree of monodispersity.

9.2.1.1 Geometrical Configurations and Applications

Different axisymmetric geometries have applied the flow focusing principle to liquid-liquid systems. The original plate-orifice configuration [18] has been implemented

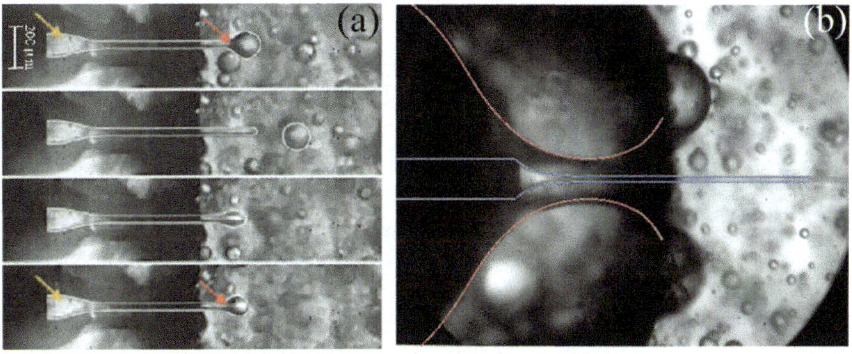

Fig. 9.4 Microdripping (**a**) and microjetting (liquid cone at the tip of an) modes of liquid-liquid flow focusing. The inner fluid is 20 cSt silicone oil, while the outer liquid is water. The white lines have been added to highlight the interface contours. The pink lines correspond to the numerical simulation (reproduced with permission, Ref. [17])

in planar silicon microfluidic chips for emulsion production. The device consisted of several orifices keeping the axisymmetric geometry per orifice [19]. Microfluidic devices with the axisymmetric flow focusing geometry have also been manufactured in PDMS [20] using stereolithography [21]. These devices have been utilized to produce double emulsions and monodisperse hydrogel beads.

Yobas et al. [22] have used an orifice with a cusp-like edge to maximize the stress exerted by the outer stream, which ensures the controlled breakup of droplets for a wide range of flow rates. The focusing effect can also be produced by locating the feeding capillary in front of another capillary, both placed inside an outer tube through which the focusing current is injected [23]. In the so-called non-embedded coflow-focusing configuration [24], a converging nozzle is placed in front of a collector tube to produce microemulsions in the microdripping and microjetting modes.

Single-step generation of multi-core double emulsion droplets can be achieved by the so-called counter-flow focusing [25]. Utada et al. [26] first assembled a counter-current double emulsion flow focusing device by aligning glass capillaries. This configuration has been examined experimentally by Nabavi et al. [27]. It can exhibit a second-order transition which leads to extremely thin jets [28], giving rise to silicone oil emulsions with a submicron particle radius. Monodispersed emulsions down to 2 μm in diameter were produced at a frequency of 20 kHz by using a similar configuration, the so-called "impinging flow-focusing" [29].

The liquid-liquid flow focusing configuration has received much attention due to its diverse applications. For the sake of illustration, we mention some additional relevant examples. Lee et al. [30] designed a flow cytometer based on the focusing effect obtained when two liquid streams flow across coaxial converging nozzles (Fig. 9.5a). The encapsulation and release of actives were achieved in the pioneering work of Utada et al. [26] and subsequent experimental studies [20, 31] using axisymmetric flow focusing. Tsuda et al. [32] fabricated an axisymmetric flow focusing device

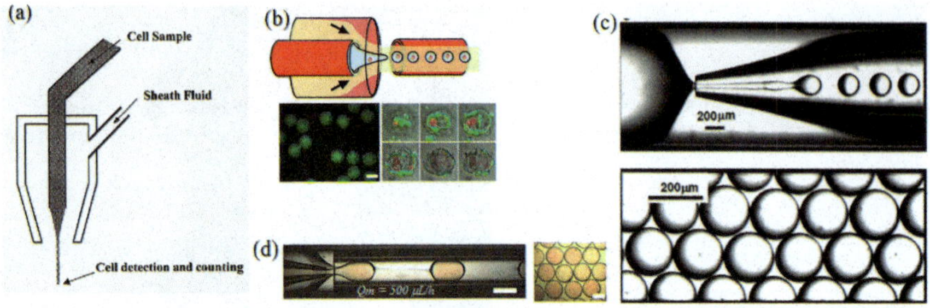

Fig. 9.5 **a** Sketch of a cytometer based on the focusing effect (reprinted with permission from Ref. [30]), **b** cell-encapsulating microgel beads (reprinted with permission from Ref. [32]. Copyright (2010) Americal Chemical Society), **c** biodegradable Poly(Lactic Acid) particles (reprinted with permission from Ref. [33]), and **d** PDMS microcapsules with tunable elastic properties [35]

to produce monodisperse cell-encapsulating microgel beads composed of a self-assembling peptide gel for 3D cell culture (Fig. 9.5b). This technique also enabled the production of Biodegradable Poly(Lactic Acid) particles (Fig. 9.5c) [33]. Multiplex coaxial flow focusing has been proposed for the single-step fabrication of multicompartment Janus microcapsules [25] and pesticide-loaded microcapsules [34].

Liquid-liquid flow focusing allows the production of microemulsions with high throughput for different purposes, such as encapsulation, chemical synthesis, and biochemical assays [36]. Monodisperse multiple-emulsions can be fabricated with the double flow focusing configuration [37, 38]. These microfluidic shapes can be used as microreactors and fine templates for synthesizing advanced particles [39].

The variety of applications of the liquid-liquid flow focusing increases when polymeric solutions are used either as the focusing current or the focused phase. For example, a variant of flow focusing allows the formation of PDMS microcapsules with tunable elastic properties [35] (Fig. 9.5d). Zhu et al. [40] have manufactured multicompartment polymeric microcapsules in an axisymmetric flow focusing device for magnetic separation and synergistic delivery. Micrometer fibers can be formed when a viscoelastic liquid is focused with a Newtonian one [41, 42] or a viscoelastic stream [43].

9.2.1.2 Microjetting Stability

Viscous stresses cause a robust focusing effect in the discharge orifice of liquid-liquid flow focusing devices. This effect favors the transition from dripping to either microdripping or microjetting. In fact, tip streaming is found in flow focusing for outer capillary numbers much smaller than those required in the equivalent coflowing configuration [44, 45]. The viscous stress in the focusing region scales as D_o^{-2}. Consequently, the droplet formation is considerably affected by the orifice diameter, while it is practically insensitive to the orifice length [45].

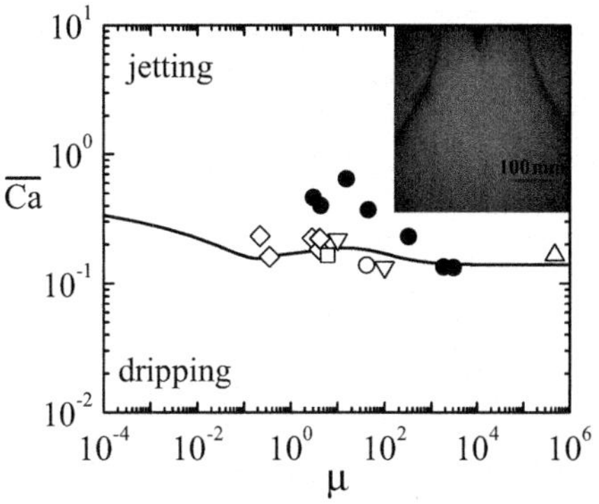

Fig. 9.6 Modified capillary number $\overline{Ca} = (\mu_i \mu_o)^{1/2} v_s/\sigma$ below which a jet becomes absolutely unstable. The line indicates the convective-to-absolute instability transition for a viscosity-dominated flow [46]. Experimental values for the jetting-to-dripping transition (symbols). The data were taken from Acero et al. [47] (•), Anna et al. [48] (□), Anna and Mayer [49] (○), Utada et al. [3] (▽), Guillot et al. [50] and Herrada et al. [51] (◇), and Gañán-Calvo [52] (△). The inset shows a liquid-liquid viscosity-dominated jetting realization in a flow focusing device [47] (reprinted from Ref. [47] with permission of IOP Publishing)

The microjetting regime of tip streaming requires two conditions: (i) the steady flow in the tapering meniscus must be stable, and (ii) the emitted jet must be *convectively* unstable (see Chap. 3). As applied to liquid-liquid flow focusing, the last condition implies that the quasi-parallel flow behind the discharge orifice must remain convectively unstable as the jet diameter decreases. The analysis of the convective-to-absolute instability transition conducted by Gañán-Calvo [46] determined the condition to verify that prerequisite in the limit of vanishing jet diameters: the capillary number $\overline{Ca} = (\mu_i \mu_o)^{1/2} v_s/\sigma$ defined in terms of the interface velocity v_s must exceed a threshold that depends on the viscosity ratio μ exclusively.

Figure 9.6 shows the theoretical curve $\overline{Ca}(\mu)$ [46] and experimental values of the critical capillary number corresponding to different coflowing and flow focusing configurations. As can be observed, the experimental data are located approximately above the theoretical prediction, which indicates that the jet convective instability is a *necessary* (not sufficient) condition for microjetting. The discrepancy between, for instance, the experimental results of Acero et al. [47] (solid circles) and the curve corresponding to the convective-to-absolute instability transition must be attributed to the instability of the steady flow in the tapering meniscus.

Most of the studies on flow focusing have been devoted to proving the applicability and versatility of this technique in diverse technological fields, such as pharmacy, the food industry, and material processing. The stability analysis of the microjetting regime of this configuration has received little attention. As in other configurations, the most relevant stability limit is that reached when the inner flow rate is reduced while the rest of the governing parameters are fixed. In fact, this limit leads to the smallest droplets produced in flow focusing.

Gañán-Calvo and Riesco-Chueca [53] studied experimentally the micro jetting-to-dripping transition taking place at the minimum inner flow rate stability limit. The experimental results were compared with the prediction obtained from the convective-to-absolute instability transition for a parallel flow. As mentioned above, this comparison is pertinent only if the instability originated in the jet, which does not occur in many experimental realizations.

In many experiments, the minimum inner flow rate is imposed by the instability of the steady flow in the tapering meniscus. Acero et al. [47] stabilized this flow by injecting the focused liquid through a hypodermic needle (see the inset in Fig. 9.6), which decreased the minimum flow rate by up to two orders of magnitude. The essential idea was to replace the tapering meniscus attached to the feeding capillary with a more stable liquid film sliding over the hypodermic needle tip.

9.2.1.3 Global Stability Analysis

The calculation of the linear global modes [54] is an adequate tool to predict the instability of the microjetting regime. As explained in Chap. 3, the idea is to assume that a long thin jet tapers from the liquid meniscus tip (microjetting). We calculate the steady basic flow corresponding to this assumption. Then, we interrogate this basic flow about its response to small-amplitude perturbations [9, 55, 56]. Microjetting becomes unstable if the largest growth rate of the eigenfrequency spectrum becomes positive.

Figure 9.7 shows experimental microjetting and dripping realizations when 20 and 100 cSt silicone oils are focused with water in a converging-diverging nozzle [17]. We regard microjetting realizations as those producing long jets compared to the discharge orifice diameter, as implicitly done in the global stability analysis. This analysis satisfactorily predicts the experimental stability limits. The discrepancies with respect to the experimental data may be due to the position of the triple contact line. In fact, the feeding capillary end is assumed to be infinitely thin in the numerical simulation, a simplification that does not apply to the experiments.

As explained in Chap. 3, the existence of dripping realizations in the globally stable parameter region may be explained in terms of the growth of linear perturbations due to the short-term superposition of stable modes [57]. This phenomenon occurs when the instability is localized in the emitted jet [58]. We will return to this point in Chap. 10.

As explained below, the global stability analysis allows one to predict whether microjetting becomes unstable due to the tapering meniscus destabilization or the emitted jet absolute instability beyond the discharge orifice. The horizontal and vertical branches of the solid line in Fig. 9.7 correspond to the former and latter cases, respectively. The experiments show the validity of the above prediction: the dripping observed below the horizontal branch was caused by the meniscus instability (Fig. 9.8a), while the microdripping arising on the left side of the vertical branch was produced by the jet breakup next to the discharge orifice (Fig. 9.8b). The horizontal branch of the stability limit is the so-called minimum flow rate stability limit,

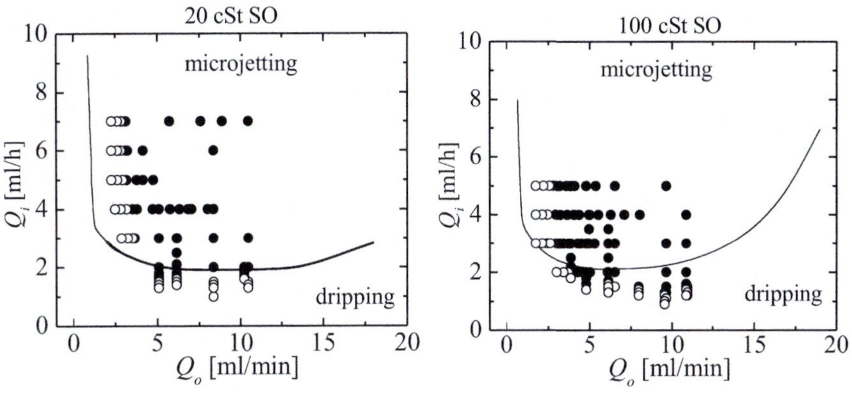

Fig. 9.7 Experimental microjetting (solid symbols) and dripping (open symbols) realizations of 20 and 100 cSt silicone oils focused with water. The solid lines are the stability limits obtained from the global linear stability analysis (reproduced from Ref. [17] with permission)

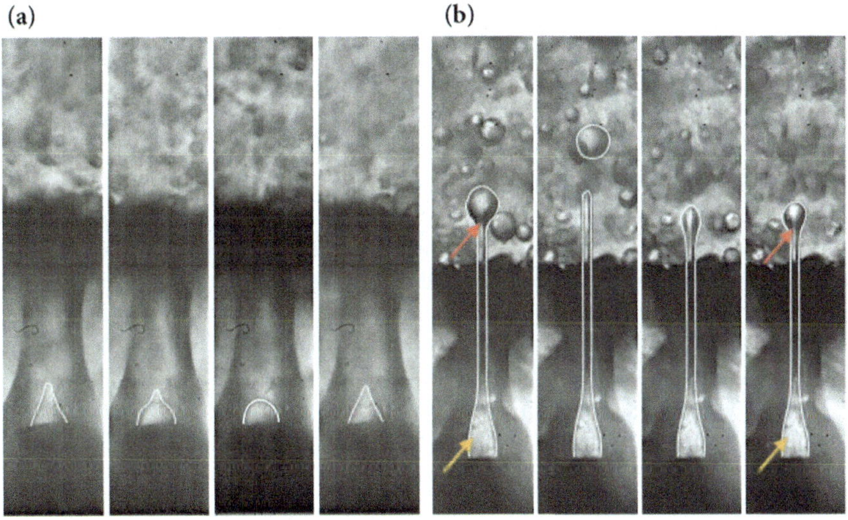

Fig. 9.8 Sequence of experimental images showing the meniscus instability (**a**) and the jet breakup (**b**). The white lines have been added to highlight the interface contours. In the right-hand image, the orange and red arrows indicate the quasi-steady meniscus and the droplet formed right after the discharge orifice, respectively. The inner fluid is 20 cSt silicone oil, while the outer liquid is water. The flow rates are $Q_i = 0.5$ ml/h and $Q_o = 5.5$ ml/min (**a**), and $Q_i = 5$ ml/h and $Q_o = 3.1$ ml/min (**b**) (reproduced from Ref. [17] with permission)

which is also characteristic of other tip streaming configurations such as gaseous flow focusing [58] and electrospray [59].

To calculate the linear global modes, we assume the following temporal dependence for the interface position (see Chap. 3):

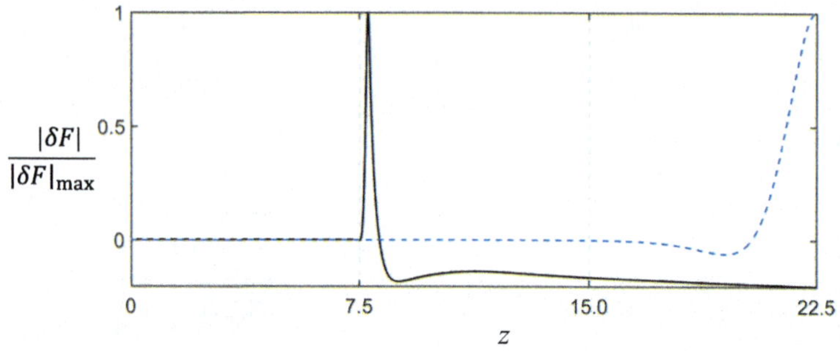

Fig. 9.9 Amplitude of the interface perturbation, $|\delta F|$, normalized with its maximum value, $|\delta F|_{\max}$. The solid and dashed lines correspond to the cases ($\mu = 0.1$, $Q_r = 585$) and ($\mu = 0.05$, $Q_r = 30$). The axial coordinate z was made dimensionless with the feeding capillary radius (reproduced from Ref. [17] with permission)

$$F(z;t) = F_0(z) + \delta F(z)\, e^{-i\omega t} + \text{c.c.} \quad (|\delta F| \ll F_0). \tag{9.2}$$

Here, $F(z;t)$ is the distance of the interface element to the symmetry axis z, $F_0(z)$ is the steady interface location in the base flow, $\omega = \omega_r + i\omega_i$ is the eigenfrequency characterizing the perturbation evolution, $\delta F(z)$ is the perturbation amplitude, and "c.c." stands for complex conjugate.

Figure 9.9 shows the magnitude of the perturbation amplitude for two marginally stable flows characterized by different values of $Q_r = Q_o/Q_i$ and $\mu = \mu_o/\mu_i$. The solid line corresponds to the horizontal instability branch in Fig. 9.7. In this case, the maximum amplitude is reached in the tapering meniscus, indicating that the perturbation originated in this region. The dashed line corresponds to the vertical branch. In this case, the perturbation grows only beyond the discharge orifice, leaving the tapering meniscus and part of the ejected liquid thread practically intact.

In the cone-jet mode of electrospray (see Chap. 8) for relatively large conductivities, the cone-jet transition is much smaller than the Taylor cone, which allows us to decouple the flows in these two regions and describe the microjetting instability in terms of the forces in the critical cone-jet transition exclusively. This large disparity between the spatiotemporal scales is not observed in the liquid-liquid flow focusing configuration for moderate values of the viscosity ratio. For this reason, one expects the microjetting stability to be affected by the flow in the tapering meniscus.

Cabezas et al. [17] proposed the following instability mechanism based on the flow in the tapering meniscus and the shape of the interface perturbation. A recirculation cell is formed right at the exit of the feeding capillary due to the relatively small value of the inner viscosity. As the inner flow rate approaches its minimum value (Q_r increases), the cell grows and gets closer to the interface, narrowing the fluid passage through which the ejected liquid crosses the meniscus (Fig. 9.10). The interface slightly deforms as this passage narrows, and an inflection point appears in the interface contour $F_0(z)$ (one of the principal curvatures vanishes). At the stability limit, the maximum of the interface perturbation amplitude $|\delta F(z)|$ is located very

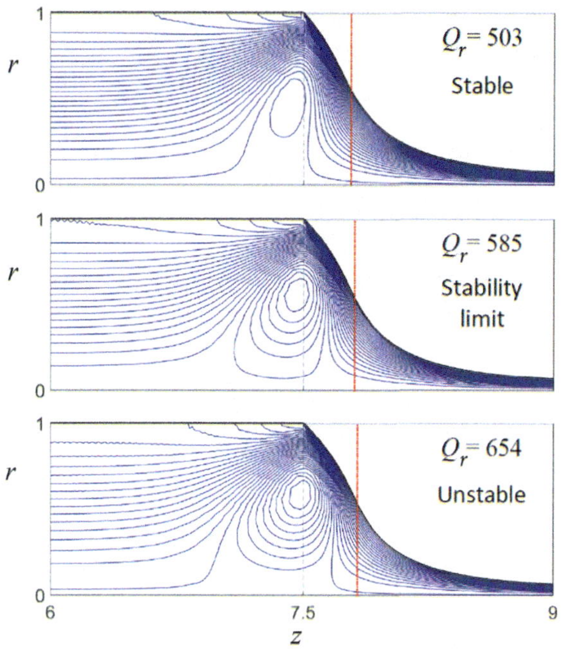

Fig. 9.10 Streamlines in the inner liquid stream for $\mu = 0.1$ and the flow rates indicated in the figure. The red vertical line indicates the location of the maximum amplitude of the interface perturbation. The axial and radial coordinates were made dimensionless with the feeding capillary radius (reproduced from Ref. [17] with permission)

close to that inflection point. This suggests that the interface does not withstand perturbations of the capillary pressure at that point. These perturbations grow on time, leading to the base flow destabilization.

Figure 9.11 shows the parameter region in the (μ, Q_r) plane within which the base flow is linearly stable, and, therefore, microjetting can be obtained. As mentioned above, the stability islands are limited by the meniscus (solid lines) and jet (dashed lines) destabilization. Decreasing the capillary number destabilizes the island's left and lower borders, showing that the axial momentum necessary to achieve microjetting becomes significantly larger. Nevertheless, the capillary number reduction has a stabilizing effect at the upper and right boundaries. The interface can withstand larger stresses due to increased interfacial tension. Therefore, the minimum inner flow rate decreases (Q_r increases), while the viscosity ratio corresponds to the stability island's upper-right vertex ($\mu = 0.115$, $Q_r = 665$) increases.

Mu et al. [60] studied both experimentally and theoretically the stability of the axisymmetric liquid-liquid flow focusing configuration. They also distinguished between the instability of the tapering meniscus and that localized in the emitted jet. This allowed them to explain their direct numerical simulations and experimental observations.

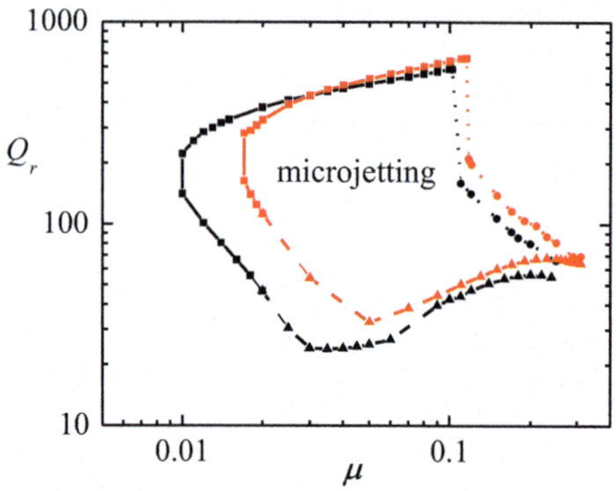

Fig. 9.11 Stability island of microjetting in the (μ, Q_r) parameter plane calculated for $\rho = 1$, $\mathrm{Re}_i = 0.02$, and $Ca_i = 0.05$ (black lines and symbols) and 0.0375 (red lines and symbols). The solid and dashed lines correspond to the tapering meniscus and jet instabilities, respectively (reproduced from Ref. [17] with permission)

9.2.1.4 Jet Diameter

The scaling law for the jet diameter can be easily deduced from mass conservation for $Q_r^{-1} \ll 1$ ($Q_i \ll Q_o$). In this case, we can assume that the thin jet and outer stream move at the same speed v_j in the nozzle neck (discharge orifice). Suppose that the velocity profiles in a nozzle cross-section are uniform (inviscid approximation). Since $d_j \ll D_o$, $Q_o \simeq v_j \pi D_o^2/4$ and, therefore,

$$\frac{d_j}{D_0} = Q_r^{-1/2}. \tag{9.3}$$

The growth of the boundary layer in the nozzle accelerates the outer stream inviscid core for a fixed outer flow rate, which enhances the transfer of axial momentum to the inner liquid jet. This reduces the jet radius with respect to that predicted by the above inviscid approximation.

Using the optimization method described by Gañán-Calvo [61], the best collapse of the data around the scaling law $d_j/D_o \sim \mu^\alpha Q_r^\beta$ is obtained for $\alpha \simeq -0.05$ and $\beta \simeq -0.45$. Figure 9.12 shows that

$$\frac{d_j}{D_o} = 0.62\, \mu^{-1/10} Q_r^{-1/2} \tag{9.4}$$

constitutes a good approximation for $\rho = 1$, $Re_i = 0.02$ and Ca_i=0.05. The minimum value of d_j/D_o was 0.038, which shows the strong focusing effect achieved with this configuration for sufficiently large viscosity ratios.

The relationship between the diameters of the jet and the droplets resulting from the jet breakup is given by the wavelength of the dominant capillary mode (see Chap. 3). If the dependence of the mode growth rate on the wavelength does not exhibit a sharp maximum, then modes with different wavelengths can affect the jet breakup, and polydisperse collections of droplets can be obtained. An external stimulus can be applied to trigger and control the jet breakup, which allows one to select the droplet diameter and narrow the size distribution. For instance, Yang

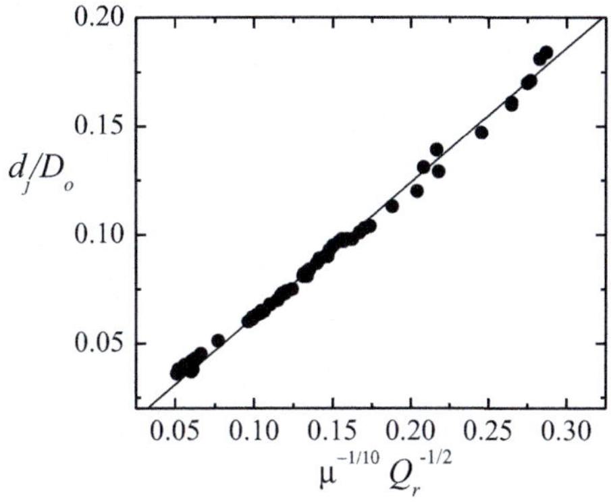

Fig. 9.12 Diameter d_j of the emitted jet calculated from simulations (symbols). The solid line is the scaling law (9.4) (reproduced from Ref. [17] with permission)

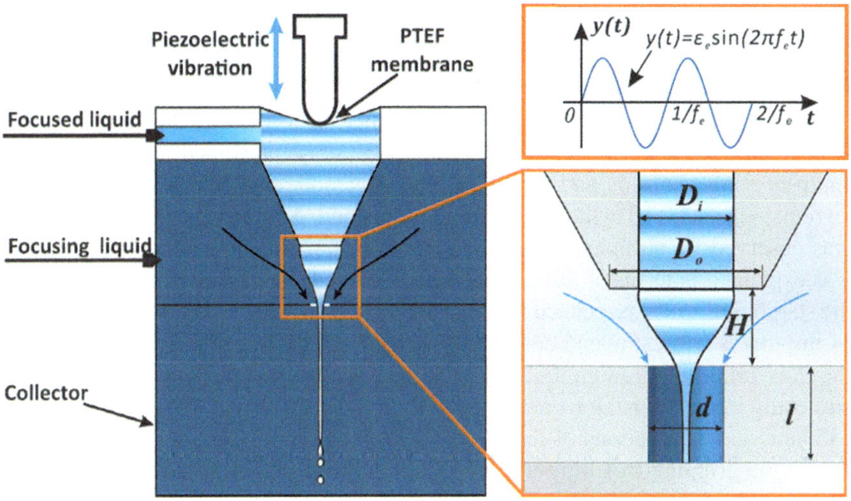

Fig. 9.13 Sketch of the mechanical vibration acting on the focused liquid (reprinted from Ref. [62] with permission of AIP Publishing)

et al. [62] employed an external sinusoidal actuation for generating monodisperse collections of droplets of the desired size from the modulated rupture of the liquid jet (Fig. 9.13).

9.2.1.5 Viscoelasticity

The 2D liquid-liquid flow focusing configuration has been used to produce micrometer jets of Boger fluids [63–68]. The thread pinch-off is initiated by the inertio-

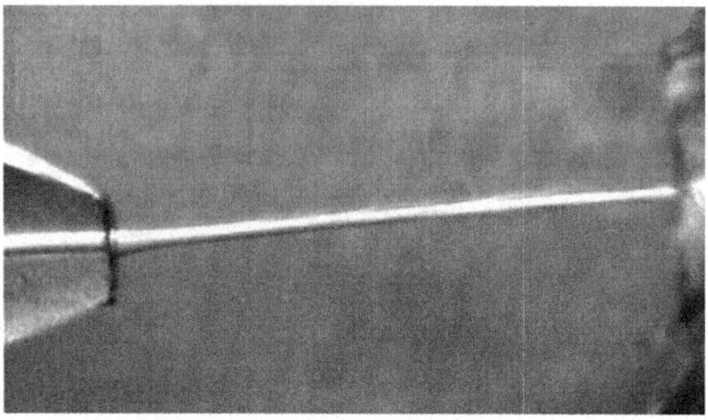

Fig. 9.14 Viscoelastic jet focused with an oil phase (reprinted with permission from Ref. [41]. Copyright (2006) Americal Chemical Society)

capillary mechanism, followed by the elastocapillary regime (see Chap. 4) [64]. As the polymer relaxation time increases, the filament dynamics become controlled primarily by the elastocapillary mechanism, and both the thread length and pinch-off time increase [64]. The flow regimes [67, 68], size of the resulting drops [63, 65], and appearance of flow asymmetries [69] have been analyzed both numerically and experimentally.

Axisymmetric flow focusing has been applied to form viscoelastic jets focused with an oil phase (Fig. 9.14) [41]. The polymeric stress prevented the formation of a sharp tapering meniscus, and thus the microjetting mode could not be established. The jet remained anchored to the edge of the capillary end. A dripping-to-jetting transition was observed for sufficiently high values of the injected flow rate. This transition was explained in terms of the balance between the destabilizing capillary force and the stabilizing stress associated with the polymer elongation [70]. An effective value of this stress was estimated considering the polymer elongation produced by both the Poiseuille-type flow in the feeding capillary [71], and the subsequent filament stretching in the focusing region.

Numerical simulations of the FENE-CR model have shown that viscoelasticity delays the dripping-to-jetting transition in liquid-liquid flow focusing [72]. One of the distinctive characteristics of viscoelastic flow focusing is that the jetting mode requires a large distance between the feeding capillary and the orifice, significantly larger than the size of the focusing region.

9.2.1.6 Surfactants

The effect of a surfactant on flow focusing has been studied on very few occasions. Moyle et al. [73] proposed a model to predict the appearance of tip streaming in the presence of a soluble surfactant. They concluded that tip streaming occurs when the surfactant adsorbed to the interface is that needed to maintain the conical interfacial shape. Wrobel et al. [74] modeled the outer velocity field in axisymmetric flow focusing by combining an imposed uniaxial extension flow at infinity with two transverse, coaxial, annular baffles placed symmetrically to either side of the drop. They

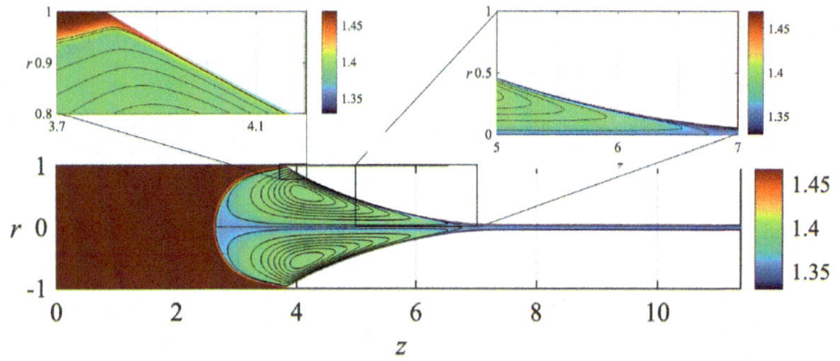

Fig. 9.15 Streamlines and surfactant volumetric concentration $c(r, z)/c_{\mathrm{cmc}}$ for $c_\infty/c_{\mathrm{cmc}} = 1.47$. The color indicates the value c/c_{cmc}. The values of the dimensionless governing parameters can be found in Ref. [77]. The spatial coordinates are made dimensionless with the capillary radius R_i

described how the focusing effect produced by the baffles collaborates with a soluble surfactant to produce tip streaming. As explained below, López et al. [75] examined experimentally the role played by a surfactant monolayer on the focusing of a liquid stream in the confined selective withdrawal configuration. The simulations of Lytra et al. [76] for a flow-focusing geometry show the development of spatially growing instabilities in the downstream region with wavelength on the order of the nozzle radius.

The global stability of flow focusing in the presence of a surfactant has been analyzed very recently [77]. The focusing effect is produced by an outer viscous stream crossing a converging nozzle which discharges into a cylindrical tube. In this way, the inner and outer stream do not slow down beyond the nozzle neck, which eliminates the unstable modes associated with the emitted jet (absolute instability) and allows one to focus on the meniscus stability. In addition, the spectrum of eigenvalues become practically insensitive to the discharge tube length for relatively small values of this parameter.

In the simulations of Rubio et al. [77], the surfactant is dissolved in the inner (dispersed) phase at concentrations larger than the CMC (Fig. 9.15). Owing to the small value of the diffusion coefficient, a thin mass boundary layer forms between the stream crossing the feeding capillary and the recirculating zone in the tapering meniscus. The surfactant is advected through a very narrow annular streamtube next to the interface. The surfactant molecules adsorb onto the interface according to the kinetic model (2.38). Adsorption occurs essentially in the region next to the feeding capillary (Fig. 9.16). The liquid flowing next to the interface is slightly emptied of surfactant. Due to the flow recirculation, this surfactant depletion is convected throughout the tapering meniscus. The parameter conditions are such as the surfactant volumetric concentration takes a relatively constant value in the entire fluid domain, including the sublayer next to the interface. This behavior was also observed by Lytra et al. [76] for a similar configuration.

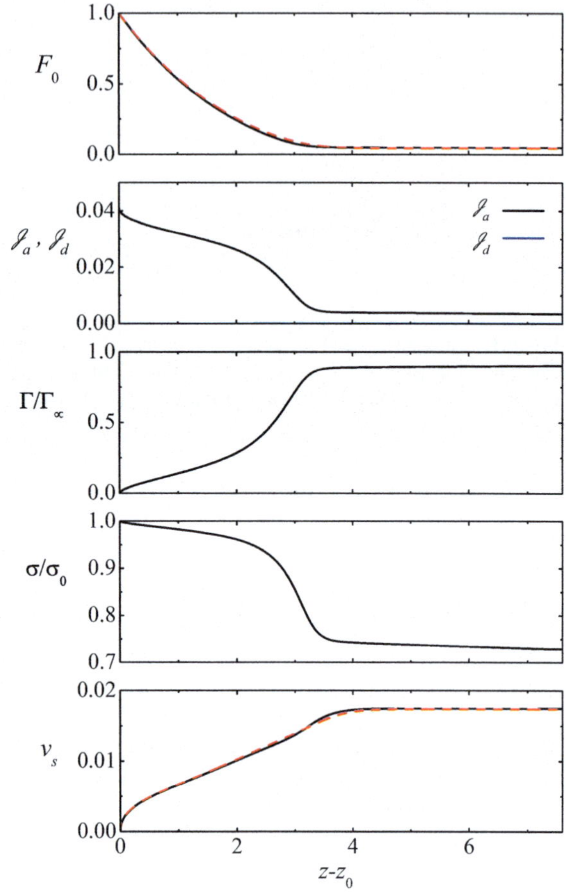

Fig. 9.16 Interface location F_0, adsorption $\mathcal{J}_a$ and desorption $\mathcal{J}_d$ surfactant flux, surface coverage Γ/Γ_∞, surface tension $\hat{\sigma}$ relative to the value σ_0 corresponding to the clean interface, and surface velocity v_s as a function of the axial coordinate $z - z_0$, where z_0 is the feeding capillary end. The dashed (red) and solid (black) lines correspond to $c_\infty/c_{\mathrm{cmc}} = 0$ and 1.47, respectively. The values of the rest of the parameters are the same as those in Fig. 9.15. The quantities have been made dimensionless as explained in Ref. [77]

The area of an interface element decreases in the meniscus tip due to the reduction of the interface radius there. The small residence time of the interface element in the meniscus tip prevents surfactant desorption even in that region. These two factors make the surfactant surface concentration increase in the meniscus tip up to values close to the maximum packing density Γ_∞, significantly decreasing the surface tension there (Fig. 9.16).

The Marangoni stress in the meniscus tip acts against the viscous drag exerted by the outer flow. However, its intensity is not large enough to reduce the surface velocity. Figure 9.16 suggests that Marangoni stress tries to immobilize the interface in the meniscus tip, where the surface tension gradient takes high values. However, the flow practically develops at the entrance of the nozzle neck due to the large viscosity forces. The Hagen-Pouiseuille parabolic profile fixes the value of v_s. The surfactant does not affect this parabolic profile, essentially determined by the flow rate ratio. Therefore, $v_s(z)$ takes practically the same value with and without surfactant

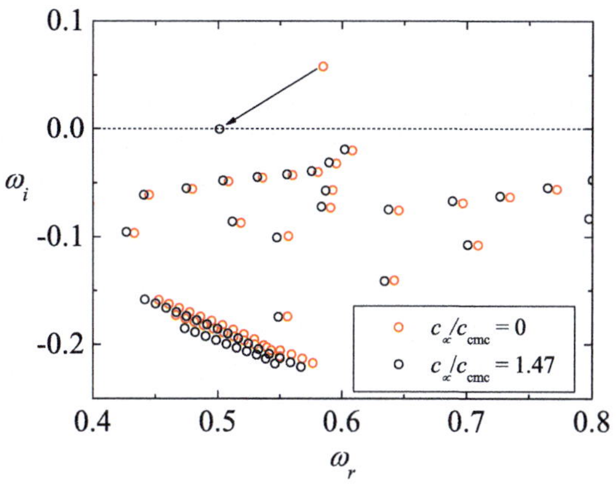

Fig. 9.17 Spectrum of eigenvalues obtained from the global stability analysis for $c_\infty/c_{\text{cmc}} = 0$ (red symbols) and 1.47 (black symbols). The values of the rest of the parameters are the same as those in Fig. 9.15. The eigenfrequency $\omega = \omega_r + i\omega_i$ has been made dimensionless as explained in Ref. [77]. The arrow indicates the displacement of the critical mode eigenfrequency in the presence of surfactant. The parameter conditions correspond to the marginal stability in the presence of surfactant

[77]. This behavior substantially differs from that observed in open flows, where the Marangoni stress reduces the interface velocity and, therefore, the intensity of the inner flow [78, 79].

The surfactant effect is localized in the meniscus tip. For this reason, the base flows with and without surfactant are similar on the scale given by the feeding capillary. Adding the surfactant slightly reduces the recirculation cell size, making the meniscus tip stagnation point move upstream. The surfactant monolayer stabilizes the flow, reducing the critical mode growth rate (Fig. 9.17). Therefore, soluto-capillarity and Marangoni stress do not significantly influence the global flow pattern but affect the evolution of the critical eigenmode.

The importance of soluto-capillarity and Marangoni stress has been assessed separately by "turning off" the corresponding terms in the interface boundary conditions [77]. Specifically, the equilibrium of tangential and normal stresses at the interface (see Chap. 2) leads to

$$\begin{aligned}
&(1 - F_z^2)(w_r^{(i)} + u_z^{(i)}) + 2F_z(u_r^{(i)} - w_z^{(i)}) \\
&= \hat{\sigma}_z(1 + F_z^2)^{1/2} + \mu[(1 - F_z^2)(w_r^{(o)} + u_z^{(o)}) + 2F_z(u_r^{(o)} - w_z^{(o)})], \qquad (9.5)
\end{aligned}$$

$$\begin{aligned}
&p^{(i)} + \hat{\sigma}\frac{FF_{zz} - 1 - F_z^2}{F(1 + F_z^2)^{3/2}} - \frac{2[u_r^{(i)} - F_z(w_r^{(i)} + u_z^{(i)}) + F_z^2 w_z^{(i)}]}{1 + F_z^2} \\
&= p^{(o)} - \frac{2\mu[u_r^{(o)} - F_z(w_r^{(o)} + u_z^{(o)}) + F_z^2 w_z^{(o)}]}{1 + F_z^2}, \qquad (9.6)
\end{aligned}$$

where r (z) is the radial (axial) coordinate, $u^{(k)}$ ($w^{(k)}$) is the radial (axial) velocity component, $F(z, t)$ is the distance of an interface element to the symmetry axis z, and $\hat{\sigma} = \sigma/\sigma_0$ is the surface tension relative to the value σ_0 corresponding to the clean interface. The subscripts t, r, and z denote the partial derivatives with respect to the corresponding variables. Equations (9.5) and (9.6) have been made dimensionless, as explained by Rubio et al. [77].

When both soluto-capillarity and Marangoni stress are present, the minimum flow rate decreases by 60%. If the term $\hat{\sigma}_z$ (the Marangoni stress) in Eq. (9.5) is set to 0, the minimum flow rate decreases by 40%. This means that the two effects significantly contribute to the microjetting stabilization. Rubio et al. [77] have shown that the decrease in the capillary pressure in the meniscus tip reduces the adverse pressure gradient in front of the stagnation point, which is a stabilizing mechanism.

9.2.2 Gaseous Stream Focused by a Liquid Current

Microbubbles occupy an important place in various biochemistry, molecular biology, and medical applications [80]. They can be used as ultrasound contrast agents, carriers for targeted drug delivery, and enhanced gene vectors, among many other applications. These applications typically demand large collections of bubbles a few microns in size with a high degree of monodispersity.

Hydrodynamic forces can be used to reduce the bubble size and increase the production frequency in T-junctions [81–84], co-flowing [85–88], and flow focusing [52, 89–103] devices. In flow focusing, a coflowing liquid stream crosses an orifice located right in front of the gas source (Fig. 9.18-left). This stream exerts both viscous and pressure forces that collaborate to stretch a gaseous meniscus attached to the feeding capillary, significantly reducing the bubble size while keeping a high degree of monodispersity. The axisymmetric version of flow focusing [52, 89, 92, 96, 97] leads to much higher production rates than its 2D counterpart [90, 91, 93–95, 98–103].

Gaseous jets injected into a quiescent pool of liquid are prone to absolute instability owing to the high values of the capillary wave velocity. For this reason, most flow focusing realizations produce bubbles in the microbubbling mode [89]. The bubble diameter can be obtained by assuming that the unsteady and convective terms of the momentum equation for the gas phase are commensurate with each other, which yields [92]

$$\frac{d_b}{D_o} = 1.1 Q_r^{-0.4}. \tag{9.7}$$

This scaling law remarkably agrees with experimental data obtained using both the plate-orifice [92] and nozzle [104] configurations (Fig. 9.18-right).

Jensen et al. [105] numerically analyzed the bubbling in an axisymmetric flow focusing device. The results for the bubble volume were consistent with those obtained experimentally by Garstecki et al. [90] for the 2D topology.

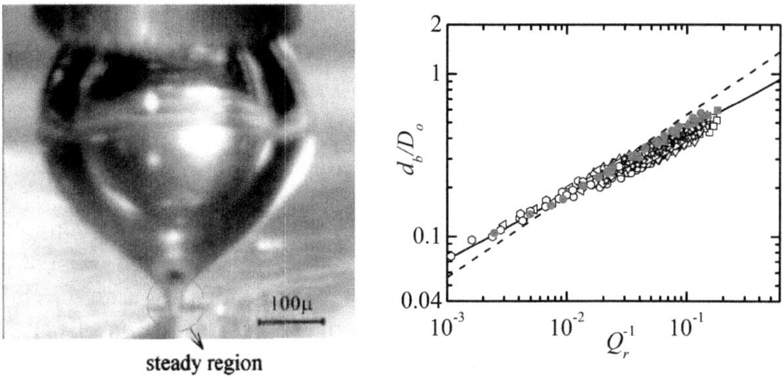

Fig. 9.18 (Left) Focusing of a gas stream with a coflowing liquid current that crosses an orifice located right in front of the gas source (reprinted with permission from Ref. [89]. Copyright (2001) by the Americal Physical Society). (Right) Diameter d_b/D_o of the bubbles produced with flow focusing as a function of the flow rate ratio Q_r. The white and grey symbols correspond to experiments conducted with the plate-orifice [92] and nozzle [104] configuration, respectively. The solid and dashed lines correspond to Eq. (9.7) and $d_b/D_o = 1.77Q_r^{-0.5}$, respectively. Results for microbubbling and quasi-microjetting are mixed (reprinted from Ref. [61] with permission of IOP Publishing)

A quasi-microjetting regime can be achieved when a low-viscosity or moderately viscous liquid stream focuses a gaseous meniscus. The tip of the stretched meniscus ejects a gaseous thread which breaks up into a quasi-monodisperse collection of bubbles with diameters smaller than that of the discharge orifice. Microbubbles were produced in this quasi-microjetting mode of flow focusing from the breakage of an air filament driven by a stream of ethanol-water [52] and glycerol water [100].

Vega et al. [104] studied the production of microbubbles in the quasi-microjetting regime realized in a converging micronozzle for moderate Reynolds numbers. The bubble diameter was approximately given by the formula

$$\frac{d_b}{D_o} = 1.77Q_r^{-\beta}, \tag{9.8}$$

where β took a value between 0.4 and 0.5 depending on the Reynolds number.

Equation (9.8) with $\beta = 0.5$ can be readily derived under two conditions: (i) neglecting both the energy dissipation in the gas meniscus and the gas kinetic energy at the capillary end, and (ii) assuming that neither the pinching location nor the proportion between the precursor jet and resulting bubble diameters significantly depends on Q_r [104].

Figure 9.18-right shows the experimental validation of Eq. (9.8), where results for microbubbling and quasi-microjetting are mixed. It is difficult to determine whether some experimental realizations should be regarded as microbubbling or quasi-microjetting owing to the short length and decreasing diameter of the precursor gaseous threads.

9.3 Confined Selective Withdrawal

9.3.1 Liquid Stream Focused by a Liquid Current

In confined selective withdrawal, the dispersed phase is injected at a constant flow rate through a capillary/needle in front of the collecting tube (see Chap. 7). The outer bath is suctioned across the discharge tube focusing the dispersed phase flow on the meniscus tip. Confined selective withdrawal can operate both in the microdripping and microjetting modes (Fig. 9.19-left). This technique has been used to produce emulsions [106], double emulsions (Fig. 9.19-right) [107], nematic shells [107], and micro-sized PDMS particles [108].

9.3.1.1 Confined Selective Withdrawal Versus Flow Focusing

There are obvious similarities between flow focusing [53] and confined selective withdrawal [106–109]. The only difference between them is the shape of the discharge orifice in front of the feeding capillary. While the original configuration of flow focusing uses an orifice located in a plate, the focusing effect in confined selective withdrawal is produced by the discharge tube.

As occurs in flow focusing, the size of the droplets produced with confined selective withdrawal [106] essentially depends on the ratio between the inner and outer

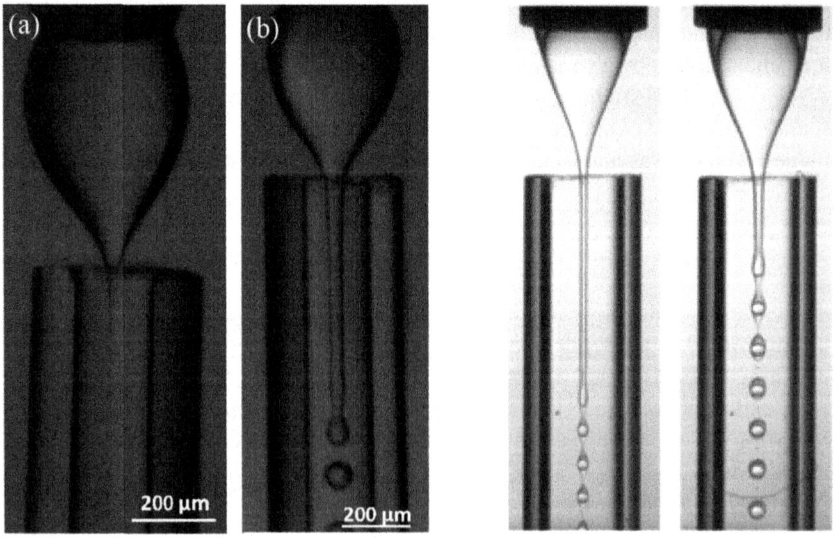

Fig. 9.19 (Left) Microdripping (**a**) and microjetting (**b**) in confined selective withdrawal [75]. (Right) Double emulsions produced with confined selective withdrawal (reproduced with permission, Ref. [106])

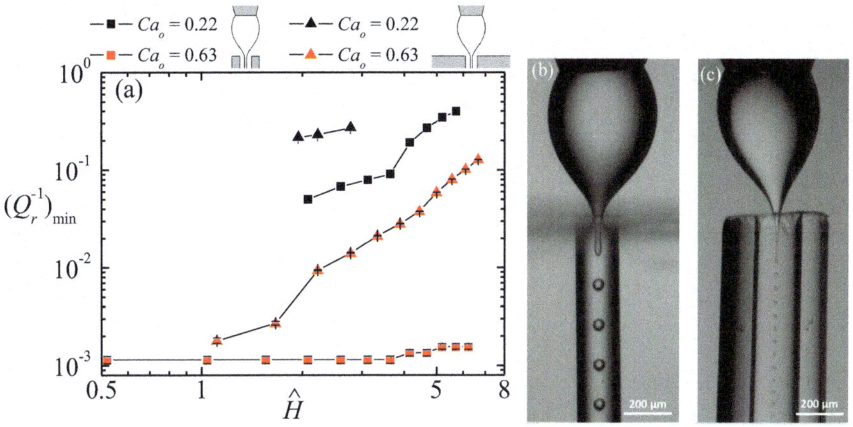

Fig. 9.20 **a** Minimum value of the flow rate ratio, $(Q_r^{-1})_{min}$, versus the dimensionless distance $\widehat{H}$ between the feeding capillary and the discharge tube [75]. **b** and **c** Images of the ejection mode in flow focusing **b** and confined selective withdrawal **c** at the corresponding minimum flow rate ratio $(Q_r^{-1})_{min} = 0.021$ and 0.012, respectively. The values of the rest of governing parameters can be found in Ref. [75]

flow rates. The smallest droplets are always produced for the minimum disperse-phase flow rate compatible with the flow stability (the minimum flow rate stability limit) [17]. This stability limit, in turn, is expected to depend on the specific geometry used to produce the hydrodynamic focusing.

The comparison between the two configurations shows that confined selective withdrawal significantly enhances the stability of the microdripping mode, which entails a significant reduction of the minimum value, $(Q_r^{-1})_{min}$, of the flow rate ratio $(Q_r)^{-1} = Q_i/Q_o$ and, consequently, of the droplet diameter (Fig. 9.20) [75, 106]. Evangelio et al. [106] found that the critical capillary number above which microjetting is produced in confined selective withdrawal is well below the corresponding value for flow focusing.

Figure 9.21 schematically compares the minimum values of the capillary numbers in viscous coflowing, flow focusing, and confined selective withdrawal. A sharp reduction is produced by the focusing effect caused by the orifice of flow focusing and confined selective withdrawal. The tip streaming stabilization is further enhanced when the plate orifice (flow focusing) is replaced with a discharge tube (confined selective withdrawal).

9.3.1.2 Global Stability Analysis

As occurs in flow focusing, the global stability analysis allows us to determine the minimum flow rate ratio, $(Q_r^{-1})_{min}$, for the microjetting mode of confined selective

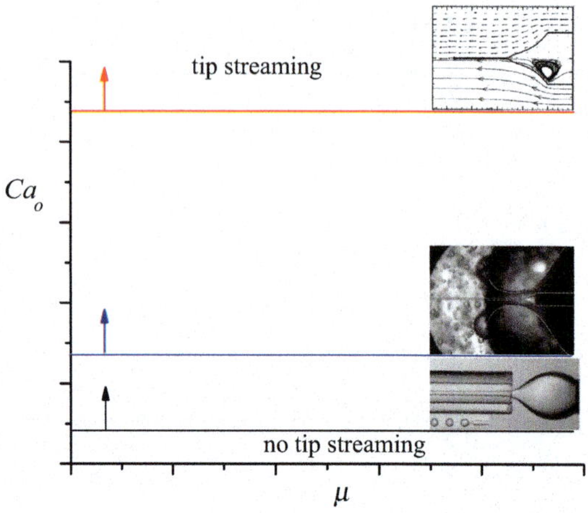

Fig. 9.21 Sketch of the minimum capillary number Ca_o for tip streaming in viscous coflowing (red line), flow focusing (blue line), and confined selective withdrawal (black line). The sketch does not consider the dependence on the viscosity ratio μ

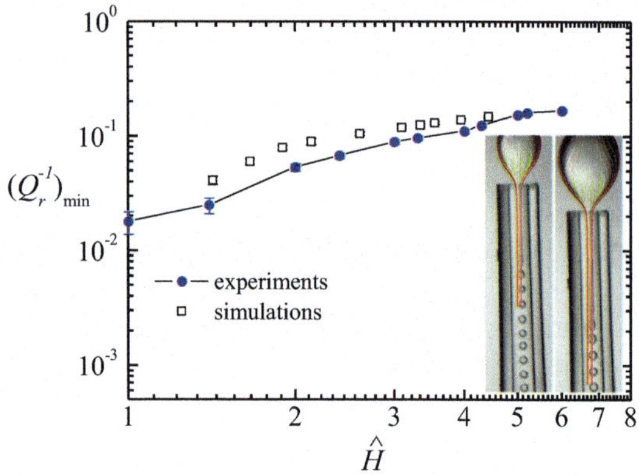

Fig. 9.22 Minimum value of the flow rate ratio, $\left(Q_r^{-1}\right)_{\rm min}$, as a function of the dimensionless distance $\widehat{H}$ between the feeding capillary and the discharge tube. The values of the rest of governing parameters were given by López et al. [75]. The inset compares the experimental images and the corresponding base flows for marginally stable states

withdrawal. Figure 9.22 shows the good agreement between the experimental values of $\left(Q_r^{-1}\right)_{\rm min}$ and the prediction of the stability analysis.

The analysis of the interface perturbation associated with the critical eigenmode allows one to anticipate whether the transition resulting from the microjetting instability leads either to microdripping or to the intermittent ejection of drops trains [75]. Specifically, if the interface perturbation takes relatively small values in the tapering meniscus, one expects the system to adopt the microdripping regime. On the contrary, if the critical eigenmode considerably perturbs the meniscus surface, the system is expected to evolve toward intermittent ejection.

Figure 9.23 shows the streamlines and the magnitude of the interface perturbation, $|\delta F(z)|$, for slightly different values of $\widehat{H}$ corresponding to stable, marginally stable, and unstable configurations. As observed, $|\delta F(z)|$ reaches its maximum value at the meniscus tip when the flow becomes unstable, indicating the transition to intermittent ejection. The figure also shows the growth of the recirculation cell as $\widehat{H}$ increases, which is assumed to be a destabilizing factor [75].

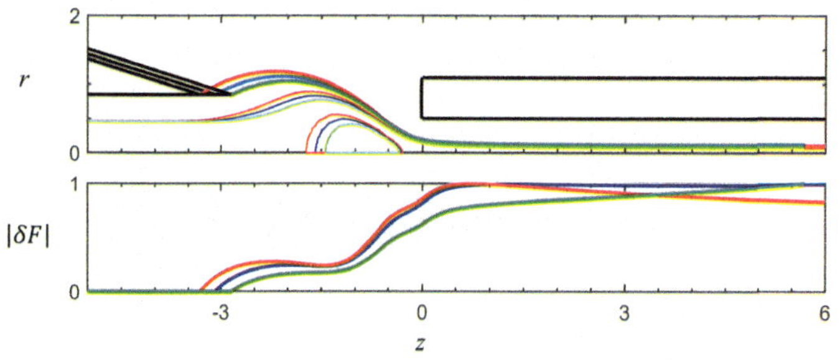

Fig. 9.23 Streamlines of the base flow and magnitude of the interface perturbation, $|\delta F(z)|$, for $\widehat{H} = 2.85$ (green, stable), 3.09 (blue, marginally stable), and 3.32 (red, unstable). The values of the rest of the governing parameters were given by López et al. [75]. The spatial coordinates have been made dimensionless with the diameter of the discharge tube, while the magnitude of the interface perturbation has been normalized with its maximum value

Microdripping was observed by López et al. [75] in all their experiments with the highest outer viscosity. The global stability analysis correctly predicted the microdripping mode in those cases and allowed the calculation of the droplet emission frequency.

9.3.1.3 Transition from Microjetting to Dripping. Oscillating Microjetting

As explained in Chap. 3, the transition from microjetting to dripping may be preceded by an intermediate oscillating jetting mode. Consider a dynamic system that adopts a steady mode when a control parameter α takes values smaller than a critical value α_c. Suppose the system undergoes a Hopf (oscillatory) bifurcation at α_c. Under certain conditions, there is a short interval of the control parameter in which the system oscillates (a limit cycle) with a frequency close to that of the critical linear eigenmode responsible for the instability transition. The non-linear terms of the governing equations become important in those oscillations and prevent the critical eigenmode from growing. The amplitude of the oscillations increases with the distance $\epsilon = (\alpha - \alpha_c)/\alpha_c$ ($\epsilon \ll 1$) from the marginal stability. These oscillations are not powered by an external actuation but by the underlying steady base flow. In tip streaming, these flow self-sustained oscillations give rise to the so called oscillating microjetting mode.

There are doubts about whether the oscillating microjetting mode truly exists or, conversely, the oscillations observed in the experiments around the marginal stability result from an uncontrolled external perturbation. The transfer function of the microjetting mode at the stability limit is expected to exhibit resonance frequencies, one of which approximately coincides with that of the critical eigenmode. If the perturbation frequency range includes that resonance frequency, uncontrolled tiny

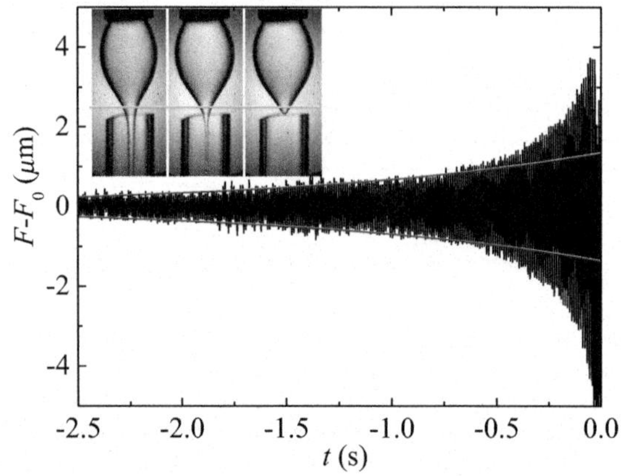

Fig. 9.24 Growth of the critical mode as a function of time. The instant $t = 0$ corresponds to the interface breakup. The black line shows the displacements $F(z, t) - F_0(z)$ of an interface point with respect to its average position $F_0(z)$. The red line is the envelope of the function $a_0 e^{\alpha t} \cos(\omega t + \phi)$ fitted to the experimental points in the time interval $(-2.5, -0.8)$ s. The insets show images at three instants. The green line in the images indicates the axial coordinate z of the interface point

external perturbations produce noticeable system oscillations. These oscillations may be erroneously interpreted as the oscillating microjetting described above.

López et al. [75] conducted careful experiments to elucidate the meaning of the oscillations observed in confined selective withdrawal close to the stability limit. When all the perturbations were eliminated, the system directly transited from the microjetting mode to the dripping regime. For the sake of illustration, Fig. 9.24 shows the growth of the critical mode as a function of time for a flow rate ratio Q_r^{-1} around 0.5% smaller than the minimum value $\left(Q_r^{-1}\right)_{\min}$. The non-linear terms did not limit the growth of the critical mode, and the oscillating microjetting regime was not established. The oscillation frequency approximately coincided with that of the critical eigenmode calculated from the global stability analysis. Non-linear effects did not significantly change the oscillation frequency but increased the exponent value.

The analysis shown in Fig. 9.24 does not allow one to rule out the existence of the oscillating microjetting regime in that experiment conclusively. In the first stage of the instability described above, the interface point oscillates with an exponentially growing amplitude. However, the exponent characterizing this growth is very small, and the oscillation is quasi-periodic (Fig. 9.25). The oscillations observed in Fig. 9.25 might correspond to the oscillating regime. Slight uncontrolled time variations of the parameter conditions could explain its short survival. However, it is difficult to admit that non-linear effects could limit the growth of such tiny oscillations, a requisite for oscillating microjetting.

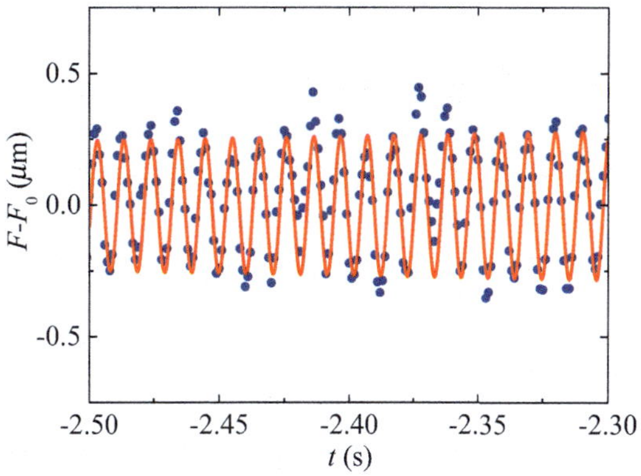

Fig. 9.25 Growth of the critical mode as a function of time. The instant $t = 0$ corresponds to the interface breakup. The symbols are the displacement $F(z, t) - F_0(z)$ of an interface point with respect to its average position $F_0(z)$. The solid line is the function $a_0 e^{\alpha t} \cos(\omega t + \phi)$ fitted to the experimental points in the time interval $(-2.5, -0.8)$ s

9.3.1.4 Droplet Diameter

Confined selective withdrawal can operate both in the microdripping and microjetting modes. Different scaling laws for the droplet diameter have been obtained in the two cases. Evangelio et al. [106] rationalized their experimental observations for microdripping using arguments based on mass conservation. They concluded that the droplet diameter d_d scales as the square root of the flow rate ratio.

The dependence of the droplet diameter on the flow rate ratio in the microdripping mode was also analyzed by López et al. [75] for a viscous outer liquid. They concluded that the power law that best fits their experimental data was $d_d/D_o = 1.14\, Q_r^{-0.45}$. The diameters were also accurately fitted by the law

$$\frac{d_d}{D_o} \sim Q_r^{-1/2} \tag{9.9}$$

previously derived by Evangelio et al. [106] (Fig. 9.26).

The droplet production frequency scales as $f \sim d_d^{-3} Q_i^{-1}$. Interestingly, the reduction of the inner flow rate does not affect the shape of the tapering meniscus (Fig. 9.26) [75]. The meniscus tip sharpens as Q_r^{-1} decreases, which lowers the size of the droplets ejected in that region.

As occurs in flow focusing, the droplet diameter in the microjetting regime can be estimated from that of the precursor jet by considering the wavelength of the most unstable capillary mode [106, 110]. The jet diameter can be estimated in terms of the flow rate ratio from the same continuity arguments used in flow focusing. This approach leads to the scaling law [106]

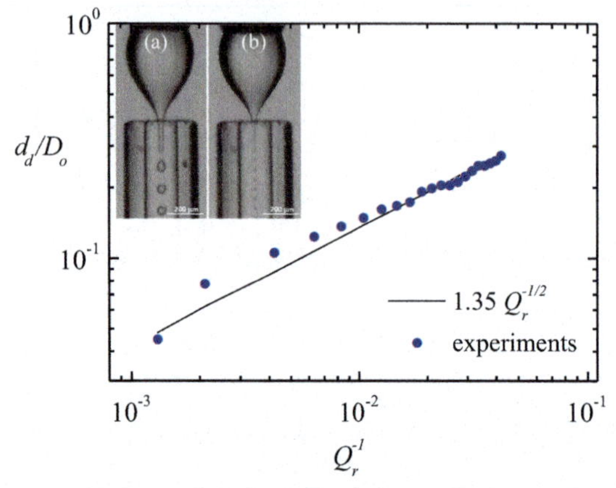

Fig. 9.26 Drop diameter d_d (normalized with the discharge tube diameter D_o) versus the flow rate ratio Q_r. The symbols and the solid line correspond to the experimental results and the scaling law $d_d/D_o = 1.35\, Q_r^{-1/2}$, respectively. The values of the governing parameters can be found in Ref. [75]. The insets are images acquired for (**a**) $Q_r^{-1} = 0.042$ and (**b**) $Q_r^{-1} = 0.001$. The yellow line in the image (**b**) is the meniscus contour in the case (**a**)

$$\frac{d_d}{D_o} \sim Q_r^{-1/3}, \tag{9.10}$$

which accurately fits the experimental data.

9.3.1.5 Surfactant Effects

López et al. [75] examined the effect of SDS and TWEEN 80 on the microdripping mode of confined selective withdrawal for concentrations close to or larger than the CMC. SDS dissolved in the inner liquid at the same concentration (0.2% wt) produced similar effects (Fig. 9.27) even though the adsorption time of SDS is much shorter than that of TWEEN. This suggests that the bulk-to-interface transfer of surfactant molecules was controlled by diffusion. When the TWEEN concentration was increased up to 3.4 CMC (0.5% wt), a significant reduction of $\left(Q_r^{-1}\right)_{\min}$ and the droplet diameter was obtained.

Figure 9.28 compares the droplet emission with and without SDS for a flow rate equal to its minimum value in the absence of surfactant. The tip of the surfactant-free meniscus enters the discharge tube. However, when a small amount of SDS is added, the tapering meniscus retracts, and the short liquid thread formed at the meniscus tip adopts a quasi-cylindrical shape in front of the discharge tube. The system reaches a quasi-jetting regime in which considerably smaller droplets are ejected at a higher frequency. Understanding the stabilization mechanism associated with the presence of a surfactant monolayer requires the addition of this factor to the global stability

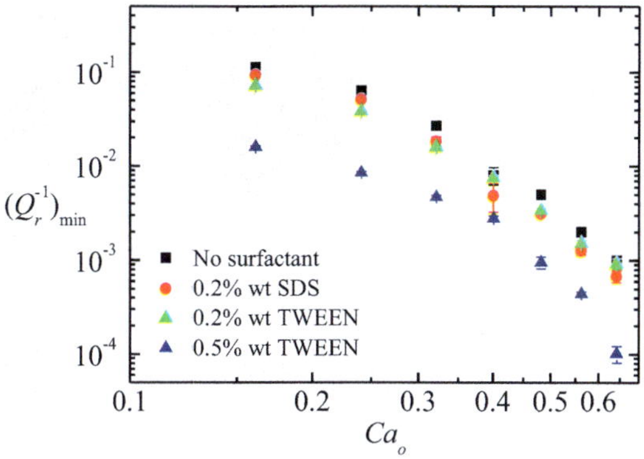

Fig. 9.27 Minimum value of the flow rate ratio, $(Q_r^{-1})_{\text{min}}$, versus the capillary number Ca_o for different surfactant concentrations. The values of the rest of governing parameters can be found in Ref. [75]

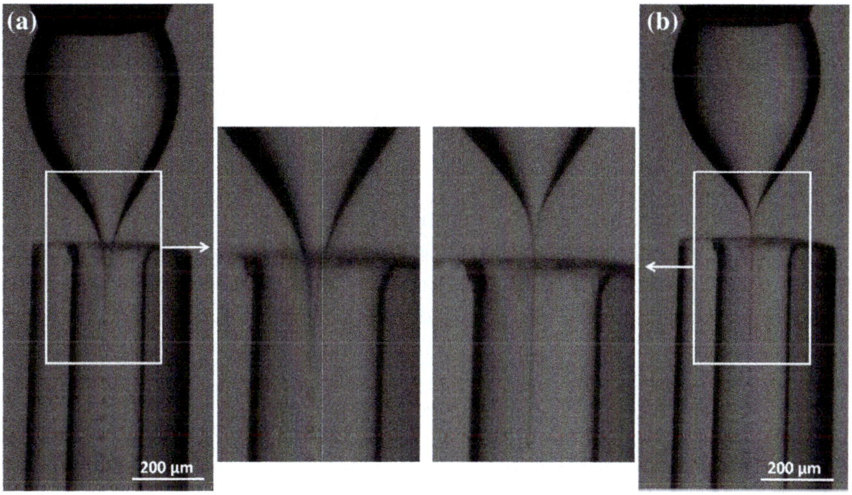

Fig. 9.28 Images of the droplet emission without surfactant (**a**) and with SDS at 0.8 CMC (**b**). The values of the governing parameters can be found in Ref. [75]

analysis. This would allow one to elucidate the effects of the local reduction of the interfacial tension and Marangoni stress.

9.3.2 Gaseous Stream Focused by a Liquid Current

When the dispersed phase is a gas, confined selective withdrawal operates in the microbubbling mode (Fig. 9.29). Evangelio et al. [109] derived and experimentally

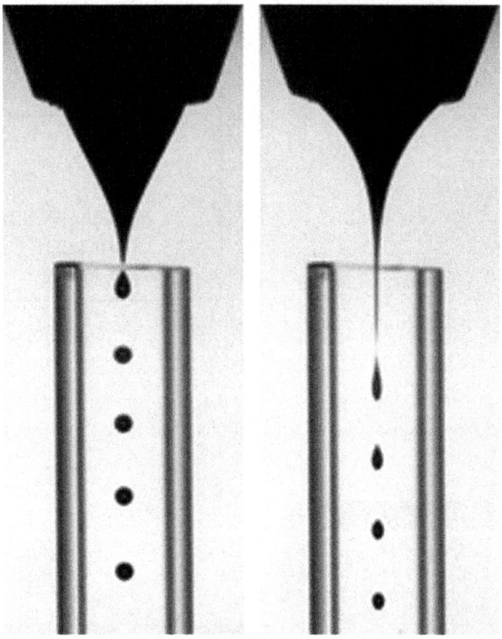

Fig. 9.29 Microbubbling in confined selective withdrawal (reproduced with permission, Ref. [109])

verified the scaling laws of the bubble diameter d_b for a fixed applied pressure to the gas flow. This quantity was shown to obey the expression $d_b/D_o \sim Q_r^{-\beta}$, where the exponent β takes a value in the interval [1/3, 1/2] depending on the gas and liquid Reynolds numbers [109]. As occurs in flow focusing, the system adopts a quasi-microjetting regime under certain experimental conditions (Fig. 9.29-right).

9.4 Selective Withdrawal

In the viscous entrainment mode of selective withdrawal (see Chap. 7), a thin jet is suctioned through a collector tube when the surrounding liquid is withdrawn at a constant flow rate (Fig. 9.30). To our knowledge, the microdripping mode has not yet been observed in this configuration. Selective withdrawal has been used to produce monodisperse emulsions of micrometer droplets and coat microparticles with polymer films [111]. This technique has proved to be flexible in terms of the chemical composition and thickness of the conformal coatings.

The steady subcritical flow giving rise to viscous entrainment has received attention over the last decades (see Chap. 6). The viscous entrainment mode *per se* has been studied on fewer occasions. Case and Nagel [112] experimentally analyzed the influence of the viscosity ratio on this mode. Pan et al. [113] showed the existence of a minimum withdrawal flow rate (Q^* in Fig. 9.31) for viscous entrainment to occur

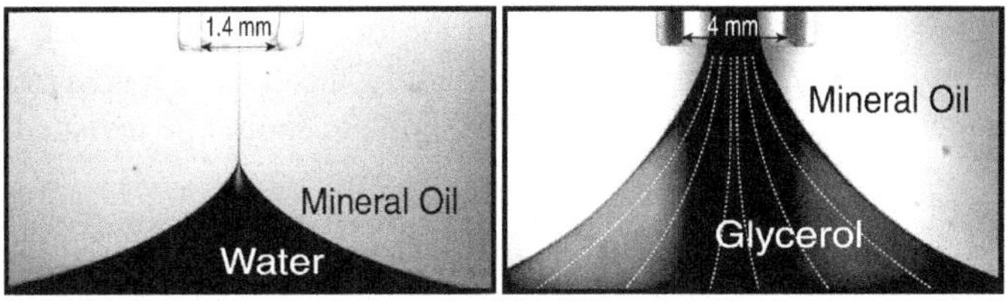

Fig. 9.30 Selective withdrawal for the viscosity ratios $\hat{\mu} = \mu_i/\mu_o = 0.0044$ (left) and 30 (right) (reprinted with permission from Ref. [112]. Copyright (2007) by the American Physical Society)

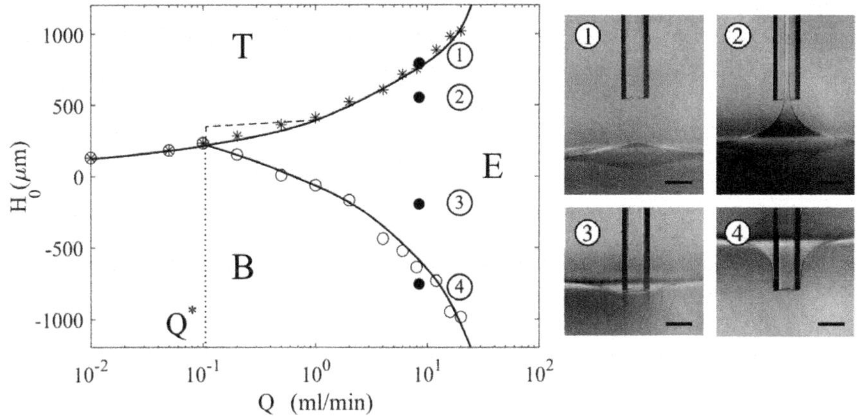

Fig. 9.31 Phase diagram of selective withdrawal using an aqueous-two-phase system determined by mixing a solution made of 25 wt% isopropanol and 15 wt% dipotassium phosphate with $D_o = 300\,\mu$m. In this figure, Q and H_0 are the withdrawal flow rate and the distance between the unperturbed interface and the capillary, respectively. The regions (T), (B), and (E) correspond to only withdrawal of the top phase, only withdrawal of the bottom phase, and entrainment, respectively. The images correspond to the experiments indicated in the figure (reprinted with permission from Ref. [113]. Copyright (2020) by the American Physical Society)

regardless of the distance between the unperturbed interface and the capillary. Their experiments revealed the effect of the Reynolds number governing the external flow around the collector tube.

There has been controversy about the mechanism that destabilizes the steady subcritical flow and gives rise to microjetting in selective withdrawal. Blanchette and Zhang [114] claimed that the instability mechanisms are not localized in the tip of the hump formed in the subcritical state. They postulated that the phenomenon is governed by a global balance between the upward force exerted by the withdrawal flow and the downward force due to interfacial tension. A similar idea was proposed by Montanero et al. [115] to explain the critical conditions for tip streaming in gaseous flow focusing operating with viscous liquids (see Chap. 10).

The theory of Blanchette and Zhang [114] assumes that the failure of the interface is insensitive to the viscosity of the entrained phase, something corroborated by the experiments of Rubio and Montanero [116] for the range $0.002 \leq \hat{\mu} \leq 0.03$ of the viscosity ratio $\hat{\mu} = \mu_i/\mu_o$. This seems to contradict the fact that there is no transition if the entrained phase is air [117, 118]. The problem should be revisited to resolve this apparent contradiction.

The steady viscous entrainment of complex fluids has been studied on very few occasions. Viscoelasticity significantly affects both the hump profiles and the stability of the subcritical mode. The presence of polymers in the withdrawn phase enhances the interface deformation, which becomes more localized [118, 119]. Extensional viscoelastic stresses enable the transition to viscous air entrainment above a polymer solution [118]. Rubio and Montanero [116] analyzed experimentally the influence of a soluble surfactant on the onset of microjetting in selective withdrawal (see Chap. 6). The steady microjetting mode was not studied.

9.5 Microjetting in a Linear Extensional Flow

The stable production of arbitrarily thin fluid jets from much bigger tubes has become the Holy Grail for many microfluidic applications that demand monodisperse collections of tiny droplets, bubbles, capsules, and emulsions. To this end, the microjetting mode of tip streaming has been the preferred method in most cases. In principle, the jet diameter can be indefinitely reduced by lowering the injected flow rate Q_i. However, all the experiments and numerical simulations have shown that the flow inevitably becomes unstable at a critical value of Q_i, which sets a minimum value of the jet diameter.

The so-called "opposed flow focusing" 2D configuration produces a continuous reduction of the jet diameter down to vanishing scales as the inner flow rate decreases. This phenomenon has been termed *second-order jetting transition* [28], alluding to the fact that microjetting producing extremely thin jets develops continuously from the jetting mode. The experimental evidence by Dong et al. [28] supported the existence of this topological transition. A similar effect was observed in experiments with an axisymmetric liquid-liquid-gas flow focusing device [120].

A droplet submerged in a linear extensional flow is a remarkable candidate for analyzing the microjetting mode of tip streaming. The simplicity of the imposed external flow allows one to combine asymptotic analysis with numerical simulation to study the steady base flow and its stability. In Sects. 9.5.1 and 9.5.2, we consider two versions of this problem.

9.5.1 *Viscous Entrainment from a Capillary*

Zhang [121] studied the viscosity-dominated microjetting mode obtained when a droplet attached to a feeding capillary of radius R_i is submerged in a linear extensional flow (Fig. 9.32). The imposed (unperturbed) external flow is given by Eqs. (6.3); i.e.,

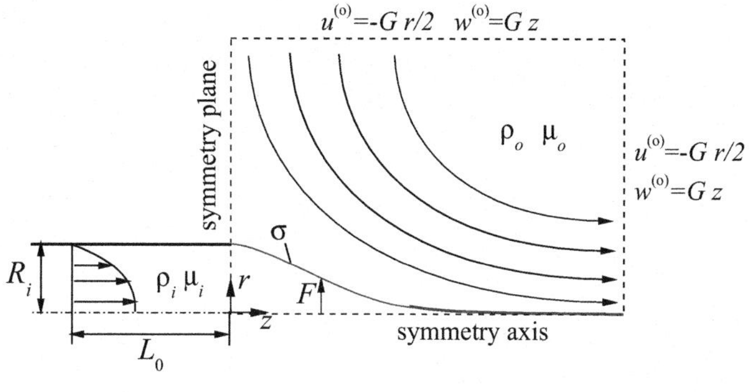

Fig. 9.32 Sketch of the microjetting mode (viscous entrainment) in a linear extensional flow for $z_N = dF/dz(z_N) = 0$

$$u^{(\text{ext})} = -Gr/2, \quad w^{(\text{ext})} = G\,z, \quad p^{(\text{ext})} = \text{const.}, \tag{9.11}$$

where $u^{(\text{ext})}$ and $w^{(\text{ext})}$ are the radial and axial components of the velocity field, G is the intensity of the extensional flow, and $p^{(\text{ext})}$ is the outer pressure field. The inner and outer viscosities are μ_i and μ_o ($\mu_i \ll \mu_o$), respectively. The parameters z_N and $dF/dz(z_N)$ are the capillary end position and interface slope at that position, respectively ($z_N = dF/dz(z_N) = 0$ in Fig. 9.32).

In this problem, the imposed pressure gradient is zero, and the viscous force exerted by the external flow drags the inner phase. The ejected volume is replaced through the feeding capillary, which is connected to a fluid reservoir where the pressure is prescribed. The reservoir pressure determines the value of $dF/dz(z_N)$. The inner phase flow rate Q_i is not fixed but results from the viscous force exerted by the outer fluid. In this sense, the problem is similar to the viscous entrainment mode of selective withdrawal. For sufficiently small values of Q_i, a stagnation point is located at the axial position z_* in front of the emitted thread. As explained below, the (unperturbed) external velocity $U_* = Gz_*$ at that position determines whether threads with arbitrarily small diameters can be produced with this flow.

A simple scaling analysis shows that the radial r_c and axial ℓ_c scales of the meniscus-jet transition verify the relation $r_c/\ell_c \sim \sqrt{\mu_i/\mu_o}$ [121]. This suggests scaling the radial and axial lengths with R_i and $R_i\sqrt{\mu_o/(4\mu_i)}$, respectively (factor 4 is introduced for convenience). The axial velocity and stresses are scaled with $U^*_{\mu\sigma} = (\sigma/\mu_o)\sqrt{\mu_o/(4\mu_i)}$ and σ/R_i, respectively. The non-dimensional parameters governing this Stokes flow are the viscosity ratio $\hat{\mu} = \mu_i/\mu_o$, the capillary number[1]

$$Ca_G = \frac{\mu_o G R_i}{\sigma}, \tag{9.12}$$

[1] Zhang [121] defined the capillary number with the prefactor 2.

the dimensionless position z_N of the capillary end, and the interface slope $\cot\theta_N/(2\hat{\mu}^{1/2})$ at that position, where θ_N is the angle formed by the interface and the capillary (the factor $2\hat{\mu}^{1/2}$ comes from the different scalings of the radial and axial coordinates).

In the slender-body theory developed by Taylor [122], the inner flow is assumed to be the sum of a plug flow induced by the external flow and the Poiseuille flow driven by the interior pressure gradient. In terms of the dimensionless variables introduced above, the model reduces to [121]

$$Q_* = 2z\,Ca_G\,F^2 - F^4\frac{dP_0}{dz}, \qquad P_0 = F^{-1} - 4\hat{\mu}\frac{d^2F}{dz^2} + 2Ca_G\,z\,F^{-1}\frac{dF}{dz}, \tag{9.13}$$

where $Q_* = 2Q_i/(\pi R_i^2 U^*_{\mu\sigma})$.

Equations (9.13) must be integrated with the downstream boundary condition $F \to \sqrt{Q_*/(2Ca_G z)}$ as $z \to \infty$, which expresses that the jet moves at the velocity imposed by the external flow sufficiently far from the capillary. This boundary condition and those at $z = z_N$ uniquely determine Q_* and $F(z)$. As mentioned above, the flow rate is not prescribed but is a function of the governing parameters.

The solution of the slender-body model shows that the steady recirculating stream in a droplet attached to the capillary evolves towards tip streaming when the capillary number Ca_G exceeds a critical value (Fig. 9.33). Starting from capillary numbers above the critical value, a continuous transition to microjetting with an arbitrarily small flow rate (jet diameter) can occur when Ca_G is decreased. For capillary numbers slightly above the critical one, the meniscus-jet transition becomes vanishingly small. Therefore, the position z_* of the stagnation point becomes practically insensitive to Ca_G in this limit. In addition, the tip of the tapering meniscus is expected to adopt a scale-invariant conical shape.

As the capillary number approaches the critical value, the solution to the slender-body model tends to the universal profile $\hat{\rho}(\eta)$ given by the equation [121]

$$1 = U_*\hat{\rho}^2 - \hat{\rho}^4\frac{d}{d\eta}\left\{\left[\frac{1}{\hat{\rho}} - 4\hat{\mu}\frac{d^2\hat{\rho}}{d\eta^2}\right] + \frac{U_*}{\hat{\rho}}\frac{d\hat{\rho}}{d\eta}\right\}, \tag{9.14}$$

where $U_* \equiv 2Ca_G\,z$, and we have introduced the scaled variables $F = \sqrt{Q_*}\hat{\rho}$ and $\eta = (z - z_*)/\sqrt{Q_*}$. The size of the region described by the universal solution $\hat{\rho}(\eta)$ vanishes around $z = z_*$, and, therefore, $U_* \simeq 2Ca_G\,z_*$.

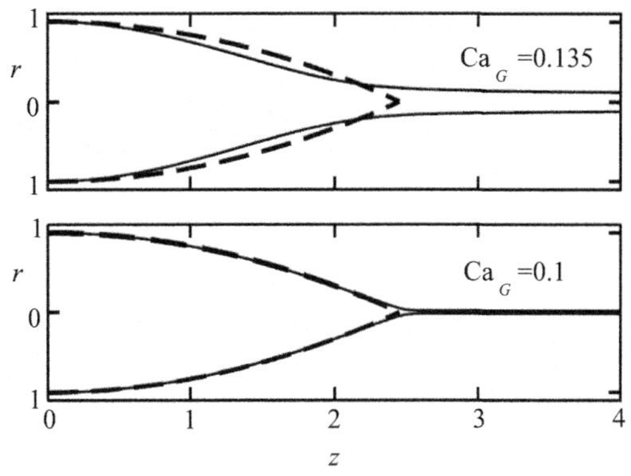

Fig. 9.33 Numerical simulation of tip streaming in a meniscus submerged in an extensional flow for $z_N = 0.05$, $\hat{\mu} = 1/80$. The solid lines represent the solutions when the capillary number Ca_G is progressively reduced. The critical capillary number is $Ca_G = 0.1$. The dashed line outlines the shape for $Ca_G = 0.1$ and zero ejected flow rate (closed droplet) (reprinted with permission from Ref. [121]. Copyright (2004) by the American Physical Society)

Equation (9.14) must be integrated with the boundary conditions $\hat{\rho} \to \sqrt{1/U_*}$ as $\eta \to \infty$ (cylindrical thread) and $\hat{\rho} = s\eta$ (conical tip) as $\eta \to -\infty$. The last condition implies that

$$s = -\frac{1}{2U_*} - \frac{1}{2}\sqrt{\frac{1}{U_*^2} - 4} \tag{9.15}$$

for a proper self-similar profile to match a conical shape upstream [121]. Equation (9.15) indicates that the velocity U_* must satisfy the condition $U^* < 1/2$ for the solution to Eq. (9.14) to verify the upstream boundary condition.

The position z_* can be calculated from (9.13) with $Q_* = 0$. The result for $\hat{\mu} = 0$ is [121]

$$z_* = 2\sqrt{\frac{1}{4Ca_G} - 1}. \tag{9.16}$$

In this case, the condition $U_* = 2Ca_G z_* < 1/2$ is equivalent to $Ca_G < 1/8$.

The solution of (9.13) matching the self-similar profile must also satisfy the boundary conditions at the capillary end. This imposes a certain relationship between z_N and θ_N. Zhang [121] concluded that if that relationship is satisfied, a continuous transition to microjetting with vanishing flow rate should take place as Ca_G approaches its critical value Ca_G^* provided that $Ca_G^* < 1/8$ (Fig. 9.33). When the variables are scaled as described above, the interface contour should converge to the universal solution of (9.14).

It should be noted that these numerical results were based on the slender body theory, which is an uncontrolled approximation for a conical interface. In addition, the stability of the numerical solutions was not determined. However, the above analysis first indicated that hydrodynamic tip streaming could steadily produce infinitely thin jets.

9.5.2 Microjetting with a Prescribed Inner Phase Flow Rate

Rubio et al. [123] conducted numerical simulations of the full hydrodynamic equations for the problem described above with $z_N = 0$ (Fig. 9.32). The flow rate was prescribed in those simulations, while the angle θ_N (the reservoir pressure) was a dependent variable. They also studied the linear stability of the base flows and determined the minimum flow rate below which those flows became unstable.

Figure 9.34 shows the stability map in the parameter plane (Ca_G, Q) for several values of the viscosity ratio $\hat{\mu} = \mu_i/\mu_o$. Here, $Q \equiv Q_i/(\pi a^2 \sigma/\mu_o)$. The symbols correspond to marginally stable cases for which the growth rate vanishes. In all the cases, the flow becomes unstable due to an oscillatory supercritical Hopf bifurcation. The growth of the interface oscillation eventually produces the interface pinching, and the system evolves from steady microjetting towards some type of dripping.

All the stability curves approximately intersect at $Q \simeq 0.031$. For $Q \lesssim 0.031$, the critical capillary number increases as $\hat{\mu}$ increases, while the opposite occurs for $Q \gtrsim 0.031$. This means that the inner viscosity destabilizes (stabilizes) the flow for $Q \lesssim 0.031$ ($Q \gtrsim 0.031$).

The stability limit $Q(Ca_G)$ exhibits two types of asymptotic behavior as $Q \to 0$. For $\hat{\mu} \geq 0.02$, $Q \sim Ca_G^{\alpha}$ with $\alpha < 0$. This implies that there is a finite value of Q below which the steady flow becomes unstable for any value of Ca_G. This expected behavior has been observed in all tip streaming configurations [61]. In contrast,

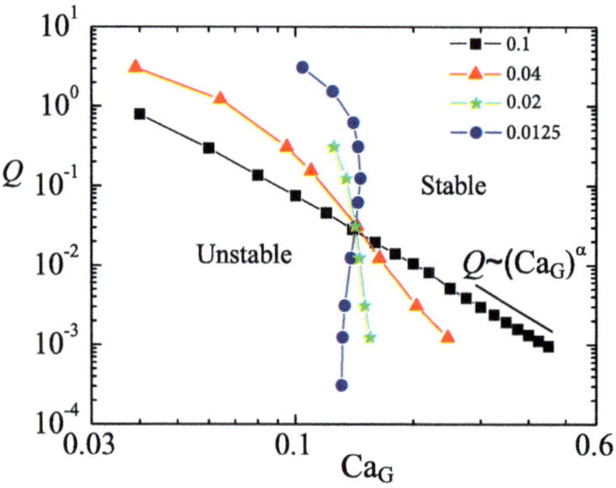

Fig. 9.34 Stability map for viscosity-dominated flow [123]. The symbols correspond to marginally stable microjetting realizations. The labels indicate the values of the viscosity ratio $\hat{\mu}$

$Q(Ca_G)$ seems to approach a vertical asymptote $Ca_G = 0.131$ for $\hat{\mu} = 0.0125$. This implies that the flow remains stable for infinitely small values of Q as long as $Ca_G \geq 0.131$.

The above conclusion has important practical consequences. As in most tip streaming configurations, the jet diameter scales as $Q^{1/2}$ at a fixed distance from the feeding capillary. Therefore, jets with arbitrarily small diameters could be emitted as $Q \to 0$ if the flow remains stable in this limit. The viscosity ratio $\hat{\mu}$ decreases and the capillary number Ca_G increases as the outer viscosity increases. This implies that, for a large enough outer viscosity, sufficiently small (large) values of $\hat{\mu}$ (Ca_G) can be obtained for the flow to remain stable in the limit $Q \to 0$. We conclude that infinitely thin jets can be steadily emitted for a large enough outer viscosity, independently of the jet viscosity and outer flow intensity. For a fixed outer viscosity, infinitely thin jets can also be produced if the jet viscosity (the viscosity ratio $\hat{\mu}$) is sufficiently small and the flow strength (the capillary number Ca_G) is sufficiently large.

The peculiar behavior of microjetting in a linear extensional flow can partially be explained as follows. For a sufficiently small viscosity ratio $\hat{\mu}$, the cone-jet transition region becomes vanishingly small as the flow rate decreases. The disparity between the size of this region and that of the meniscus suggests that the microjetting stability is determined by the flow in the cone-jet transition region. The meniscus stretches as the flow rate decreases. Therefore, the meniscus tip is exposed to larger outer fluid velocities as the flow rate decreases, which favors the convection of the perturbation downstream and stabilizes the flow. A definitive explanation of the singular stability of microjetting in a linear extensional flow is yet to be given.

These numerical results can constitute a useful guide for experimentalists in their search for new microjetting techniques. Gañán-Calvo et al. [120] designed a coaxial flow focusing configuration for producing extremely thin liquid jets. In this configuration, the inner liquid was dragged by the viscosity force exerted by an outer liquid stream within a very small region. The local flow in that region may approximately correspond to the linear extensional flow analyzed here.

References

1. Christopher GF, Anna SL (2007) Microfluidic methods for generating continuous droplet streams. J Phys D Appl Phys 40:R319–R336
2. Cramer C, Fischer P, Windhab EJ (2004) Drop formation in a co-flowing ambient fluid. Chem Engin Sci 59:3045–3058
3. Utada AS, Fernandez-Nieves A, Stone HA, Weitz DA (2007) Dripping to jetting transitions in coflowing liquid streams. Phys Rev Lett 99(094):502
4. Tomotika S (1935) On the instability of a cylindrical thread of a viscous liquid surrounded by another viscous fluid. Proc R Soc Lond 150:322–337
5. van Hoeve W, Dollet B, Gordillo JM, Versluis M, van Wijngaarden L, Lohse D (2011) Bubble size prediction in co-flowing streams. EPL 94(64):001
6. Utada AS, Fernandez-Nieves A, Gordillo JM, Weitz DA (2008) Absolute instability of a liquid jet in a coflowing stream. Phys Rev Lett 100(014):502

7. Suryo R, Basaran OA (2006) Tip streaming from a liquid drop forming from a tube in a co-flowing outer fluid. Phys Fluids 18(082):102
8. Castro-Hernández E, Gundabala V, Fernández-Nieves A, Gordillo JM (2009) Scaling the drop size in coflow experiments. New J Phys 11(075):021
9. Gordillo JM, Sevilla A, Campo-Cortés F (2014) Global stability of stretched jets: conditions for the generation of monodisperse micro-emulsions using coflows. J Fluid Mech 738:335–357
10. Oliveira JE, Moraes EA, Costa RGF, Afonso AS, Mattoso LHC, Orts WJ, Medeiros ES (2011) Nano and submicrometric fibers of poly(d, l-lactide) obtained by solution blow spinning: Process and solution variables. J Appl Polym Sci 122:3396–3405
11. Daristotle JL, Behrens AM, Sandler AD, Kofinas P (2016) A review of the fundamental principles and applications of solution blow spinning. ACS Appl Mater Interfaces 8:34,951–34,963
12. Lauricella M, Succi S, Zussman E, Pisignano D, Yarin AL (2020) Models of polymer solutions in electrified jets and solution blowing. Rev Mod Phys 92(035):004
13. Medeiros ES, Glenn GM, Klamczynski AP, Orts WJ, Mattoso LHC (2009) Solution blow spinning: a new method to produce micro- and nanofibers from polymer solutions. J Appl Polym Sci 113:2322–2330
14. Lou H, Han W, Wang X (2014) Numerical study on the solution blowing annular jet and its correlation with fiber morphology. Ind Eng Chem Res 53:2830–2838
15. Benavides RE, Jana SC, Reneker DH (2012) Nanofibers from scalable gas jet process. ACS Macro Lett 1:1032–1036
16. Zhmayev E, Cho D, Joo YL (2010) Nanofibers from gas-assisted polymer melt electrospinning. Polymer 51:4140–4144
17. Cabezas MG, Rubio M, Rebollo-Muñoz N, Herrada MA, Montanero JM (2021) Global stability analysis of axisymmetric liquid-liquid flow focusing. J Fluid Mech 909:A10
18. Gañán-Calvo AM (1998) Generation of steady liquid microthreads and micron-sized monodisperse sprays in gas streams. Phys Rev Lett 80:285–288
19. Luque A, Perdigones FA, Esteve J, Montserrat J, Gañán-Calvo AM, Quero JM (2007) Silicon microdevice for emulsion production using three-dimensional flow focusing. J Microelectromech Syst 16:1201–1208
20. Takeuchi S, Garstecki P, Weibel DB, Whitesides GM (2005) An axisymmetric flow-focusing microfluidic device. Adv Mater 17:1067–1072
21. Morimoto Y, Tan WH, Takeuchi S (2009) Three-dimensional axisymmetric flow-focusing device using stereolithography. Biomed Microdevices 11:369–377
22. Yobas L, Martens S, Onga W, Ranganathan N (2006) High-performance flow-focusing geometry for spontaneous generation of monodispersed droplets. Small 6:1073–1079
23. Zhu P, Kong T, Lei L, Tian X, Kang Z, Wang L (2016) Droplet breakup in expansion-contraction microchannels. Sci Rep 6(21):527
24. Dewandre A, Rivero-Rodriguez J, Vitry Y, Sobac B, Scheid B (2020) Microfluidic droplet generation based on non-embedded co-flow-focusing using 3d printed nozzle. Sci Rep 10(21):616
25. Wu Q, Yang CY, Liu GL, Xu WH, Zhu ZQ, Si T, Xu RX (2017) Multiplex coaxial flow focusing for producing multicompartment Janus microcapsules with tunable material compositions and structural characteristics. Lab Chip 17:3168–3175
26. Utada AS, Lorenceau E, Link DR, Kaplan PD, Stone HA, Weitz DA (2005) Monodisperse double emulsions generated from a microcapillary device. Science 308:537–541
27. Nabavi SA, Vladisavljevic GT, Manovic V (2017) Mechanisms and control of single-step microfluidic generation of multi-core double emulsion droplets. Chem Eng J 322:140–148
28. Dong J, Meissner M, Faers MA, Eggers J, Seddon AM, Royall CP (2018) Opposed flow focusing: evidence of a second order jetting transition. Soft Matter 14:8344–8351
29. Wu T, Luo Z, Ding W, Cheng Z, He L (2017) Monodisperse droplets by impinging flow-focusing. Microfluid Nanofluid 21:129
30. Lee GB, Hung CI, Ke BJ, Huang GR, Hwei BH, Lai HF (2001) Hydrodynamic focusing for a micromachined flow cytometer. J Fluids Eng 123:672–679

31. Sun BJ, Shum HC, Holtze C, Weitz DA (2010) Microfluidic melt emulsification for encapsulation and release of actives. Appl Mat Interf 2:3411–3416
32. Tsuda Y, Morimoto Y, Takeuchi S (2010) Monodisperse cell-encapsulating peptide microgel beads for 3D cell culture. Langmuir 26:2645–2649
33. Vladisavljevic GT, Henry J, Duncanson WJ, Shum HC, Weitz DA (2012) Fabrication of biodegradable poly(lactic acid) particles in flow-focusing glass capillary devices. Progr Colloid Polym Sci 139:111–114
34. Zhong F, Yang C, Wu Q, Wang S, Cheng L, Dwivedi P, Zhu Z, Si T, Xu RX (2020) Preparation of pesticide-loaded microcapsules by liquid-driven coaxial flow focusing for controlled release. Int J Polym Mater 69:840–847
35. do Nascimento DF, J A Avendaño AM, Moura MJB, Carvalho MS, Duncanson WJ, (2017) Flow of tunable elastic microcapsules through constrictions. Sci Rep 7(11):898
36. Gu H, Duits MHG, Mugele F (2011) Droplets formation and merging in two-phase flow microfluidics. Int J Mol Sci 12:2572–2597
37. Chen H, Li J, Shum HC, Stone HA, Weitz DA (2011) Breakup of double emulsions in constrictions. Soft Matter 7:2345–2347
38. Chen Y, Wub L, Zhang L (2015) Dynamic behaviors of double emulsion formation in a flow-focusing device. Int J Multiphase Flow 82:42–50
39. Wang JT, Wang J, Han JJ (2011) Fabrication of advanced particles and particle-based materials assisted by droplet-based microfluidics. Small 4:1728–1754
40. Zhu Z, Wu Q, Han S, Xu W, Zhong F, Yuan S, Dwivedi P, Si T, Xu RX (2018) Rapid production of single- and multi-compartment polymeric microcapsules in a facile 3D microfluidic process for magnetic separation and synergistic delivery. Sens Actuators B Chem 275:190–198
41. Edmond KV, Schofield AB, Marquez M, Rothstein JP, Dinsmore AD (2006) Stable jets of viscoelastic fluids and self-assembled cylindrical capsules by hydrodynamic focusing. Langmuir 22:9052–9056
42. Madurga R, Guinea GV, Elices M, Pérez-Rigueiro J, Gañán-Calvo AM (2017) Straining flow spinning: simplified model of a bioinspired process to mass produce regenerated silk fibers controllably. Eur Polymer J 97:26–39
43. Jeong W, Kim J, Kim S, Lee S, Mensingc G, Beebe DJ (2004) Hydrodynamic microfabrication via "on the fly" photopolymerization of microscale fibers and tubes. Lab Chip 4:S76–S80
44. Liu X, Wu L, Zhao Y, Chen Y (2017) Study of compound drop formation in axisymmetric microfluidic devices with different geometries. Colloids Surf A 533:87–98
45. Wu L, Liu X, Zhao Y, Chen Y (2017) Role of local geometry on droplet formation in axisymmetric microfluidics. Chem Eng Sci 163:56–67
46. Gañán-Calvo AM (2008) Unconditional jetting. Phys Rev E 78(026):304
47. Acero AJ, Rebollo-Muñoz N, Montanero JM, Gañán-Calvo AM, Vega EJ (2013) A new flow focusing technique to produce very thin jets. J Micromech Microeng 23(065):009
48. Anna SL, Bontoux N, Stone HA (2003) Formation of dispersions using flow focusing in microchannels. Appl Phys Lett 82:364–366
49. Anna SL, Mayer HC (2006) Microscale tipstreaming in a microfluidic flow focusing device. Phys Fluids 18(121):512
50. Guillot P, Colin A, Utada AS, Ajdari A (2007) Stability of a jet in confined pressure-driven biphasic flows at low Reynolds numbers. Phys Rev Lett 99(104):502
51. Herrada MA, Gañán-Calvo AM, Guillot P (2008) Spatiotemporal instability of a confined capillary jet. Phys Rev E 78(046):312
52. Gañán-Calvo AM, Herrada MA, Garstecki P (2006) Bubbling in unbounded coflowing liquids. Phys Rev Lett 96(124):504
53. Gañán-Calvo AM, Riesco-Chueca P (2006) Jetting-dripping transition of a liquid jet in a lower viscosity co-flowing immiscible liquid: the minimum flow rate in flow focusing. J Fluid Mech 553:75–84
54. Theofilis V (2011) Global linear instability. Annu Rev Fluid Mech 43:319–352
55. Sauter US, Buggisch HW (2005) Stability of initially slow viscous jets driven by gravity. J Fluid Mech 533:237–257

56. Tammisola O, Lundell F, Soderberg LD (2012) Surface tension-induced global instability of planar jets and wakes. J Fluid Mech 713:632–658
57. Schmid PJ (2007) Nonmodal stability theory. Annu Rev Fluid Mech 39:129–162
58. Cruz-Mazo F, Herrada MA, Gañán-Calvo AM, Montanero JM (2017) Global stability of axisymmetric flow focusing. J Fluid Mech 832:329–344
59. Ponce-Torres A, Rebollo-Muñoz N, Herrada MA, Gañán-Calvo AM, Montanero JM (2018) The steady cone-jet mode of electrospraying close to the minimum volume stability limit. J Fluid Mech 857:142–172
60. Mu K, Ding H, Si T (2018) Instability analysis of the cone-jet flow in liquid-driven flow focusing. Microfluid Nanofluidics 22:138
61. Montanero JM, Gañán-Calvo AM (2020) Dripping, jetting and tip streaming. Rep Prog Phys 83(097):001
62. Yang C, Qiao R, Mu K, Zhu Z, Xu RX, Si T (2019) Manipulation of jet breakup length and droplet size in axisymmetric flow focusing upon actuation. Phys Fluids 31(091):702
63. Zhou C, Yue P, Feng JJ (2006) Formation of simple and compound drops in microfluidic devices. Phys Fluids 18(092):105
64. Steinhaus B, Shena AQ, Sureshkumar R (2007) Dynamics of viscoelastic fluid filaments in microfluidic devices. Phys Fluids 19(073):103
65. Miller E, Rotea M, Rothstein JP (2010) Microfluidic device incorporating closed loop feedback control for uniform and tunable production of micro-droplets. Lab on a Chip 10:1293–1301
66. Lee W, Walker LM, Anna SL (2011) Competition between viscoelasticity and surfactant dynamics in flow focusing microfluidics. Macromol Mater Eng 296:203–213
67. Derzsi L, Kasprzyk M, Plog JP, Garstecki P (2013) Flow focusing with viscoelastic liquids. Phys Fluids 25(092):001
68. Duboin A, Middleton R, Malloggi F, Monti F, Tabeling P (2013) Cusps, spouts and microfiber synthesis with microfluidics. Soft Matter 9:3041–3049
69. Oliveira MSN, Pinho FT, Poole RJ, Oliveira PJ, Alves MA (2009) Purely elastic flow asymmetries in flow-focusing devices. J Non-Newtonian Fluid Mech 160:31–39
70. Goldin M, Yerushalmi J, Pfeffer R, Shinnar R (1969) Breakup of a laminar capillary jet of a viscoelastic fluid. J Fluid Mech 38:689–711
71. Bird RB, C R, Armstrong, Hassager O (1987) Dynamics of polymeric liquids volume I: fluid mechanics; volume II: kinetic theory. Wiley, New York
72. Nooranidoost M, Izbassarov D, Muradoglu M (2016) Droplet formation in a flow focusing configuration: effects of viscoelasticity. Phys Fluids 28(123):102
73. Moyle TM, Walker LM, Anna SL (2012) Predicting conditions for microscale surfactant mediated tipstreaming. Phys Fluids 24(082):110
74. Wrobel JK, Booty MR, Siegel M, Wang Q (2018) Simulation of surfactant-mediated tip-streaming in a flow-focusing geometry. Phys Rev Fluids 3(114):003
75. López M, Cabezas MG, Montanero JM, Herrada MA (2022) On the hydrodynamic focusing for producing microemulsions via tip streaming. J Fluid Mech 934:A47
76. Lytra A, Vlachomitrou M, Pelekasis N (2023) Numerical study of the steady core-annular flow in a focusing geometry in the presence of soluble surfactants. Int J Multiphase Flow 170:104652
77. Rubio M, Montanero JM, Cabezas MG, Herrada MA (2024) Global stability analysis of hydrodynamic focusing in the presence of a soluble surfactant. J Fluid Mech
78. Frumkin A, Levich V (1947) On surfactants an interfacial motion (in Russian). Zhur Fiz Khim 21:1183
79. Herrada MA, Ponce-Torres A, Rubio M, Eggers J, Montanero JM (2022) Stability and tip streaming of a surfactant-loaded drop in an extensional flow. influence of surface viscosity. J Fluid Mech 934:A26
80. Stride E, Edirisinghe M (2008) Novel microbubble preparation technologies. Soft Matter 4:2350–2359

81. Thorsen T, Roberts RW, Arnold FH, Quake SR (2001) Dynamic pattern formation in a vesicle-generating microfluidic device. Phys Rev Lett 86:4163–4166
82. Gunther A, Khan SA, Thalmann M, Trachsel F, Jensen KF (2004) Transport and reaction in microscale segmented flow. Lab Chip 4:278–286
83. Garstecki P, Fuerstman MJ, Stone HA, Whitesides GM (2006) Formation of droplets and bubbles in a microfluidic T-junction-scaling and mechanism of break-up. Lab Chip 6:437–446
84. Herrada MA, Gañán-Calvo AM, Montanero JM (2013) Theoretical investigation of a technique to produce microbubbles by a microfluidic T-junction. Phys Rev E 88(033):027
85. Smith CS (1949) On blowing bubbles for Bragg's dynamic crystal model. J Appl Phys 20:631–632
86. Bhunia A, Pais SC, Kamotani Y, Kim IH (1998) Bubble formation in a coflow normal and reduced configuration in gravity. AIChE 44:1499–1509
87. Serizawa A, Feng Z, Kawara Z (2002) Two-phase flow in microchannels. Exp Thermal Fluid Sci 26:703–714
88. Xiong R, Bai M, Chung JN (2007) Formation of bubbles in a simple co-flowing micro-channel. J Micromech Microeng 17:1002–1011
89. Gañán-Calvo AM, Gordillo JM (2001) Perfectly monodisperse microbubbling by capillary flow focusing. Phys Rev Lett 87(274):501
90. Garstecki P, Gitlin I, DiLuzio W, Whitesides GM, Kumacheva E, Stone HA (2004) Formation of monodisperse bubbles in a microfluidic flow-focusing device. Appl Phys Lett 85:2649–2651
91. Cubaud T, Ho CM (2004) Transport of bubbles in square microchannels. Phys Fluids 16:4575–4585
92. Gañán-Calvo AM (2004) Perfectly monodisperse microbubbling by capillary flow focusing: An alternate physical description and universal scaling. Phys Rev E 69(027):301
93. Garstecki P, Stone HA, Whitesides GM (2005) Mechanism for flow-rate controlled breakup in confined geometries: A route to monodisperse emulsions. Phys Rev Lett 94(164):501
94. Garstecki P, Fuerstman MJ, Whitesides GM (2005) Nonlinear dynamics of a flow-focusing bubble generator: an inverted dripping faucet. Phys Rev Lett 94(234):502
95. Raven JP, Marmottant P (2006) Periodic microfluidic bubbling oscillator: insight into the stability of two-phase microflows. Phys Rev Lett 97(154):501
96. Lorenceau E, Sang YYC, Hohler R, Cohen-Addad S (2006) A high rate flow-focusing foam generator. Phys Fluids 18(097):103
97. Talu E, Lozano MM, Powell RL, Dayton PA, Longo ML (2006) Long-term stability by lipid coating monodisperse microbubbles formed by a flow-focusing device. Langmuir 22:9487–9490
98. Dollet B, van Hoeve W, Raven JP, Marmottant P, Versluis M (2008) Role of the channel geometry on the bubble pinch-off in flow-focusing devices. Phys Rev Lett 100(034):504
99. Marmottant P, Raven JP (2009) Microfluidics with foams. Soft Matter 5:3385–3388
100. Fu T, Ma Y, Funfschilling D, Li HZ (2009) Bubble formation and breakup mechanism in a microfluidic flow-focusing device. Chem Engin Sci 64:2392–2400
101. Fu T, Funfschilling D, Ma Y, Li HZ (2010) Scaling the formation of slug bubbles in microfluidic flow-focusing devices. Microfluid Nanofluid 8:467–475
102. Fu T, Ma Y, Funfschilling D, Li HZ (2011) Gas-liquid flow stability and bubble formation in non-Newtonian fluids in microfluidic flow-focusing devices. Microfluid Nanofluid 10:1135–1140
103. Castro-Hernández E, van Hoeve W, Lohse D, Gordillo JM (2011) Microbubble generation in a co-flow device operated in a new regime. Lab Chip 11:2023–2029
104. Vega EJ, Acero AJ, Montanero JM, Herrada MA, Gañán-Calvo AM (2014) Production of microbubbles from axisymmetric flow focusing in the jetting regime for moderate Reynolds numbers. Phys Rev E 89(063):012
105. Jensen MJ, Stone HA, Bruus H (2006) A numerical study of two-phase Stokes flow in an axisymmetric flow-focusing device. Phys Fluids 18(077):103

106. Evangelio A, Campo-Cortés F, Gordillo JM (2016) Simple and double microemulsions via the capillary breakup of highly stretched liquid jets. J Fluid Mech 804:550–577
107. He K, Campo-Cortés F, Goral M, López-León T, Gordillo JM (2019) Micron-sized double emulsions and nematic shells generated via tip streaming. Phys Rev Fluids 4(124):201
108. Muñoz-Sánchez BN, Silva SF, Pinho D, Vega EJ, Lima R (2016) Generation of micro-sized PDMS particles by a flow focusing technique for biomicrofluidics applications. Biomicrofluidics 10(14):122
109. Evangelio A, Campo-Cortés F, Gordillo JM (2015) Pressure gradient induced generation of microbubbles. J Fluid Mech 778:653–668
110. Tomotika S (1936) Breaking up of a drop of viscous liquid immersed in another viscous fluid which is extending at a uniform rate. Proc Roy Soc Lond 153:302–318
111. Cohen I, Li H, Hougland JL, Mrksich M, Nagel SR (2001) Using selective withdrawal to coat microparticles. Science 292:265–267
112. Case SC, Nagel SR (2007) Spout states in the selective withdrawal of immiscible fluids through a nozzle suspended above a two-fluid interface. Phys Rev Lett 98(114):501
113. Pan Z, Nunes JK, Stone HA (2020) Regime map and triple point in selective withdrawal. Phys Rev Lett 125(264):502
114. Blanchette F, Zhang WW (2009) Force balance at the transition from selective withdrawal to viscous entrainment. Phys Rev Lett 102(144):501
115. Montanero JM, Rebollo-Muñoz N, Herrada MA, Gañán-Calvo AM (2011) Global stability of the focusing effect of fluid jet flows. Phys Rev E 83(036):309
116. Rubio M, Montanero JM (2023) Influence of a soluble surfactant on the transition to tip streaming. Exp Therm Fluid Sci 141(110):776
117. Eggers J, Courrech du Pont S (2010) Comment on force balance at the transition from selective withdrawal to viscous entrainment. Phys Rev Lett 105(089):401
118. Zhoua D, Feng JJ (2010) Selective withdrawal of polymer solutions: experiments. J Non-Newtonian Fluid Mech 165:829–838
119. Zhoua D, Feng JJ (2010) Selective withdrawal of polymer solutions: computations. J Non-Newtonian Fluid Mech 165:839–851
120. Gañán-Calvo AM, González-Prieto R, Riesco-Chueca P, Herrada MA, Flores-Mosquera M (2007) Focusing capillary jets close to the continuum limit. Nat Phys 3:737–742
121. Zhang W (2004) Viscous entrainment from a nozzle: singular liquid spouts. Phys Rev Lett 93(184):502
122. Taylor G (1964) Disintegration of water drops in electric field. Proc R Soc Lond A 280:383–397
123. Rubio M, Montanero JM, Eggers J, Herrada MA (2024) Stable production of fluid jets with vanishing diameters via tip streaming. J Flui Mech 893: A4

Chapter 10
Gaseous Flow Focusing I

Abstract We describe the geometries that produce the flow focusing effect and their applications. The forces, flow patterns, and instability mechanisms in the low-viscosity and viscous cases are discussed. We show results on the popular gas dynamic virtual nozzle (GDVN) geometry. Global stability analysis is used to gain insight into the axisymmetric and whipping instabilities limiting the applicability of this technique. Results of the microdripping mode are also presented.

Keywords Gaseous flow focusing · Microjetting · Microdripping · Global stability analysis

Gaseous flow focusing is a particular example of gas-liquid tip streaming flow, with characteristics qualitatively different from those of the liquid-liquid configuration described in Chap. 9. Liquid streams have been focused with high-speed gaseous currents practically always using the axisymmetric configuration. The 2D counterpart of this technique has been implemented on very few occasions. The gaseous flow focusing technique typically operates in the microjetting mode, although there have also been some experimental realizations in the microdripping regime.

Gaseous flow focusing is a genuine example of tip streaming because the scale of the flow in the tapering meniscus tip can be two orders of magnitude smaller than the feeding capillary size. Gaseous flow focusing enables the ejection of Newtonian and viscoelastic jets down to very few microns in diameter by purely hydrodynamic means. The jets can reach speeds higher than 100 km/h. They can be discharged in the air at atmospheric pressure and in a vacuum.

The versatility and simplicity of gaseous flow focusing have considerably increased its range of applications. This technique has been widely used in different technological fields, such as pharmacy, materials engineering, and crystallography. We consider three versions of gaseous flow focusing: incompressible, transonic, and viscoelastic. This chapter reviews some relevant results about the stability and outcome of this technique in the incompressible regime. Transonic and viscoelastic flow focusing are analyzed in the next chapter.

J. M. Montanero, *Tip Streaming of Simple and Complex Fluids*, Fluid Mechanics and Its Applications 137, https://doi.org/10.1007/978-3-031-52768-5_10

We describe the geometries that produce the flow focusing effect and their applications. The forces, flow patterns, and instability mechanisms in the low-viscosity and viscous cases are discussed. We show results on the popular gas dynamic virtual nozzle (GDVN) geometry. Global stability analysis is used to gain insight into the axisymmetric and whipping instabilities limiting the applicability of this technique. Results of the microdripping mode are also presented.

10.1 Geometrical Configurations and Applications

In gaseous flow focusing [1], the capillary feeding the liquid phase is located in front of an orifice in a plate or inside a nozzle. The liquid is injected at a constant flow rate through that capillary. The gas coflows with the liquid across the orifice/nozzle driven by a constant pressure drop, stretching the liquid meniscus attached to the feeding capillary until a thin jet or tiny droplets are ejected from the meniscus tip (see Chap. 7) [1]. In the incompressible version of this technique, the gas current is driven by a relatively small pressure drop so that the Mach number does not exceed, say, 0.4. Therefore, the gas density can be regarded as a constant, and the pressure drop across the orifice/nozzle is the only parameter controlling the action of the gas stream.

The original plate-orifice configuration was proposed by Gañán-Calvo [1]. Around ten years later, DePonte et al. [2] focused the liquid current inside a converging nozzle. This configuration is called Gas Dynamic Virtual Nozzle (GDVN). As explained below, the GDVN configuration has become the main tool for sample delivery in Serial Femtosecond Crystallography, which has conferred great popularity on this technique. Gaseous flow focusing has practically always been implemented in the axisymmetric configuration, but there are some realizations in 2D and 3D geometries as well [3]. Experimenters have utilized glass [2, 4, 5] or ceramic [6] nozzles and have used two-photon polymerization to manufacture the microfluidic device [7] (Fig. 7.3). Different injection methods have been proposed to enhance the tip streaming stability. One of the essential ideas is to reduce the liquid volume dragged by the gas stream and eliminate the recirculation pattern arising in the meniscus of the original configuration, which stabilizes the flow [8–10].

The incompressible version of gaseous flow focusing has been used to produce microparticles of complex structures [11–13] by injecting coaxially two immiscible liquid streams [11]. Si et al. [14] have developed a gaseous co-flow focusing process to produce stimuli-responsive microbubbles that comprise perfluorocarbon suspension of silver nanoparticles in a lipid. The sketch on the left side of Fig. 10.1 depicts the flow focusing configuration used to produce compound capillary jets. The core (shell) fluid is injected through an inner (outer) capillary. The resulting compound meniscus is stretched by a focusing gas stream driven by a pressure drop until a thin jet is emitted. Herrada et al. [15] examined the dripping-to-jetting transition in this configuration from the spatiotemporal linear stability analysis (see Chap. 3).

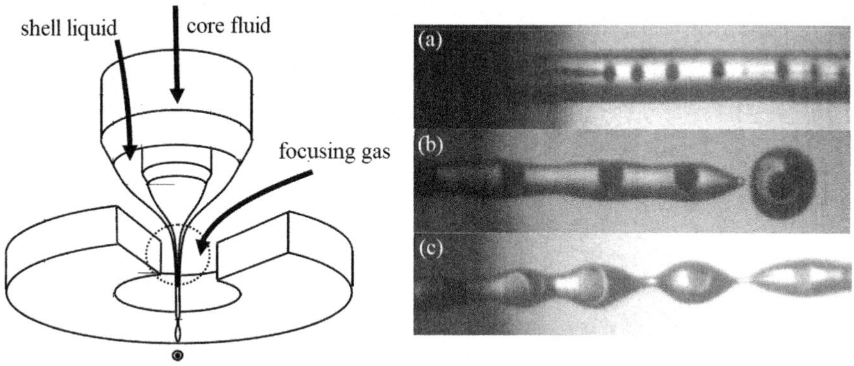

Fig. 10.1 (Left) Flow focusing device for producing compound jets. (Right) The applied pressure drop and injected flow rates can be selected to produce different outcomes: multi-vesicle capsules (**a**), mixed capsules of one and two cores (**b**), and single-core capsules (**c**) (reproduced from Ref. [15] with permission)

Gaseous flow focusing offers several attractive advantages over other atomization methods. One of them is that the collimated droplet streams acquire the velocity of the outer gas current in times of the order of $t_s = \rho_l d_d^2/(18\mu_g)$, where ρ_l is the liquid density, d_d is the droplet diameter, and μ_g the gas viscosity. This means that a micrometer water droplet in a co-flowing subsonic air stream moving at the speed, say, $U_g \sim 10^2$ m/s can reach very high kinetic energies (per unit volume) within distances from the source as small as $t_s U_g \sim 1$ mm.

Electro-flow focusing [16] is a combination of electrospray and flow focusing in which a relatively small voltage is applied to the focused liquid to charge the droplets produced via flow focusing. These droplets, with the proper combination of size, speed, and electrical charge, are perfect candidates for applications such as surface sample desorption (Fig. 10.2) [17, 18]. In electro-flow focusing, the driving force is exerted by the gas stream, while the mechanical effect of the electric stresses is negligible. Therefore, this technique cannot be generally regarded as a version of the cone-jet mode of electrospray.

Gañán-Calvo [19] studied an application of electro-flow focusing in which Maxwell stresses do collaborate with the gas stream in propelling the liquid jet. He showed how strong electric fields applied to the gas-focused liquid stream affect the jet's diameter and the convective-to-absolute instability transition.

Bisterov et al. [20] have recently increased the resolution of electrochemical jet machining by reducing the jet diameter using gaseous flow focusing. Resolution is also a central feature of high-density circuits. The so-called aerosol jet printing [21] is a version of electro-flow focusing for the fabrication of printed electronics that uses the gaseous flow focusing principle to reduce the deposit size (Fig. 10.3). It has been used in various applications, such as the printing of transistors, electrode arrays, interconnects, solid oxide fuel cells, strain gauges, and solar cells.

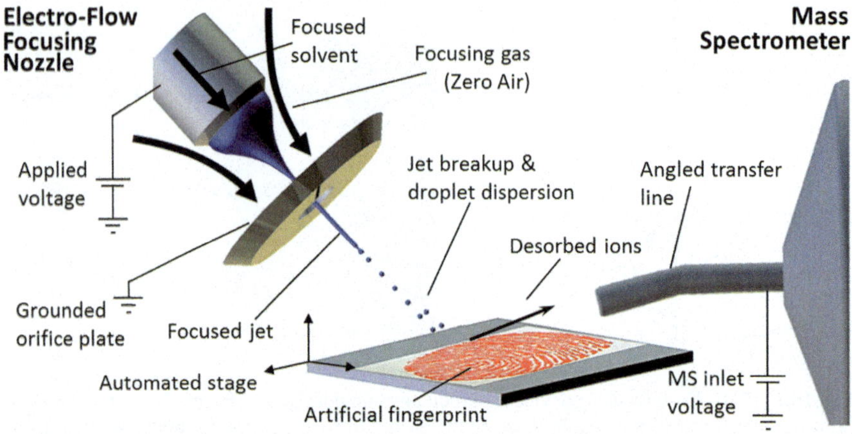

Fig. 10.2 Electro-flow focusing for surface sample desorption (reprinted with permission from Ref. [17])

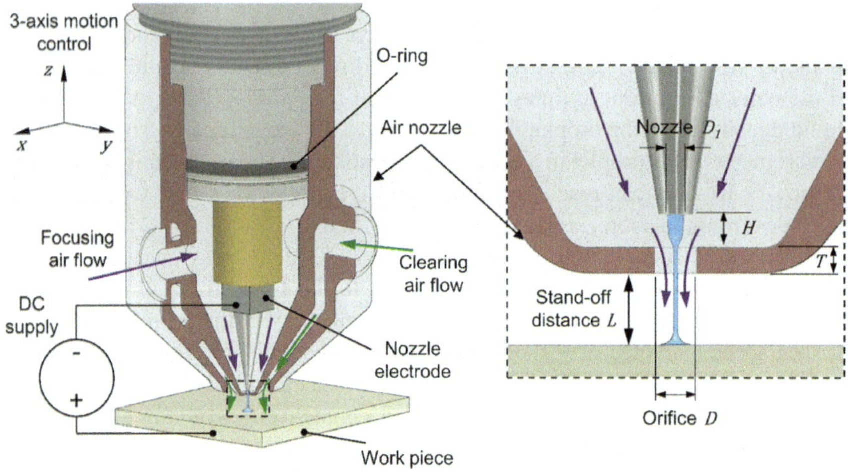

Fig. 10.3 Aerosol jet printing for the fabrication of printed electronics [21]

10.2 The Microjetting Mode

Three regimes of gaseous flow focusing can be readily identified [22, 23]: (i) the dripping regime, where the tapering liquid meniscus is unstable and large droplets or liquid threads are intermittently emitted; (ii) the microdripping regime, where the liquid meniscus is stable, but the emitted jet is absolutely unstable so that small droplets are produced in the meniscus tip; and (iii) the microjetting regime, where the liquid meniscus is stable, and the emitted jet is convectively unstable so that an almost steady liquid ligament can be found close to the orifice.

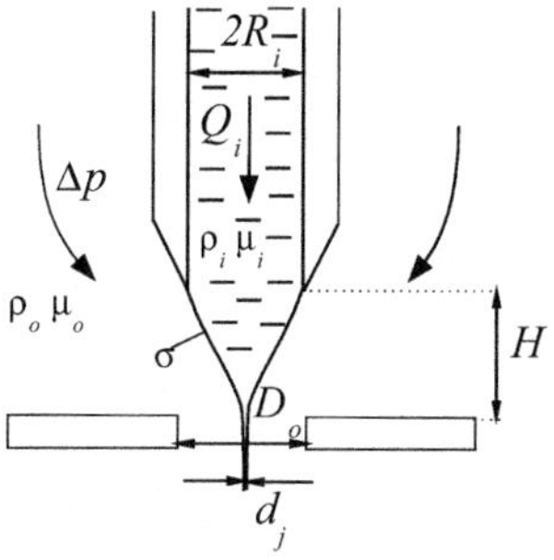

Fig. 10.4 The microjetting regime in the original plate-orifice flow focusing configuration

Figure 10.4 depicts the microjetting regime in the original plate-orifice configuration. A liquid jet of density ρ_i, viscosity μ_i, and diameter d_j tapers from the tip of an axisymmetric liquid meniscus. The meniscus is formed by injecting a constant flow rate Q_i through a capillary of radius R_i. The meniscus is held by the surface tension σ. The energy necessary to expel the jet essentially comes from the pressure drop Δp applied to a gas stream of density ρ_o and viscosity μ_o, which coflows with the liquid jet across an orifice of diameter D_o. The main geometrical parameters of this fluid configuration are the diameter D_o and the capillary-to-orifice distance H.

Because of the importance of the surface tension in the meniscus, one must compare the viscous-capillary velocity $U_{\mu\sigma} = \sigma/\mu_i$ with the inertio-capillary speed $U_{\rho\sigma} = [\sigma/(\rho_i R_i)]^{1/2}$ based on the feeding capillary radius R_i. The ratio $U_{\rho\sigma}/U_{\mu\sigma}$ yields the Ohnesorge number $\mathrm{Oh}_i = \mu_i/(\rho_i \sigma R_i)^{1/2}$ (see Chap. 7). In what follows, we shall refer to low (high) viscosity liquids as those for which $\mathrm{Oh}_i \ll 1$ ($\mathrm{Oh}_i \gg 1$). We devote the next section to the low-viscosity case.

10.3 Microjetting for Low Viscosities

10.3.1 The Jet Diameter

In low-viscosity flow focusing realizations, the energy sinks associated with the surface tension and viscosity can be neglected. Thus, the energy injected into the system per unit time, $Q_i \Delta p$, is mainly invested in the jet kinetic energy, $Q_i \rho_i v_j^2/2$, where $v_j = 4Q_i/(\pi d_j^2)$ is the jet mean velocity [24]. Therefore, the jet diameter d_j can be estimated from the exact balance between the applied pressure and the resulting kinetic energy:

$$d_j = d_j^{\mathrm{FF}} \equiv \left(\frac{8\rho_i Q_i^2}{\pi^2 \Delta p} \right)^{1/4} . \tag{10.1}$$

Equation (10.1) is in remarkable agreement with experiments, as reported by numerous studies going back to 1998 [1, 25–28].

10.3.2 Microjetting Instability

The injected flow rate Q_i is the main governing parameter determining the jet size and microjetting stability. For a small pressure drop Δp, the jet becomes absolutely unstable as Q_i decreases, and microjetting yields microdripping. For sufficiently large values of Δp, the steady flow in the tapering meniscus destabilizes when Q_i falls below a critical value independent from Δp, and the dripping mode arises [22, 23].

For the sake of illustration, Fig. 10.5 shows the stability regions for water. The data are projected onto the parameter plane defined by the jet Reynolds and Weber numbers

$$\mathrm{Re}_j^{\mathrm{FF}} = \frac{\rho_i v_j^{\mathrm{FF}} d_j^{\mathrm{FF}}}{\mu_i}, \quad \mathrm{We}_j^{\mathrm{FF}} \equiv \frac{\rho_i (v_j^{\mathrm{FF}})^2 d_j^{\mathrm{FF}}}{\sigma}, \tag{10.2}$$

where $v_j^{\mathrm{FF}} = 4Q_i/[\pi(d_j^{\mathrm{FF}})^2]$ is the jet velocity corresponding to the diameter d_j^{FF}.

For $10 \lesssim \Delta p \lesssim 65$ mbar, the microjetting regime becomes unstable and gives rise to microdripping due to the convective-to-absolute instability transition in the emitted jet. This transition is explained in terms of the intrinsic dynamics of the jet [29] and, therefore, is not affected by the meniscus slenderness H/D_o (Fig. 10.6). The experimental data agreed reasonably well with the prediction obtained from the spatiotemporal linear stability analysis of the uniform basic flow [29] (see Chap. 3). The theoretical prediction slightly overestimates the transitional Weber number due to the simplicity of the basic flow considered [22, 24]. Owing to the co-flowing gas stream, the jet's free surface moves with a velocity greater than that of the bulk, which naturally favors the jet to sweep growing surface waves downstream, displacing the onset of microdripping to lower flow rates (Weber numbers).

For $\Delta p \simeq 65$ mbar, there is a turning point in the microjetting/microdripping-to-dripping transition line. This line almost coincides with the curve $Q_i = 2.9$ ml/h beyond that point. This means that, for $\Delta p \gtrsim 65$ mbar, there is a threshold value $Q_{i\min}$ of the flow rate below which flow focusing becomes unstable independently of Δp [22, 24]. As may be expected, the critical flow rate $Q_{i\min}$ does depend on the capillary-to-orifice distance H because it is related to the flow pattern in the tapering meniscus (Fig. 10.5-right).

Figure 10.7-left shows the minimum flow rate $Q_{i\min}$ as a function of H for three capillary radii R_i [23]. In all the cases, there is a value of H for which $Q_{i\min}$ diverged. Below that value, flow focusing could not be attained and yielded to the flow blurring phenomenon [30]. There is an optimum value H_{opt} of the distance H at which $Q_{i\min}$ was minimum. The best collapse of the experimental data is obtained when those data are projected onto the plane defined by $(Q_{i\min}/Q_D)(2R_i/D_o)^{-1/3}$ and $(H/D_o)(2R_i/D_o)^{-1/4}$ (Fig. 10.7-right). Here, we have introduced the characteristic flow rate $Q_D \equiv D_o\mu_i/\rho_i$, whose meaning will be explained below. The above result leads to the scaling laws [23]

$$Q_{i\text{opt}}/Q_D \simeq 2.5(2R_i/D_o)^{1/3}\,, \quad H_{\text{opt}}/D_o \simeq 0.6(2R_i/D_o)^{1/4}\,. \tag{10.3}$$

The vertical dashed line in Fig. 10.7-right corresponds to the value of H/D_o estimated by Gañán-Calvo [30] at which the transition from flow focusing to flow blurring takes place.

The minimum flow rate $Q_{i\min}$ can be estimated from simple mass conservation arguments [26]. The viscous tangential stress exerted by the gas current on the meniscus surface makes the low-viscosity liquid move toward the tip. The flow rate dragged in the meniscus tip, Q_d, can be estimated as $Q_d \sim v_s[R_s^2 - (R_s - \delta_l)^2]$, where v_s and R_s are the free surface velocity and radius at the meniscus tip, respectively, while $\delta_l \sim [D_o\mu_i/(\rho_i v_s)]^{1/2}$ is the boundary layer thickness on the liquid side (Fig. 10.8). We assume that, for $Q_i \sim Q_{i\min}$, the flow rate evacuated by the jet is approximately that dragged by the outer stream Q_d (the dragged stream occupies the whole jet section), which suggests that $\delta_l \sim R_s$ at this stability limit (Fig. 10.8). In this case, $Q_d \sim v_s\delta_l^2$, the surface velocity cancels out, and $Q_d \sim Q_D$. The flow rate evacuated by the jet must coincide with that injected through the feeding capillary. Thus,

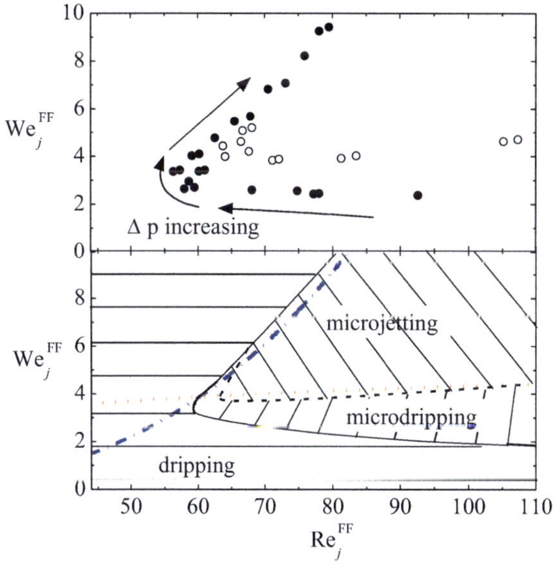

Fig. 10.5 Stability regions in experiments conducted with water. The upper graph shows the original experimental data. The solid and open symbols correspond to the transitions to dripping and microdripping, respectively. The lower graph shows the stability regions plotted from the experimental data shown in the upper graph. The dash-dot (blue) line is the curve $Q_i = 2.9$ ml/h. The dotted (red) line shows the Leib and Goldstein prediction [29] for the convective/absolute instability transition (reprinted from Ref. [23] with the permission of AIP Publishing)

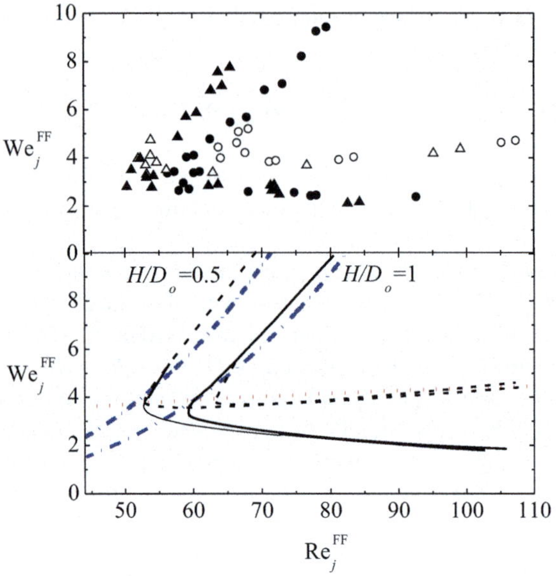

Fig. 10.6 Stability regions in experiments conducted with water. The upper graph shows the original experimental data. The solid and open symbols correspond to the transitions to dripping and microdripping, respectively. The triangles and the circles are the results obtained for $H/D_o = 0.5$ and $H/D_o = 1$, respectively. The lower graph shows the regions occupied by the three regimes mentioned above. The left and right dash-dot (blue) lines are the curves $Q_i = 1.9$ and 2.9 ml/h, respectively. The dotted (red) line shows the Leib and Goldstein prediction [29] for the convective/absolute instability transition (reprinted from Ref. [23] with the permission of AIP Publishing)

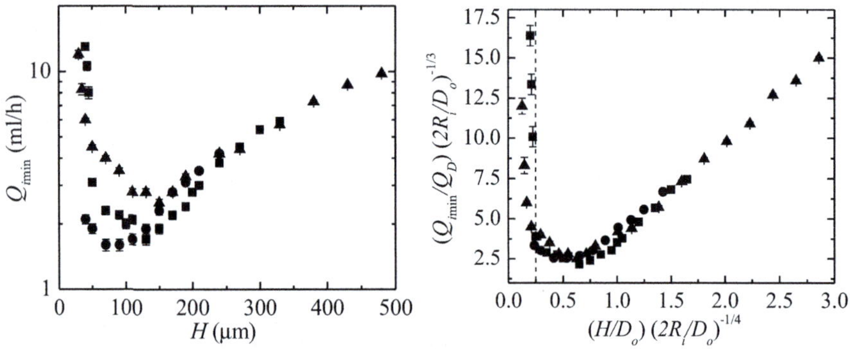

Fig. 10.7 (Left) Minimum flow rate $Q_{i\min}$ as a function of H for $R_i = 50\,\mu$m (circles), 100 μm (squares), and 200 μm (triangles). The experiments were conducted with water and $\Delta p = 250$ mbar. (Right) Dimensionless representation of the experimental data plotted in the left-hand graph (reprinted from Ref. [23] with the permission of AIP Publishing)

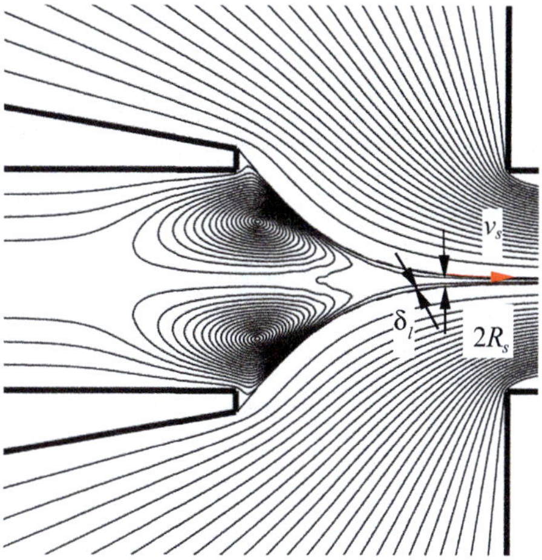

Fig. 10.8 The recirculation cell in gaseous flow focusing. The symbols v_s and R_s are the free surface velocity and radius at the meniscus tip, respectively, while δ_l is the boundary layer thickness on the liquid side (reprinted with permission from Ref. [31]. Copyright (2011) by the American Physical Society)

we conclude that $Q_{i\min} \sim Q_D$ in the low-viscosity case [26]. The ratio Q_i/Q_D is essentially the Reynolds number Re_i defined in Chap. 7.

The major assumption made in the above scaling analysis is that, for $Q_i \sim Q_{i\min}$, the flow rate evacuated by the jet is approximately that dragged by the outer stream. A recirculation cell must grow in the tapering meniscus when this occurs (Fig. 10.8). For this reason, the growth of that cell has been linked to the meniscus instability in low-viscosity gaseous flow focusing. It has been hypothesized that the penetration of the recirculation cell into the feeding capillary for $Q_i \sim Q_{i\min}$ produces an extra viscous dissipation of energy that cannot be compensated for by the outer stream. This energy imbalance might originate the microjetting instability [23, 26, 31]. It is not clear whether the recirculation cell growth is the cause of instability or an indicator that the stability limit has been reached.

10.3.3 The Role of the Viscosity Ratio

The flow rate $Q_{i\min}$ at the meniscus instability is a function of the geometrical parameters and the fluid properties $(\rho_i, \mu_i, \rho_o, \mu_o, \sigma)$. As explained above, the applied pressure drop Δp is irrelevant to this instability. Besides, the density ratio takes an approximately constant value in most experimental realizations, and its influence can be neglected. For a fixed geometry, dimensional analysis dictates that $Q_{i\min}$ obeys the formal relationship

$$Q_{i\min}/Q_D = f(\text{Oh}_i, \mu). \tag{10.4}$$

The results presented in the previous section were obtained obviating the influence of viscosity ratio. Rubio et al. [32] have re-visited this problem by examining experimentally the effect of μ on the meniscus instability in the low-viscosity case.

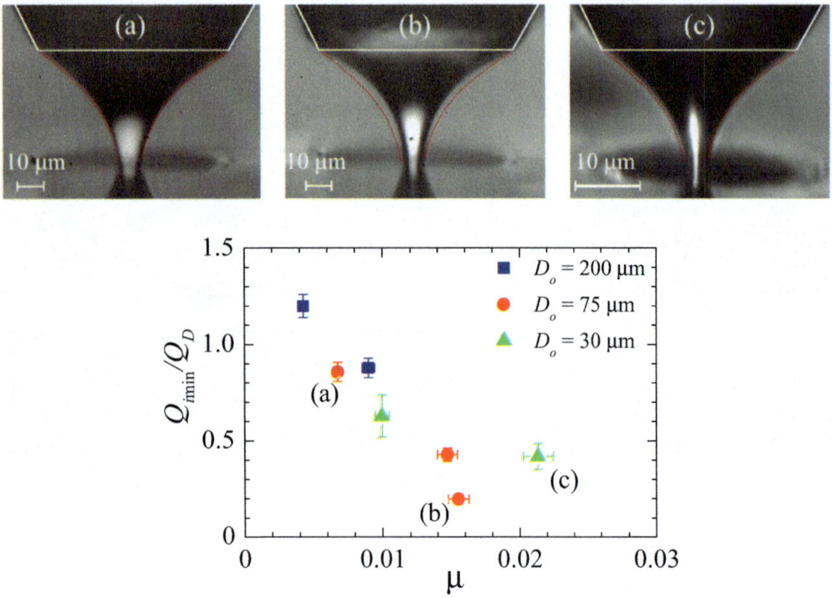

Fig. 10.9 $Q_{i\text{min}}/Q_D$ as a function of the viscosity ratio μ for $\text{Oh}_i = 0.055 \pm 0.005$. The upper images show the meniscus shape for the experiments indicated in the graph. The images have been scaled so that the orifice has approximately the same diameter. The red line indicates the shape of the meniscus for case (a) (reprinted from Ref. [32] with permission of Elsevier)

The meniscus Reynolds number $Q_{i\text{min}}/Q_D$ exhibits a non-monotonous behavior as μ varies (Fig. 10.9). The dependence of the meniscus shape on the viscosity ratio reflects this non-monotonous behavior: the maximum liquid (dimensionless) acceleration is achieved for the optimum viscosity ratio (image (b) in Fig. 10.9).

Rubio et al. [32] have refined the scaling analysis presented in the previous section to include the effect of the viscosity ratio μ. The free surface radius at the meniscus tip can be estimated from the energy balance equation

$$\frac{1}{2}\rho_i v_j^2 \simeq \Delta p(1 - \Phi_v) + \mathcal{C}\,\frac{\mu_o v_g}{\delta_g}, \tag{10.5}$$

where v_j is the mean velocity of the liquid jet in front of the discharge orifice, $\mathcal{C}$ is a dimensionless positive constant accounting for the influence of the free surface shape, $v_g \sim (\Delta p/\rho_o)^{1/2}$ is the gas outer stream velocity in the orifice, and $\delta_g \sim (D_o\mu_o/\rho_o v_g)^{1/2}$ is the gas viscous boundary layer thickness (Fig. 10.10).

The term $\Phi_v > 0$ in Eq. (10.5) represents the pressure loss in the discharge orifice due to the gas viscosity. In other words, $\Delta p(1 - \Phi_v)$ is the pressure drop in the liquid meniscus. As can be observed, Eq. (10.5) includes the small effects of the gaseous viscous dissipation and shear stress, both neglected in the classical formula of flow focusing (10.1). This formula is recovered for $\Phi_v = \mu_o = 0$.

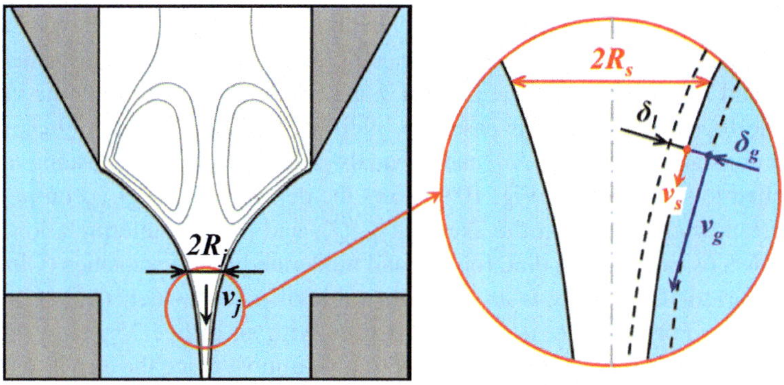

Fig. 10.10 Sketch of the liquid and gas boundary layers in the orifice entrance. The symbols v_s and R_s are the free surface velocity and radius at the meniscus tip, respectively, while δ_l and δ_g are the boundary layer thickness on the liquid and gas side, respectively. In addition, R_j is the jet radius in the orifice entrance, and v_j and v_g are the jet and gas velocities, respectively (reprinted from Ref. [32] with permission of Elsevier)

The Taylor series around $\mu = 0$ of the solution to the energy equation (10.5) is [32]

$$d_j \simeq d_j^{\rm FF}\left(1 - \alpha\,\mu^{1/2} + \frac{5}{2}\alpha^2\mu + \frac{1}{4}\mathcal{B}\,\mu + \cdots\right), \tag{10.6}$$

where $d_j = (4Q_i/\pi v_j)^{1/2}$ is the jet diameter and $\alpha = \mathcal{C}\mu_i^{1/2}/(4D_o^{1/2}\Delta p^{1/4}\rho_o^{1/4})$. To derive Eq. (10.6), we have assumed that the dissipation in the orifice scales as μ_o, i.e., $\Phi_v \sim \mathcal{B}\mu$, where $\mathcal{B}$ is a dimensionless function of the governing parameters.

Assuming that $R_s \simeq d_j/2$, Eq. (10.6) can be written as

$$\frac{Q_d}{Q_D} \simeq \frac{d_j^{\rm FF}}{\delta_l}\left(1 - \alpha\,\mu^{1/2} + \frac{5}{2}\alpha^2\mu + \frac{1}{4}\mathcal{B}\mu\right) - 1. \tag{10.7}$$

As observed, the liquid flow rate Q_d dragged within the boundary layer first decreases and then increases as the viscosity ratio increases. This result explains why $Q_{i\min}/Q_D$ exhibits a non-monotonous dependence on μ (Fig. 10.9-left).

The non-monotonous dependence of the dragged flow rate Q_d on the gas viscosity results from the competition between two opposite effects as the gas viscosity increases: (i) the decrease in the gas velocity for a fixed applied pressure due to the viscous dissipation in the orifice, and (ii) the increase in the viscous shear stress for a fixed gas velocity.

The present analysis does not rely on the hypothesis $\delta_l \simeq d_j/2$ at the stability limit, as assumed in the original scaling analysis presented in Sect. 10.3.2. For $\delta_l \simeq d_j^{\rm FF}/2$ ($\delta_l \simeq R_s$) and $\mu \simeq 0$, the previous result $Q_d \simeq Q_D$ is recovered [31].

The meniscus Reynolds number $Q_{i\min}/Q_D$ and viscosity ratio μ are the only parameters involving the minimum flow rate $Q_{i\min}$ and the gas viscosity μ_o, respectively. This allows us to interpret Fig. 10.9 in dimensional terms as follows. Consider a flow focusing realization with a given ejector and focused liquid. The orifice diameter D_o and liquid properties ρ_i, μ_i, and σ are fixed. We change only the viscosity μ_o of the focusing gas. In this case, Oh_i is fixed, $Q_{i\min}/Q_D$ equals $Q_{i\min}$ divided by by a constant value $D_o\mu_i/\rho_i$, and μ equals μ_o divided by the constant value μ_i. As applied to this example, Fig. 10.9 shows the dependence of $Q_{i\min}$ on μ_o except for constant scale factors in the axes $Q_{i\min}/Q_D$ and μ. This interpretation of Fig. 10.9 allows us to conclude that, for low and moderate liquid viscosities (Ohnesorge numbers up to 10^{-1}), there is an optimum value of μ_o for which $Q_{i\min}$ reaches its lowest value. For instance, if water is focused with an orifice 75 µm in diameter, the optimum gas viscosity is around 0.035 mPa s, almost twice the air viscosity. The optimum gas viscosity would reduce around 100% of the minimum flow rate [32].

10.4 Microjetting for Viscous Liquids

In incompressible gaseous flow focusing, the primary energy source generally is the pressure drop Δp caused by the outer current. If we neglect the interfacial and viscous sinks of energy, the jet's diameter is approximately given by Eq. (10.1) [1]. The interfacial tension has a negligible effect on the energy balance in practically all the flow focusing realizations. However, the liquid viscosity can dissipate a significant amount of energy for large values of the Ohensorge number. Equation (10.1) is no longer valid in this case.

Gañán-Calvo [33] found that the experimental results for the jet diameter $d_j(z)$ approximately match the universal solution of the 1D momentum equation for a constant driving force [34], which includes both the drop of hydrostatic pressure and the viscous shear stress exerted by the outer stream (Fig. 10.11). This solution can be approximated by the simple expression [33]

$$v = \left[\frac{1}{(2x)^{1/2}} + \frac{2}{x^2}\right]^{-1}, \tag{10.8}$$

where

$$v = 2\left(\frac{6\sqrt{2}R_\mu}{D_o}\right)^{-1/3}\left(\frac{d_j^{\text{FF}}}{d_j}\right)^2, \quad x = 2\left(\frac{6\sqrt{2}R_\mu}{D_o}\right)^{-2/3}\frac{z}{D_o}, \tag{10.9}$$

are the scaled velocity and section coordinate, respectively, and

$$R_\mu = \left(\frac{\mu_i^2}{\rho_i \Delta p}\right)^{1/2} \tag{10.10}$$

is a viscous characteristic length.

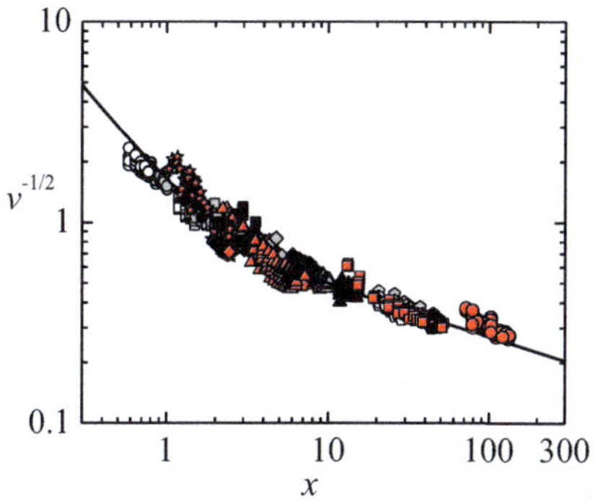

Fig. 10.11 Universal jet shape (velocity) $v^{-1/2}$ as a function of the scaled axial coordinate x. The symbols correspond to different liquids. The white, grey, and black symbols were measured with the plate-orifice configuration [33], while the red ones were obtained by focusing the liquid stream in a glass nozzle [4]. The solid and dashed lines correspond to the exact solution of the 1D momentum equation and (10.8), respectively (they overlap) (reprinted from Ref. [35] with permission of IOP Publishing)

Fig. 10.12 $Q_{i\min}$ as a function of the applied pressure drop Δp (reprinted with permission from Ref. [31]. Copyright (2011) by the American Physical Society)

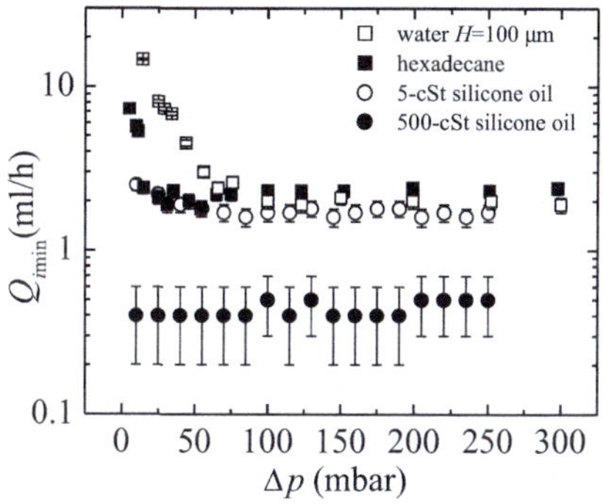

For $R_\mu \ll 1$ ($x \gg 1$) and $z = D_o$, Eq. (10.1) is recovered. For $R_\mu \gg 1$ ($x \ll 1$) and $z \sim D_o \sim H$, the viscous scaling law [36]

$$d_j \sim \left(\frac{\mu_i Q_i}{\Delta p H}\right)^{1/2} \tag{10.11}$$

is obtained. This equation can be derived from the balance between the driving pressure drop Δp and the resistance extensional viscous stress $\mu_i(Q_i/d_j^2)/H$.

In the high-viscosity case, viscosity stabilizes the emitted jet, delaying the growth of the capillary waves and preventing the transition from microjetting to microdripping. Therefore, the microjetting instability is caused exclusively by the destabilization of the steady flow in the tapering meniscus at $Q_i = Q_{i\min}$, independently of the applied pressure drop [31]. As observed in Fig. 10.12, $Q_{i\min}$ becomes independent from Δp as the liquid viscosity (Ohnesorge number) increases.

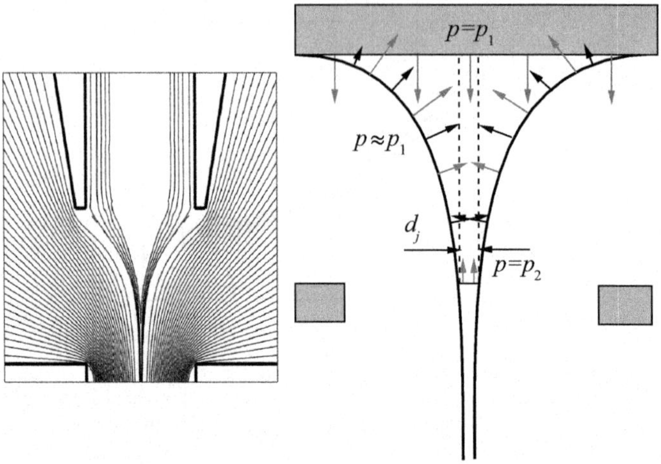

Fig. 10.13 (Left) Streamlines calculated from a numerical simulation of the Navier-Stokes equations for silicone oil of 100 cSt ($Q_i = 2$ ml/h and $\Delta p = 250$ mbar). (Right) Sketch of the main forces acting on the viscous liquid meniscus. The fluid surface delimiting the meniscus is surrounded by a fluid medium with a practically constant pressure p_1 except in the region close to the orifice, where the outer gas stream experiences a sharp pressure drop $\Delta p = p_1 - p_2$. The gray arrows represent the pressure forces. In addition, the liquid meniscus is subjected to the action of the capillary forces represented by the black arrows at the free surface. The projection of the forces onto the meniscus axis results in two main opposite forces: the pressure $F_p \sim \Delta p d_j^2$ and surface tension $F_\sigma \sim \sigma \kappa R_i^2$ forces (reprinted with permission from Ref. [31]. Copyright (2011) by the American Physical Society)

Numerical simulations show that the flow pattern in highly viscous menisci is entirely different from that described in the previous section for almost inviscid liquids [4]. Viscous stresses "arrange" the streamlines and "direct" the flow in the meniscus tip, which prevents the recirculation cell from being formed even for very small injected flow rates and very large applied pressure drops (Fig. 10.13-left). The meniscus stretches progressively mainly due to the action of the pressure drop Δp transmitted across the free surface. The flow rate dragged by the outer gas stream is much smaller than in the low-viscosity case. Therefore, the instability mechanism is entirely different from that in low-viscosity menisci.

The instability of a viscous meniscus can be explained as follows. Consider the liquid meniscus sketched in Fig. 10.13-right. The grey arrows represent the pressure distribution imposed by the gas stream over the free surface, at the capillary end, and in front of the jet. At this point, we do not consider the effect of the surface tension. Therefore, the hydrostatic pressure is regarded as a continuous field across the free surface. The hydrostatic pressure approximately equals the upstream value p_1 over most of the free surface and decreases to p_2 in front of the jet. Then, the axial pressure force F_p driving the fluid motion scales as $F_p \sim \Delta p d_j^2$, where $\Delta p = p_1 - p_2$ is the pressure drop in the gas domain.

Numerical simulations show that F_p is much larger than the axial viscous forces both at the capillary end and in front of the jet. The viscous tangential force exerted by the gas stream on the free surface also plays a secondary role. Thus, these forces can be neglected. Inertia and, therefore, the net flux of axial momentum across the meniscus are subdominant too.

Now we consider the effect of the surface tension force (black arrows in Fig. 10.13-right). It is worth noting that the meniscus shape resembles that formed in selective withdrawal [37]. As in that case, the contribution to the capillary force corresponding to the interface next to the capillary end can be estimated as $F_\sigma \sim \sigma\kappa R_i^2$, where $\kappa \sim H/R_i^2$ is the free surface curvature in that region. This contribution is much greater than that of the meniscus tip, $\sigma/d_j\, d_j^2 \sim \sigma d_j$. Therefore, the latter can be neglected, and the capillary force scales as $F_\sigma \sim \sigma H$.

Montanero et al. [31] hypothesized that the meniscus becomes unstable when $F_\sigma \sim F_p$. A similar hypothesis was postulated by Blanchette and Zhang [37] for the viscous entrainment mode of selective withdrawal (see Chap. 9). At the onset of instability, $\Delta p d_j^2 \sim \sigma H$ ($F_p \sim F_\sigma$), thus $\Delta p \sim \sigma/d_j\; H/d_j$, and therefore $\Delta p \gg \sigma/d_j$. This implies that the pressure drop across the liquid meniscus, Δp_l, is commensurate with that in the gas domain Δp ($\Delta p_l \sim \Delta p$).

The Stokes equations for negligible inertia yield $\Delta p_l \sim \mu_i v_j/H$, where v_j is the jet's velocity at the discharge orifice. Therefore, $F_p \sim \Delta p_l d_j^2 \sim \mu_i v_j d_j^2/H \sim \mu_i Q_i/H$. Since $R_i \simeq H$ in the configurations of interest, the condition $F_\sigma \sim F_p$ becomes $\sigma R_i \sim \mu_i Q_i/R_i$, which can also be expressed as $Q_i \sim Q_v \equiv R_i^2\sigma/\mu_i$ [4].

The ratio Q_i/Q_v is essentially the capillary number Ca_i defined in Chap. 7. We conclude that the capillary number becomes of the order of unity at the onset of the microjetting instability for viscous liquids.

The surface tension plays a critical role in the stability of the viscous meniscus. The meniscus destabilizes when the retraction force F_σ exerted by the surface tension on the meniscus base becomes comparable with the driving pressure force F_p exerted by the gas stream. However, the surface tension does not affect the energy balance in the meniscus, and therefore it does not significantly alter the jet diameter in most experimental realizations.

Equation (10.4) applies to both the low-viscosity and high-viscosity limits. This equation can be re-written as

$$g(Q_{i\min}/Q_D, Q_{i\min}/Q_v, \mu) = 0, \tag{10.12}$$

where we have considered that

$$\mathrm{Oh}_i = \left(\frac{Q_{i\min}}{Q_D}\right)^{1/2} \left(\frac{Q_{i\min}}{Q_v}\right)^{-1/2} \left(\frac{R_i}{D_o}\right)^{1/2} \tag{10.13}$$

and the ratio R_i/D_o is fixed. In the high-viscosity case (low values of $Q_{i\min}/Q_D$), the liquid inertia becomes negligible, the influence of $Q_{i\min}/Q_D$ can be neglected, and the viscosity ratio takes such small values that its effect on $Q_{i\min}/Q_v$ can be ignored as well. Then, one concludes that $Q_{i\min}/Q_v \simeq$ const. According to the scaling analysis presented above, this constant should be of order unity.

10.5 Stability Map

The solid symbols of Fig. 10.14 shows the projection onto the parameter plane $(Q_{i\min}/Q_D, Q_{i\min}/Q_v)$ of experimental data obtained when the liquid stream tapers from a capillary located in front of a plate orifice [31] or inside a glass nozzle [4]. In this projection, the viscosity ratio is not considered, although this parameter has a noticeable effect on the stability limit in the low-viscosity case (large values of $Q_{i\min}/Q_D$) (Fig. 10.9) [32]. As predicted in the scaling analysis, $Q_{i\min}/Q_v$ becomes a constant of order unity for sufficiently small values of $Q_{i\min}/Q_D$ (high-viscosity case). Interestingly, the stability limit is not significantly affected by the geometry that focuses the liquid stream (either an orifice in a plate or a converging-diverging nozzle).

Figure 10.14 also shows the strong stabilizing effect produced when the feeding capillary is replaced with a hypodermic needle. As explained in Chap. 7, the meniscus is substituted by a liquid film sliding over the needle tip, which drastically reduces the minimum flow rate and, therefore, the minimum jet diameter [9]. Using this idea, Rebollo-Muñoz et al. [38] designed and built a hybrid flow focusing nozzle to produce micron and sub-micron capillary jets. The same concept was employed in transonic flow focusing to reduce the jet's diameter down to hundreds of nanometers and increase the jet's speed beyond 170 m/s [39]. The hypodermic needle partially

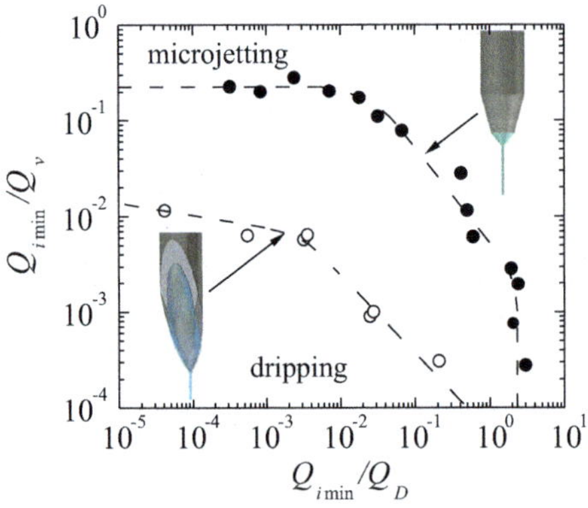

Fig. 10.14 Minimum flow rate $Q_{i\min}$ in gaseous flow focusing. The solid symbols are the results obtained when the liquid stream tapers from a capillary located in front of a plate orifice [31] or inside a converging-diverging glass nozzle [4]. The open symbols are the results obtained when the liquid is injected through a hypodermic needle inside a glass nozzle [9]. The dashed lines are to guide the eye (reprinted from Ref. [9] with permission of AIP Publishing)

suppresses the mechanism destabilizing the meniscus in the low-viscosity and high-viscosity cases. Specifically, the needle reduces the liquid flow rate dragged by the gas stream and removes the recirculation cell for low-viscosity liquids. In addition, it eliminates the surface tension retraction force in the high-viscosity case.

10.6 Focusing Microjets with Nozzles

DePonte et al. [2] first used a flame-shaped glass micro-nozzle for gaseous flow focusing. The nebulizer was built by introducing a feeding capillary inside a converging-diverging nozzle concentrically. One injects the liquid through the capillary to obtain the flow focusing effect, while a gas stream crosses the nozzle driven by an applied pressure drop. As in the original plate-orifice configuration [1], the gas stream suctions the liquid meniscus attached to the capillary. A microjet tapers from the meniscus tip pushed by the pressure gradient and dragged by the viscous stress. This configuration is referred to as Gas Dynamic Virtual Nozzle (GDVN).

The flame polishing method commonly used to fabricate the nozzles can be summarized as follows. When a flame is applied to the tip of a clean-cut millimeter glass tube, the glass melts and flows, driven by the surface tension and viscosity forces. This flow narrows the inner duct formed in the tube. Suppose the heat source is swiftly withdrawn before the glass's inner free surface pinches. In that case, the glass solidifies into various hourglass shapes, which depend on the flame's intensity, focus, and location [40]. The manufacturing process has been considerably improved to increase reproducibility and control the nozzle shape [41, 42]. The alternative pulling method produces slightly asymmetric nozzles with needle-like shapes, significantly increasing the length and decreasing the outer section. These nozzles are not appropriate for flow focusing.

The nozzles offer several advantages over the plate-orifice configuration: they are smooth even at the micrometer scale, demand gas flow rates (energy) significantly smaller than those required by the plate-orifice configuration, and are transparent. In addition, one can readily obtain nozzles with neck diameters in the range of a few tens of microns. Glass micro-nozzles have been used to flow focusing hydrated streams of nanocrystals with transonic gaseous currents in Serial Femtosecond Crystallography [43], which have conferred them on great popularity.

Significant differences exist between the gas flow patterns in the plate-orifice configuration and the converging-diverging nozzle. Specifically, the axial momentum of the gas stream close to the feeding capillary end is significantly larger in the nozzle than in the plate-orifice configuration. This enhances the transfer of momentum from the gas to the liquid via viscosity, which reduces the jet diameter. However, the difference between the liquid and gas velocities on the two sides of the meniscus surface can trigger the whipping (bending) instability in front of the discharge orifice, a phenomenon not observed in the plate-orifice configuration. This is the so-called *absolute whipping* (Fig. 10.15), which was first observed by Acero et al. [4].

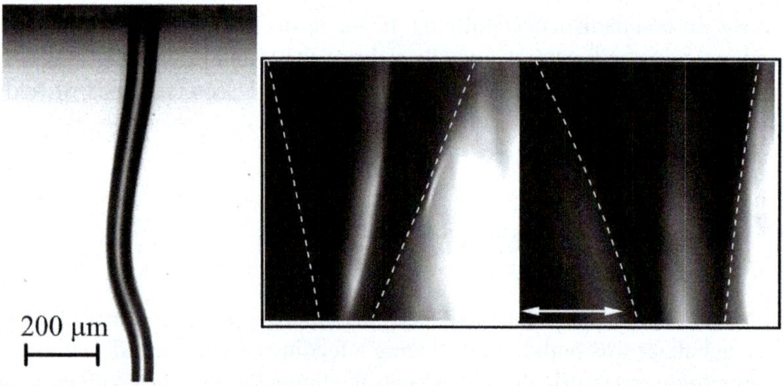

Fig. 10.15 (Left) Convection downstream of whipping oscillations over a glycerol jet produced by a plate-orifice flow-focusing device. (Right) Images of a 5 cSt solicone oil meniscus oscillating inside the glass nozzle when absolute whipping appears (reprinted from Ref. [4] with permission of AIP Publishing)

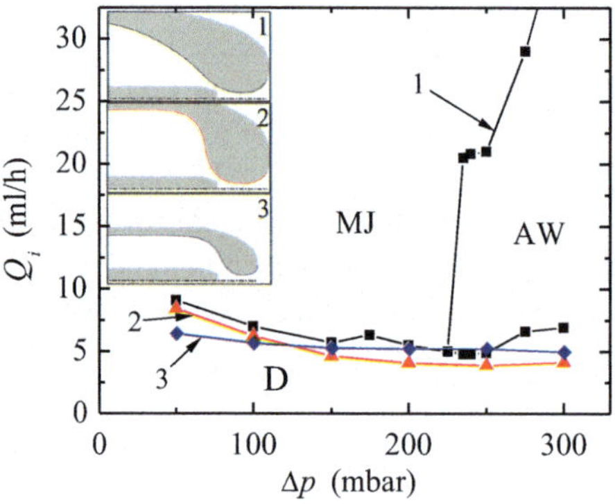

Fig. 10.16 Experimental stability regions for water focused with air using Nozzles 1, 2, and 3, whose necks have the same diameter. In all the cases, $H=450\,\mu$m. Microjetting (MJ), dripping (D), and absolute whipping (AW) were found in the experiments conducted with Nozzle 1. Nozzles 2 and 3 produced neither absolute nor convective whipping. The insets show the nozzle shape and the feeding capillary position (reprinted from Ref. [44] with permission of Elsevier)

More recently, Blanco-Trejo et al. [44] examined the influence of the ejector geometry on the microjetting stability. To this end, nozzles that differed only in their convergence rate were fabricated [42]. This parameter is the inverse of the length over which the nozzle diameter reduction takes place. The increase of the nozzle convergence rate eliminated the absolute whipping instability (Fig. 10.16), which explains why this phenomenon is not found in the plate-orifice configuration.

When the bending of the jet emitted in flow focusing is adequately controlled, whipping can become helpful. Narayanasamy et al. [45] have recently found a unique whipping instability that emanates a multi-monodisperse whipping spray jet with a 2D profile. A high-aspect-ratio non-axisymmetric microfluidic device produces this instability. Some potential applications of this phenomenon are the preparation for static single-particle cryo-EM studies, electrospray and electrosonic spray ionization, and drug delivery [45].

10.7 Global Stability Analysis

10.7.1 Axisymmetric Instability

Experimental studies have separately analyzed the instability in the tapering meniscus and the jet. Microjetting is stable if and only if both fluid domains are stable. The global linear stability analysis allows one to determine the stability limits of the microjetting mode without splitting the system into those two parts.

As explained in Chap. 3, one calculates the axisymmetric linear modes by assuming the temporal dependence

$$U(r, z; t) = U_0(r, z) + \delta U(r, z)\, e^{-i\omega t} + \text{c.c.}, \tag{10.14}$$

where $U(r, z; t)$ represents any hydrodynamic quantity, $U_0(r, z)$ and $\delta U(r, z)$ stand for the base (steady) solution and the spatial dependence of the eigenmode, respectively, and $\omega = \omega_r + i\omega_i$ is the eigenfrequency. The dominant eigenmode is that with the largest growth rate ω_i. If that rate is positive, the base flow is asymptotically unstable [46]. The meaning of the term "asymptotically" was explained in Chap. 3. We will come back to this point below.

The theoretical stability limits are compared with the corresponding experimental data in Fig. 10.17. The linear stability analysis satisfactorily captures the influence of the viscosity in all the cases. The differences between the numerical and experimental geometries may explain the discrepancies between the theoretical predictions and experiments. In fact, the ratio H/D_o was close to its maximum possible value; therefore, slight variations of the capillary position H considerably influence the critical flow rate [31].

Any small-amplitude perturbation introduced into the system can be seen as the superposition of the infinite set of eigenmodes. For sufficiently large time (within the linear regime), the mode with the largest growth rate (lowest damping rate) dominates the system dynamics, while the contribution of the rest of the eigenmodes can be neglected. Therefore, only the dominant mode determines the system stability *asymptotically* (for sufficiently large time). If the dominant mode growth rate ω_i is positive, that mode grows exponentially, and the system is *asymptotically* unstable. In contrast, $\omega_i < 0$ implies that the dominant mode decays and the system is *asymptotically* stable.

As explained in Chap. 3, asymptotic linear stability does not necessarily imply linear stability. If the linearized Navier-Stokes operator is normal, then the energy of the perturbation introduced into an asymptotically stable system decreases monotonously

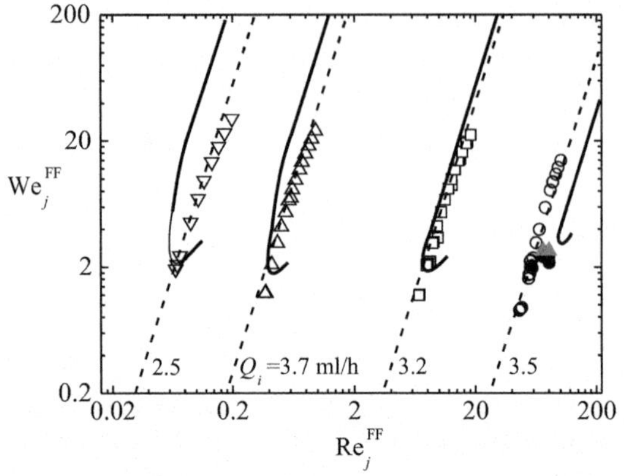

Fig. 10.17 Stability map for the flow focusing configuration analyzed by Cruz-Mazo et al. [47]. From right to left, the solid lines approximately correspond to experimental microjetting-to-dripping transitions for $Oh_i = 0.0136$ (water), 0.127 (5-cSt silicone oil), 2.47 (100-cSt silicone oil), and 12.6 (500-cSt silicone oil), respectively. The open symbols are the corresponding transitions from asymptotically stable to unstable base flows. The figure shows the minimum flow rates estimated from these theoretical predictions (dashed lines). The solid triangles and circles are stable and unstable direct numerical simulations for water, respectively. (Ref. [47], reproduced with permission)

not only for a sufficiently large time but also during the system's short-term response. On the contrary, if this last condition does not apply, there can be a transient growth of the perturbation energy before the asymptotic exponential regime is reached [48, 49]. This short-term, non-exponential growth of small-amplitude perturbations can be responsible for the instability of asymptotically stable systems. This behavior has been observed in, for instance, convective instabilities commonly arising in problems with inflow and outflow conditions [50].

Cruz-Mazo et al. [47] integrated the Navier-Stokes equations linearized around the base flow solution to examine its short-term response to a perturbation. The equations were the same as those solved in the asymptotic global stability analysis. The perturbation consisted in a deformation of the free surface (the velocity and pressure fields were not perturbed) given by the Dirac delta function at $t = 0$. Figure 10.18 shows the free surface deformation at three instants for an asymptotically stable base flow. Due to the small magnitude of the perturbation, the free surface deformation is hardly noticeable at $t = 0$. Owing to the non-normal character of the linearized Navier-Stokes operator, the superposition of decaying eigenmodes gives rise to the free surface pinch-off within the numerical domain. This kind of short-term convective instability does not cause oscillations of the liquid meniscus. This is consistent with both numerical simulations [26] and experiments [22, 23] for the plate-orifice configuration.

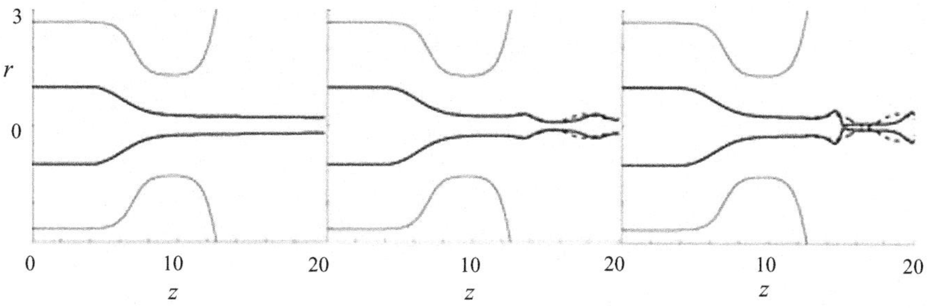

Fig. 10.18 Free surface deformation calculated with the linearized (dash line) and nonlinear (solid line) hydrodynamic equations at three instants for $\mathrm{Re}_j^{\mathrm{FF}} = 80.2$, $\mathrm{We}_j^{\mathrm{FF}} = 2.42$, and $\mathrm{Oh}_i = 0.0136$ (reproduced with permission, Ref. [47])

10.7.2 Whipping Instability

As explained above, the gas kinetic energy can trigger the whipping instability in the emitted jet and the tapering meniscus formed in a converging-diverging nozzle. The global linear stability analysis also predicts the condition under which whipping arises. To this end, one assumes the temporal dependence

$$U(r, \theta, z; t) = U_0(r, z) + \delta U(r, z)\, e^{i(m\theta - \omega_m t)} + \text{c.c.}, \tag{10.15}$$

where m is the azimuthal wave number. If the growth rate ω_{mi} of the dominant mode is positive, the base flow is unstable under the corresponding axisymmetric ($m = 0$) or non-axisymmetric ($m \neq 0$) perturbation.

Figure 10.19 shows the stability map when the eigenmodes $m = 0$ and $m = 1$ are considered. Whipping arises as Δp increases due to the gas velocity increase. The value of the capillary-to-orifice distance H considerably affects the instability transition for $m = 0$, while it hardly alters the flow stability under the $m = 1$ perturbation. The theoretical predictions for different nozzles were consistent with the experimental observations [44]. Specifically, the flow becomes more stable under $m = 1$ perturbations as the nozzle convergence rate increases.

10.8 The Microdripping Mode

Gaseous flow focusing can also operate in the microdripping mode under certain specific conditions [51]. In this mode, drops are ejected right behind the discharge orifice with sizes smaller than the discharge orifice (Fig. 10.20). The existence of a filter (the feeding capillary-discharge orifice arrangement) for the breakup wavenumber and the absence of both satellite droplets and whipping instabilities enable the production of perfectly monodisperse collections of droplets. Their diameters are

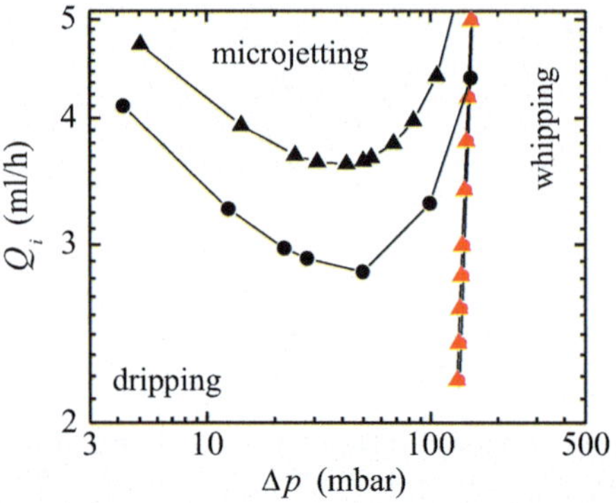

Fig. 10.19 Marginally stable flows under $m = 0$ (black symbols) and $m = 1$ (red symbols) perturbations calculated for 5-cSt silicone oil focused with the nozzle used by Acero et al. [4]. The circles and triangles correspond to $H = 350$ and $440\,\mu$m, respectively (reprinted from Ref. [44] with permission of Elsevier)

much smaller than those obtained when a liquid simply drips from a faucet, making this mode a good candidate for various technological applications. The Weber number of the droplets produced in the microdripping mode is much smaller than in the microjetting regime, which may suppress the splashing on solid surfaces and favor the coalescence with other droplets.

Owing to the pulsatile character of this regime, the pressure drop applied to the gas stream is essentially balanced by the fluid particle local acceleration, which constitutes the fundamental difference with respect to the steady microjetting mode. Based on this idea, Cruz-Mazo et al. [51] derived the scaling law for the droplet diameter as a function of the governing parameters.

First, we project the momentum equation onto the symmetry axis z. The result is

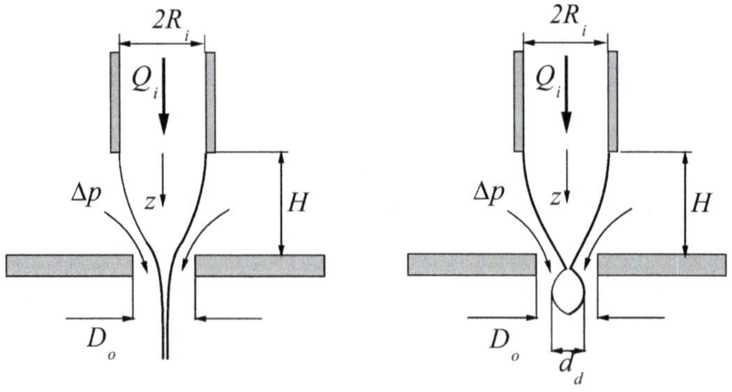

Fig. 10.20 Sketch of the microjetting (left) and microdripping (right) modes of gaseous flow focusing

$$\frac{\partial w}{\partial t} + w\frac{\partial w}{\partial z} = -\frac{1}{\rho_i}\frac{\partial p}{\partial z} + \frac{\mu_i}{\rho_i}\frac{\partial^2 w}{\partial z^2}, \tag{10.16}$$

where w is the z-component of the velocity, and p is pressure in the liquid domain, respectively. We have retained only the dominant contribution to the viscosity term and have neglected the gravitational force due to the smallness of the fluid configuration.

In the microdripping mode, the orders of magnitude of the terms in Eq. (10.16) are

$$\frac{\partial}{\partial t} \sim \frac{Q_i}{d_d^3}, \quad w \sim \frac{Q_i}{d_d^2}, \quad w\frac{\partial w}{\partial z} \sim \frac{Q_i^2}{d_d^4 R_i}, \quad \frac{\partial p}{\partial z} \sim \frac{\Delta p_l}{R_i}, \quad \frac{\mu_i}{\rho_i}\frac{\partial^2 w}{\partial z^2} \sim \frac{\mu_i}{\rho_i}\frac{Q_i}{d_d^2 R_i^2}, \tag{10.17}$$

where d_d is the droplet diameter, and Δp_l is the pressure drop across the liquid meniscus. In addition, we have taken into account that $R_i \sim H$.

In the steady microjetting regime, the pressure gradient $\partial p/\partial z$ induced by the outer gaseous stream essentially transforms into the convective term $w\partial w/\partial z$ (the jet's kinetic energy). On the contrary, the convective term is at most comparable to $\partial w/\partial t$ in the pulsatile microdripping mode. For this reason, the drop inflation time d_d^3/Q_i is selected as the characteristic time of the process.

The Womersley number $\alpha = Q_i R_i^2 \rho_i/(\mu_i d_d^3)$ measures the relative importance of local acceleration versus viscosity force. This parameter takes large values for small and moderate liquid viscosities. Therefore, viscous stresses can be neglected. Considering the dominant terms in Eq. (10.16), one gets

$$\frac{Q_i^2}{d_d^5} \sim \frac{\Delta p_l}{\rho_i R_i}. \tag{10.18}$$

As mentioned above, the convective term is at most comparable to the unsteady term, i.e.,

$$\frac{Q_i^2}{d_d^4 R_i} \lesssim \frac{Q_i^2}{d_d^5}, \tag{10.19}$$

which implies that $d_d \lesssim R_i$. Equations (10.18) and (10.19) lead to

$$\frac{Q_i^2}{d_d^4 R_i} \lesssim \frac{\Delta p_l}{\rho_i R_i}. \tag{10.20}$$

We assume that the droplet forms during a time $t_f \sim d_d^3/Q_i$ of the order of that of the free surface breakup t_b. For sufficiently low viscosities, t_b is of the order of the inertio-capillary time defined in terms of the droplet diameter $\tilde{t}_{\rho\sigma} = (\rho_i d_d^3/\sigma)^{1/2}$. The condition $t_f \sim \tilde{t}_{\rho\sigma}$ leads to

$$Q_i \sim \sigma^{1/2}\rho_i^{-1/2} d_d^{3/2}, \tag{10.21}$$

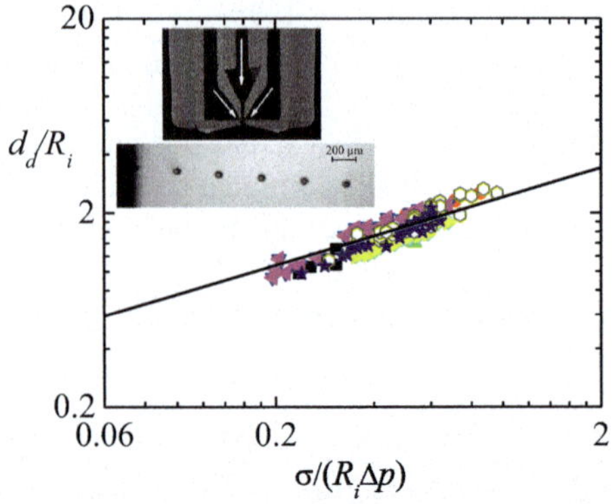

Fig. 10.21 Diameters of the drops produced in the microdripping mode of gaseous flow focusing. The solid line is the function $d_d/R_i = 2.4[\sigma/(R_i\Delta p)]^{1/2}$. The inset shows an image of the flow focusing device and the ejected droplets (reprinted with permission from Ref. [51]. Copyright (2016) by the American Physical Society)

which allows one to eliminate the flow rate Q_i from Eq. (10.20).

Equations (10.20) and (10.21) allow one to conclude that $\sigma/d_d \lesssim \Delta p_l$. Thus, surface tension triggers the instability pinching the free surface and stopping the droplet formation, but it does not provide the energy necessary for that process. That energy comes from the pressure drop in the outer gaseous stream, which is of the order of that across the liquid meniscus, i.e., $\Delta p_l \sim \Delta p$.

Taking into account the above considerations, Eqs. (10.18) and (10.21) yield the scaling law

$$\frac{d_d}{R_i} \sim \left(\frac{\sigma}{R_i\Delta p}\right)^{1/2} \tag{10.22}$$

for the droplet diameter. The scaling law (10.22) was validated experimentally by Cruz-Mazo et al. [51] (Fig. 10.21). The scattering of the experimental data around the function $d_d/R_i = 2.4[\sigma/(R_i\Delta p)]^{1/2}$ can be attributed to viscosity effects. [51] also determined the conditions for the microdripping mode to appear.

References

1. Gañán-Calvo AM (1998) Generation of steady liquid microthreads and micron-sized monodisperse sprays in gas streams. Phys Rev Lett 80:285–288
2. DePonte DP, Weierstall U, Schmidt K, Warner J, Starodub D, Spence JCH, Doak RB (2008) Gas dynamic virtual nozzle for generation of microscopic droplet streams. J Phys D: Appl Phys 41(195):505
3. Tirandazi P, Tomic G, Hidrov CH (2018) An ultra-high-throughput flow-focusing microfluidic device for creation of liquid droplets in air. In: Proceedings of the ASME 2017 15th international

conference on nanochannels, microchannels, and minichannels, Cambridge, Massachusetts, USA, pp 1–6
4. Acero AJ, Ferrera C, Montanero JM, Gañán-Calvo AM (2012) Focusing liquid microjets with nozzles. J Micromech Microeng 22(065):011
5. Vladisavljevic GT, Henry J, Duncanson WJ, Shum HC, Weitz DA (2012) Fabrication of biodegradable poly(lactic acid) particles in flow-focusing glass capillary devices. Progr Colloid Polym Sci 139:111–114
6. Beyerlein KR, Adriano L, Heymann M, Kirian R, Knoska J, Wilde F, Chapman HN, Bajt S (2015) Ceramic micro-injection molded nozzles for serial femtosecond crystallography sample delivery. Rev Sci Instrum 86(125):104
7. G Nelson et al (2016) Three-dimensional-printed gas dynamic virtual nozzles for X-ray laser sample delivery. Opt Express 24:11,515–11,530
8. Acero AJ, Montanero JM, Ferrera C, Herrada MA, Gañán-Calvo AM (2012) Enhancement of the stability of the flow focusing technique for low-viscosity liquids. J Micromech Microeng 22(115):039
9. Acero AJ, Rebollo-Muñoz N, Montanero JM, Gañán-Calvo AM, Vega EJ (2013) A new flow focusing technique to produce very thin jets. J Micromech Microeng 23(065):009
10. Muñoz-Sánchez BN, Silva SF, Pinho D, Vega EJ, Lima R (2016) Generation of micro-sized PDMS particles by a flow focusing technique for biomicrofluidics applications. Biomicrofluidics 10(14):122
11. Gañán-Calvo AM, González-Prieto R, Riesco-Chueca P, Herrada MA, Flores-Mosquera M (2007) Focusing capillary jets close to the continuum limit. Nat Phys 3:737–742
12. Gañán-Calvo AM, Montanero JM, Martín-Banderas L, Flores-Mosquera M (2013) Building functional materials for health care and pharmacy from microfluidic principles and Flow Focusing. Adv Drug Deliv Rev 65:1447–1469
13. Gañán-Calvo AM, Castro-Hernández E, Flores-Mosquera M, Martín-Banderas L (2014) Massive, generic, and controlled microencapsulation by flow focusing: some physicochemical aspects and new applications. J Flow Chem
14. Si T, Li GB, an Z Q Zhu QW, Luo XS, Xu RX (2016) Optical droplet vaporization of nanoparticle-loaded stimuli-responsive microbubbles. Appl Phys Lett 108(111):109
15. Herrada MA, Montanero JM, Ferrera C, Gañán-Calvo AM (2010) Analysis of the dripping-jetting transition in compound capillary jets. J Fluid Mech 649:523–536
16. Gañán-Calvo AM, López-Herrera JM, Riesco-Chueca P (2006) The combination of electrospray and flow focusing. J Fluid Mech 566:421–445
17. Forbes TP, Sisco E (2014) Chemical imaging of artificial fingerprints by desorption electro-flow focusing ionization mass spectrometry. Analyst 139:2982
18. Forbes TP, Sisco E (2014) Mass spectrometry detection and imaging of inorganic and organic explosive device signatures using desorption electro-flow focusing ionization. Anal Chem 86:7788–7797
19. Gañán-Calvo AM (2007) Electro-flow focusing: the high-conductivity low-viscosity limit. Phys Rev Lett 98(134):503
20. Bisterov I, Abayzeed S, Speidel A, Magnini M, Zubayr M, Clare AT (2023) Adapting 'tool' size using flow focusing: a new technique for electrochemical jet machining. J Mater Process Technol 311(117):807
21. Mahajan A, Frisbie CD, Francis LF (2013) Optimization of aerosol jet printing for high-resolution, high-aspect ratio silver lines. ACS Appl Mater Interfaces 5:4856–4864
22. Si T, Li F, Yin XY, Yin XZ (2009) Modes in flow focusing and instability of coaxial liquid-gas jets. J Fluid Mech 629:1–23
23. Vega EJ, Montanero JM, Herrada MA, Gañán-Calvo AM (2010) Global and local instability of flow focusing: the influence of the geometry. Phys Fluids 22(064):105
24. Gañán-Calvo AM, Montanero JM (2009) Revision of capillary cone-jet physics: electrospray and flow focusing. Phys Rev E 79(066):305
25. Gordillo JM, Gañán-Calvo AM, Pérez-Saborid M (2001) Monodisperse microbubbling: absolute instabilities in coflowing gas-liquid jets. Phys Fluids 13:3839–3842

26. Herrada MA, Gañán-Calvo AM, Ojeda-Monge A, Bluth B, Riesco-Chueca P (2008) Liquid flow focused by a gas: jetting, dripping, and recirculation. Phys Rev E 78(036):323
27. Martín-Banderas L, Flores-Mosquera M, Riesco-Chueca P, Rodríguez-Gil A, Cebolla A, Chávez S, Gañán-Calvo AM (2005) Flow focusing: a versatile technology to produce size-controlled and specific-morphology microparticles. Small 1:688–692
28. Rosell-Llompart J, Gañán-Calvo AM (2008) Turbulence in pneumatic flow focusing and flow blurring regimes. Phys Rev E 77(036):321
29. Leib SJ, Goldstein ME (1986) Convective and absolute instability of a viscous liquid jet. Phys Fluids 29:952–954
30. Gañán-Calvo AM (2005) Enhanced liquid atomization: from flow-focusing to flow-blurring. Appl Phys Lett 86(214):101
31. Montanero JM, Rebollo-Muñoz N, Herrada MA, Gañán-Calvo AM (2011) Global stability of the focusing effect of fluid jet flows. Phys Rev E 83(036):309
32. Rubio A, Muñoz-Sánchez BN, Cabezas MG, Montanero JM (2023) Influence of the gas viscosity on the stability of flow focusing. Exp Therm Fluid Sci 141(110):788
33. Gañán-Calvo AM, Ferrera C, Montanero JM (2011) Universal size and shape of viscous capillary jets: application to gas-focused microjets. J Fluid Mech 670:427–438
34. Clarke NS (1966) A differential equation in fluid mechanics. Mathematika 12:51–53
35. Montanero JM, Gañán-Calvo AM (2020) Dripping, jetting and tip streaming. Rep Prog Phys 83(097):001
36. Gañán-Calvo AM, Pérez-Saborid M, López-Herrera JM, Gordillo JM (2004) Steady high viscosity liquid micro-jet production and fiber spinning using co-flowing gas conformation. Eur Phys J B 39:131–137
37. Blanchette F, Zhang WW (2009) Force balance at the transition from selective withdrawal to viscous entrainment. Phys Rev Lett 102(144):501
38. Rebollo-Muñoz N, Acero AJ, Marcos de León JZ, Montanero J, Gañán-Calvo A (2016) A hybrid flow focusing nozzle design to produce micron and sub-micron capillary jets. Int J Mass Spectrom 403:32–38
39. Nazari R, Zaare S, Alvarez RC, Karpos K, Engelman T, Madsen C, Nelson G, Spence JCH, Weierstall U, Adrian RJ, Kirian RA (2020) 3D printing of gas-dynamic virtual nozzles and optical characterization of high-speed microjets. Opt Express 28:21,749–21,765
40. Montanero JM, Gañán-Calvo AM, Acero AJ, Vega EJ (2010) Micrometer glass nozzles for flow focusing. J Micromech Microeng 20(075):035
41. Muñoz-Sánchez BN, Cabezas MG (2018) Borosilicate nozzles manufactured by reproducible fire shaping. J Mater Proces Tech 261:173–183
42. Muñoz-Sánchez BN, Gañán-Calvo AM, Cabezas MG (2019) A new fire shaping approach to produce highly axisymmetric and reproducible nozzles. J Mater Process Tech 270:241–253
43. Chapman HN et al (2011) Femtosecond X-ray protein nanocrystallography. Nature 470:73–79
44. Blanco-Trejo S, Herrada MA, Gañán-Calvo AM, Rubio A, Cabezas MG, Montanero JM (2020) Whipping in gaseous flow focusing. Int J Multiphase Flow 130(103):367
45. Narayanasamy SR, RVasireddi, Holman HYN, Trebbin M (2023) A sui generis whipping-instability-based self-sequencing multi-monodisperse 2D spray from an anisotropic microfluidic liquid jet device. Cell Rep Phys Sci 4(101):221
46. Theofilis V (2011) Global linear instability. Annu Rev Fluid Mech 43:319–352
47. Cruz-Mazo F, Herrada MA, Gañán-Calvo AM, Montanero JM (2017) Global stability of axisymmetric flow focusing. J Fluid Mech 832:329–344
48. Chomaz J (2005) Global instabilities in spatially developing flows. Annu Rev Fluid Mech 37:357–392
49. Schmid PJ (2007) Nonmodal stability theory. Annu Rev Fluid Mech 39:129–162
50. de Luca L, Costa M, Caramiello C (2002) Energy growth of initial perturbations in two-dimensional gravitational jets. Phys Fluids 14:289–299
51. Cruz-Mazo F, Montanero JM, Gañán-Calvo AM (2016) Monosized dripping mode of axisymmetric flow focusing. Phys Rev E 94(053):122

Chapter 11
Gaseous Flow Focusing II

Abstract Serial femtosecond crystallography (SFX) is the most remarkable application of gaseous flow focusing. In this application, the gaseous stream driving the jet is normally discharged into a vacuum chamber, which makes the gas current reach the sound velocity in the nozzle orifice. We devote the first part of this chapter to analyzing this transonic version of flow focusing.

Gaseous flow focusing can be employed to produce fibers and films when the liquid phase is a viscoelastic fluid. The second part of this chapter describes these interesting applications of flow focusing.

The large strain rates produced in the tip of the tapering meniscus of flow focusing can trigger the coil-stretch transition of the polymers dissolved in the liquid, which leads to the building of large axial viscoelastic stresses. This can fundamentally change the behavior of weakly viscoelastic flow focusing, increasing the stability of both the tapering meniscus and the emitted jet. This chapter closes by analyzing this novel and interesting phenomenon. Attention is paid to this viscoelastic transition and the superstability of the jets emitted when that transition occurs.

Keywords Transonic flow focusing · Strongly viscoelastic flow focusing · Weakly viscoelastic flow focusing

11.1 Transonic Flow Focusing

11.1.1 Serial Femtosecond Crystallography

In SFX, single flash diffraction patterns of many individual protein crystals are recorded to analyze complex biochemical species [1] (Fig. 11.1). X-ray free-electron lasers produce flashes sufficiently bright to generate resolvable diffraction patterns from sub-micron crystals. Those flashes destroy the crystals. This implies that the data must be collected from a row of crystals streaming across the X-ray beam.

J. M. Montanero, *Tip Streaming of Simple and Complex Fluids*, Fluid Mechanics and Its Applications 137, https://doi.org/10.1007/978-3-031-52768-5_11

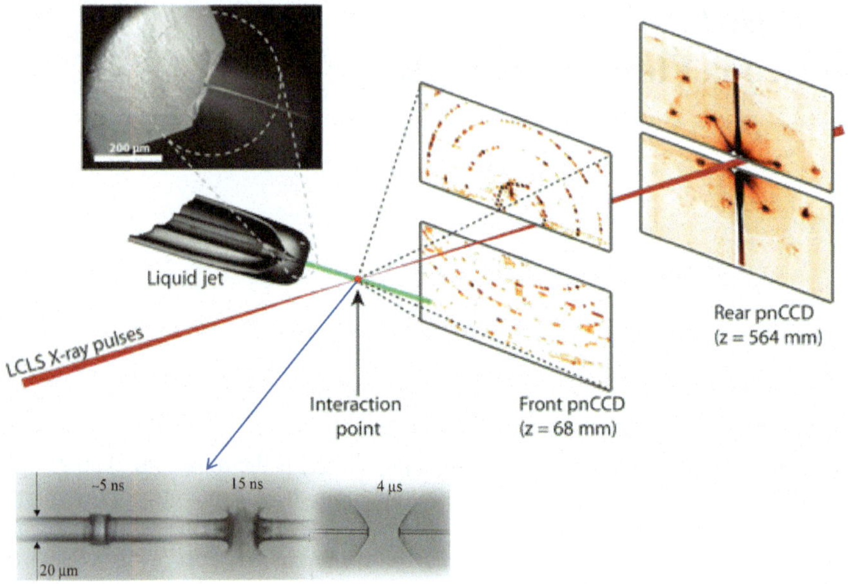

Fig. 11.1 The serial femtosecond crystallography technique. A thin liquid jet places the sample into the beam focus (reprinted with permission from Ref. [1]). The interaction with the X-ray pulse produces an explosion (reprinted with permission from Ref. [2]. Copyright (2019) by the American Physical Society)

In most SFX experiments, the sample is placed into the beam focus using a thin liquid jet. SFX demands severe conditions for the emission of the jet. The jet must be, at most, very few microns in diameter to reduce the background diffraction signal. The jet must be perfectly steady to interact consistently with the X-ray beam. The jet portion interacting with the X-ray pulse must be located sufficiently far away from the discharge orifice to avoid a rapid collection of sputtered material from the explosion (Fig. 11.1). These last two conditions imply that the so-called jet intact region must be much larger than the orifice. Finally, the jet must be sufficiently fast for the exposed sample to exit the interaction region before the next flash strikes the jet [3, 4]. To summarize, liquid jets as thin, long, steady, and fast as possible are required for the proper functioning of SFX.

Gaseous flow focusing [5–7] has been the preferred method to produce jets fulfilling the above stringent constraints. The search for the best ejection conditions has led to the use of transonic streams of helium to focus the liquid jet. The nozzle geometry has been optimized to improve the quality of the jet [4, 8, 9].

In transonic flow focusing, the capillary feeding the liquid is introduced concentrically inside a converging nozzle which discharges the gas stream into a vacuum. The gaseous flow exceeds the sound speed beyond the orifice due to the sharp gas expansion. This configuration does not suffer from whipping instability owing to the low density of the gas stream beyond the discharge orifice. Due to the high speed

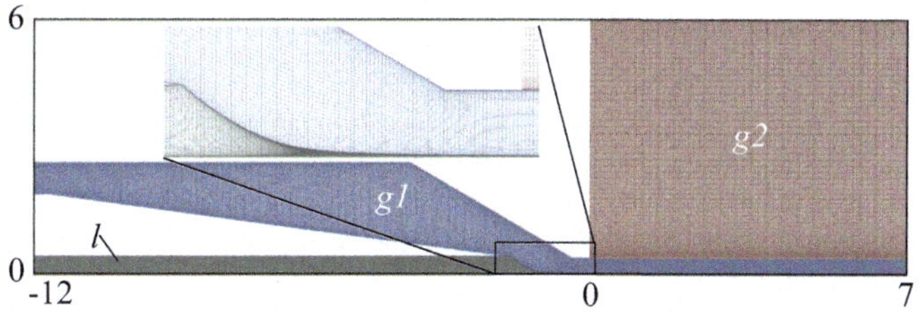

Fig. 11.2 Geometry of transonic flow focusing [12]. Detail of the mesh used in the numerical simulations

reached by the liquid jet, the capillary waves growing on the free surface are convected downstream. Consequently, a straight, perfectly steady, long, thin, and very fast liquid jet is emitted from the nozzle orifice.

Different geometrical configurations have been studied to minimize/maximize the jet's diameter/speed. For instance, jets with diameters down to 325 nm and speeds up to 170 m/s have been produced by injecting the liquid with a beleved nozzle [10] using the method proposed by Acero et al. [11] for incompressible flow focusing. Despite the advances mentioned above, optimizing liquid ejection remains an open problem. One of the primary goals is to reduce the jet diameter to the minimum possible value to suppress the background diffraction signal.

11.1.2 Formulation of the Problem

We consider a flow focusing realization in which a gas stream (the outer phase) is injected from a chamber pressurized at a pressure p_0 and temperature T_0. The gas enters into a converging nozzle with an exit orifice of diameter D_o and discharges into a chamber connected to that orifice (Fig. 11.2). The pressure of the gas stream at the discharge chamber exit is p_e.

The gas properties entering the problem are the viscosity μ_o, heat coefficients c_{vo} and c_{po}, and thermal conductivity κ_{To}. The dilatational coefficient of viscosity $\lambda_{\mu o}$ is taken as $\lambda_{\mu o} = -2/3\,\mu_o$ (the dilatational viscosity vanishes) [13]. To simplify the analysis, we neglect the influence of both the pressure and temperature on the gas properties. This approximation is acceptable for our purposes because (i) the gas viscosity and thermal conductivity are hardly affected by the pressure [14], and (ii) the temperature of the gas layer in contact with the liquid is practically the liquid temperature T_0, except at the nozzle exit where variations of around 10% are found [15].

The liquid stream (the inner phase) is injected at the flow rate Q_i and temperature T_0. The liquid properties involved in the problem are the density ρ_i, viscosity μ_i, heat coefficient c_{vi}, and thermal conductivity κ_{Ti}. Finally, the surface tension is σ.

For a fixed geometry, the transonic flow focusing is governed by the set of dimensional parameters

$$\{D_o;\, \mu_o, c_{vo}, c_{po}, \kappa_{To};\, \rho_0, p_0, p_e;\, \rho_i, \mu_i, c_{vi}, \kappa_{Ti}, \sigma;\, Q_i\}, \tag{11.1}$$

where the upstream stagnation temperature T_0 has been replaced by the corresponding density ρ_0 calculated from the ideal gas law. The dimensionless jet diameter can be expressed as

$$d_j^* \equiv \frac{d_j}{d_j^{\mathrm{FF}}} = f(p_r, \rho, Q), \tag{11.2}$$

where d_j^{FF} is the jet diameter given by Eq. (10.1) for incompresible flow focusing, $p_r = p_0/p_e$, $\rho = \rho_0/\rho_i$, $Q = Q_i/Q_D$, and $Q_D = D_o\mu_i/\rho_i$ is the characteristic liquid flow rate defined in Chap. 10. There are eight additional dimensionless parameters that are fixed in most experimental and numerical analyses.

When the ratio p_0/p^* between the stagnation pressure p_0 and the hydrostatic pressure p^* at the nozzle orifice exceeds the critical value $p_0/p^* = [(\gamma+1)/2]^{\gamma/(\gamma-1)}$ (γ is the heat capacity ratio), the quasi-1D isentropic gas flow becomes chocked [13]; i.e., the gaseous flow in the nozzle cannot be accelerated by increasing p_0 (and keeping T_0 constant). The increase of the stagnation pressure increases the mass flow rate by raising the gas density, but it does not affect the velocity field inside the nozzle. In addition, the temperature field and, therefore, the gas viscosity do not change when the stagnation pressure is increased. This implies that the viscous stress exerted by the outer stream on the liquid current inside the nozzle remains practically constant when p_0 is increased under choking flow conditions. This occurs in most experiments, even for the small values of the upstream stagnation pressure.

As mentioned above, the gas flow is controlled by the parameters $\{T_0, p_0, p_e\}$. As a first approximation, one can state that these parameters determine the force driving the liquid ejection, while the flow rate Q_i controls the response of the liquid to that force. In a second approximation, one must also consider the influence of Q_i on the gaseous flow. The liquid current moves much slower than the gas stream, acting as a solid boundary for the gaseous flow. Since the liquid flow rate Q_i determines the free surface contour, it controls the shape of the "nozzle" crossed by the gas stream, thus affecting all the properties of that stream.

11.2 Global Stability

As in the incompressible regime, the liquid flow rate is the major factor determining the jet diameter. The specific energy (energy per unit volume) transferred to the liquid does not considerably depend on the liquid flow rate. Therefore, the jet speed remains practically constant, and the diameter decreases as the flow rate decreases. However, the flow rate cannot be reduced without limit. The microjetting regime

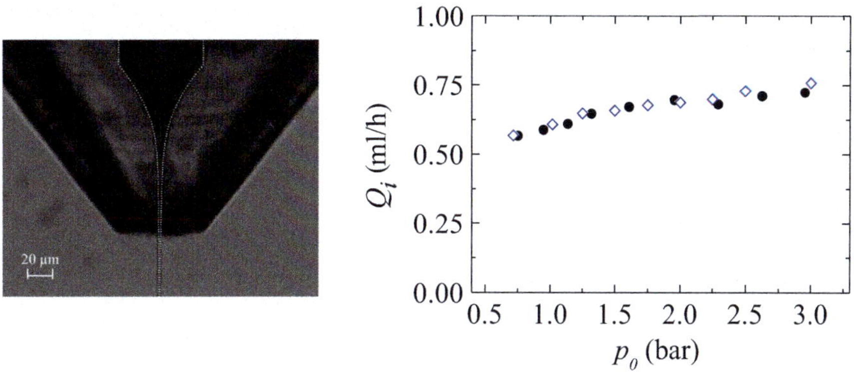

Fig. 11.3 (Left) Experimental image and free surface position calculated numerically (dotted line) for $p_0 = 1.75$ bar and $Q_i = 0.68$ ml/h [12]. (Right) Liquid flow rate Q_i at the stability limit as a function of the upstream stagnation pressure p_0 obtained experimentally (open symbols) and numerically (solid symbols) [12]

produced by flow focusing becomes unstable for flow rates below a minimum value, which depends on the rest of the governing parameters. Therefore, the search for the optimum liquid ejection conditions necessarily involves two aspects of the problem: (i) the maximization of the energy transmitted by the gas stream to the liquid jet and (ii) the stabilization of the flow producing the liquid ejection.

As explained throughout this text, the global linear stability analysis is an adequate tool to determine the stability limit of the microjetting regime. Again, one calculates the axisymmetric linear global modes assuming the temporal dependence

$$U(r, z; t) = U_0(r, z) + \delta U(r, z)\, e^{-i\omega t} + \text{c.c.} \tag{11.3}$$

$$F(z; t) = F_0(z) + \delta F(z)\, e^{-i\omega t} + \text{c.c.} \tag{11.4}$$

for any hydrodynamic field U and the free surface location F. The base flow is (asymptotically) unstable when the largest growth rate ω_i becomes positive.

Rubio et al. [12] examined both numerically and experimentally the transonic flow focusing when a helium stream focuses a water current. Figure 11.3-left shows the excellent agreement between the free surface shape in the experiment and that calculated from the simulation. The parameter conditions correspond to a quasi-marginally stable realization [12].

The minimum values of liquid flow rate Q_i obtained in the experiments are practically the same as those calculated from the global stability analysis (Fig. 11.3-right). Besides, the critical flow rate does not considerably depend on the upstream stagnation pressure for the range of values analyzed in Fig. 11.3-right. As occurs in incompressible gaseous flow focusing, this flow rate is commensurate with that dragged by the gaseous viscous force. As mentioned in Sect. 11.1.2, this force and, therefore,

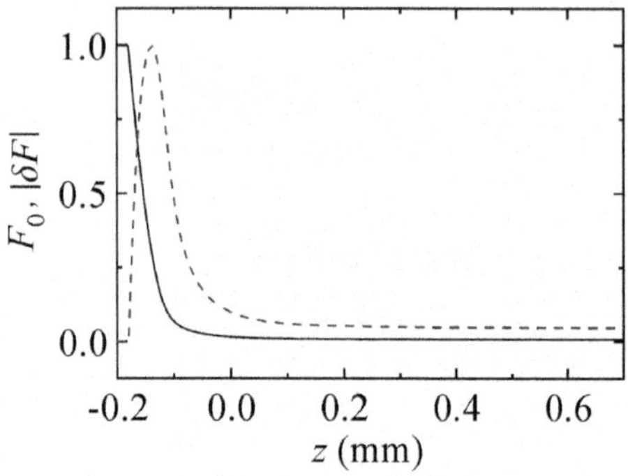

Fig. 11.4 Steady free surface position $F_0(z)$ (black solid line) and magnitude of the free surface perturbation, $|\delta F(z)|$ (dashed blue line), for $p_0 = 1.6$ bar and $Q_i = 0.65$ ml/h [12]. The curves have been normalized by dividing by the corresponding maximum value

the dragged liquid flow rate are hardly influenced by the stagnation pressure. This explains why the critical flow rate is not considerably affected by p_0.

One can gain insight into the instability mechanism by examining the critical eigenmode. The magnitude of the interface perturbation $\delta F(z)$ peaks at the meniscus tip (Fig. 11.4), which suggests that flow becomes unstable due to the destabilization of the liquid meniscus (not the emitted jet). Given the low viscosity of the focused liquid, the meniscus destabilizes because the injected flow rate becomes commensurate with the dragged one, as occurs in the incompressible regime (see Chap. 10).

11.2.1 Jet Diameter

In the numerical simulations of Rubio et al. [12], the values of $\{p_r, \rho, Q\}$ were changed while the rest of the parameters were fixed. For incompressible flow focusing, the dependence of d_j on the gas pressure is approximately the same as that given by Eq. (10.1) for $d_j^{\rm FF}$. Therefore, and as a first approximation, the ratio $d_j^* = d_j/d_j^{\rm FF}$ is independent of p_r. According to Eq. (11.2), the jet diameter approximately obeys the expression $d_j^* = f(\rho, Q)$.

Rubio et al. [12] looked for a scaling law of the form

$$d_j^* = C Q^\alpha \rho^\beta. \tag{11.5}$$

Using the optimization method described in Chap. 4, the best collapse of the numerical data around (11.5) was obtained for $C = 0.433$, $\alpha = 0.308$, and $\beta = -0.06$ (Fig. 11.5). As expected, $\alpha \gg |\beta|$ shows the dominant role of Q. The scaling law remarkably fits the numerical data in the range of values $0.549 \lesssim d_j \lesssim 10.9$ μm calculated in the simulations.

The scaling law (11.5) with $\alpha = 0.308$ and $\beta = -0.06$ implies that the jet diameter also scales as

$$d_j \sim Q_i^{0.808} p_0^{-0.31} \tag{11.6}$$

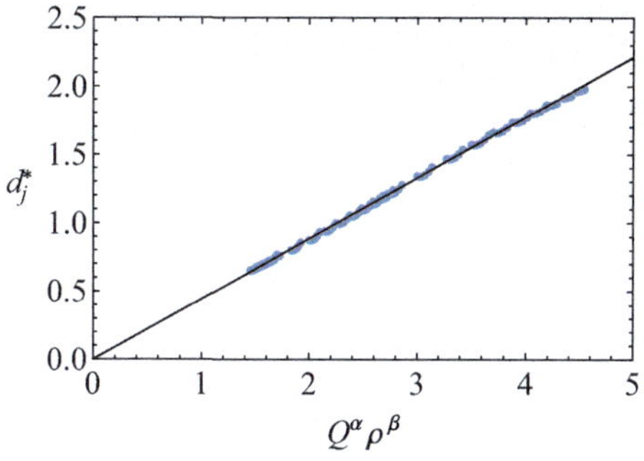

Fig. 11.5 d_j^* versus $Q^\alpha \rho^\beta$ for $\alpha = 0.308$ and $\beta = -0.06$. The solid line is the function $y = 0.443x$

when the ejector geometry, properties of the fluids, and stagnation temperature are fixed [12]. The minimum liquid flow rate decreases with the stagnation pressure (Fig. 11.3-right) approximately as $Q_i \sim p_0^{0.18}$. Taking into account the scaling law (11.6), $d_j \sim p_0^{-0.17}$ at the stability limit. This implies that thinner jets can be produced as p_0 increases even though the minimum flow rate slightly increases with p_0.

Direct numerical simulations of transonic flow focusing [16] have been conducted from the time integration of the Navier-Stokes equations in the laminar regime with the VoF method [17]. These simulations are a valuable tool for analyzing different aspects of the problem. Among the geometrical parameters analyzed in the simulations, the nozzle outlet diameter most significantly influences the jet diameter. In contrast, the feeding capillary angle and the capillary-to-orifice distance have a very small effect on the liquid flow [15].

The influence of the capillary-to-orifice distance on the minimum liquid flow rate is approximately given by the scaling law empirically derived by for the incompressible regime (see Chap. 10) [18]. Sarler et al. [19] have recently shown that a change in the angle of the nozzle outlet orifice significantly affects jet emission. The influence of the gas [20] and liquid [21] properties on the jet's characteristics has also been examined from direct numerical simulations. Zahoor et al. [16] found a qualitative agreement between the scaling law for the incompressible regime and their numerical results.

11.3 Strongly Viscoelastic Flow Focusing

The response of a viscoelastic solution to a straining flow is generally characterized by the so-called extensional relaxation time λ_e, measured in filament thinning rheometers such as CaBER [22] and FiSER [23] (see Chap. 4). This parameter, linked to the polymer relaxation time, increases not only with the polymer molecular weight but also with the concentration of a dilute solution due to the hydrodynamic (flow-mediated) interaction among polymers.

Fig. 11.6 Strongly-viscoelastic flow focusing. The image shows a jet of Poly(acrylic acid) (PAA) (Mw= 18×10^6 g/mol) dissolved in water at the concentration $c = 1000$ ppm and formed between the feeding capillary and the discharge orifice. The extensional relaxation time is $\lambda_e = 250$ ms. The inner diameter of the feeding capillary is 200 μm (reprinted from Ref. [24] with permission of Elsevier)

Two categories of viscoelastic flow focusing can be distinguished according to the extensional relaxation time: (i) strongly-viscoelastic flow focusing in which this parameter takes values larger than, say, 100 μs, and (ii) weakly-viscoelastic (quasi-Newtonian) flow focusing characterized by relaxation times of the order of tens of microseconds.

In the strongly-viscoelastic version of flow focusing [24], a cylindrical feeding capillary, long as compared to its diameter, is located in front of the discharge orifice, whose diameter is of the same order of magnitude as that of the capillary. The viscoelastic liquid is injected at a constant flow rate Q_i through the feeding capillary. A gas stream crosses the discharge orifice driven by a prescribed pressure drop Δp. If the distance H between the feeding capillary and the discharge orifice lies within the appropriate range, then a steady liquid jet is attached to the capillary end's edge (Fig. 11.6). The gas stream sucks and drags the liquid, progressively reducing the jet's diameter to values much smaller than that of the orifice. This jetting realization cannot be regarded as true tip streaming because the decrease in the jet's diameter does not take place at the tip of a tapering meniscus.

Because of the rheological nature of the liquid, this focusing phenomenon considerably differs from the Newtonian one in several regards [24]. The aerodynamic focusing effect is confined within a region much smaller than the capillary-to-orifice distance. Thus, the liquid jet flows freely in front of the discharge orifice, with no significant interaction with the ambient. The axial stress caused by the focusing effect spreads upstream due to the polymer stretching, which pulls the liquid from the feeding capillary. The large value of the extensional relaxation time delays the macromolecule recoiling. The viscous shear stress exerted by the gas stream on the emitted jet beyond the discharge orifice increases this effect. The unrelaxed axial stress results in a large extensional viscosity, which inhibits the breakup process. Thus, very long jets with virtually zero Weber numbers are produced.

For given values of the polymer concentration, the injected flow rate, and the applied pressure drop, there is an interval of the capillary-to-orifice distance for which the jetting regime is reached (Fig. 11.7). Outside that interval, the jet either suffers from pull-out instability or breaks up before reaching the discharge orifice. The pull-out instability is caused by an excess of polymer stress occurring for small values of H. This instability makes the triple contact line detach from the capillary end and climb over the capillary inner wall [24].

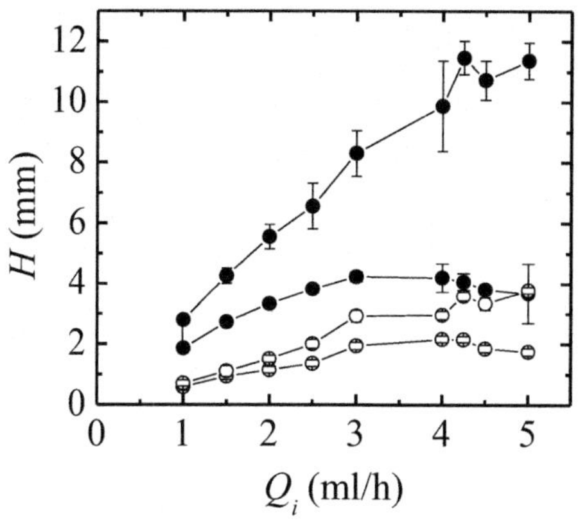

Fig. 11.7 The minimum and maximum values, $H_{\min}$ and $H_{\max}$, of the capillary-to-orifice distance H at which the triple contact line anchors to the capillary edge as a function of the flow rate Q_i. The open circles, open triangles, and solid circles correspond to PAA aqueous solutions with $c = 250$, 500, and 1000 ppm, respectively. The experiments were conducted with $D_o = 200$ μm and $\Delta p = 250$ mbar (reprinted from Ref. [24] with permission of Elsevier)

Significant free surface oscillations can be observed in most of the jetting realizations. This oscillatory behavior is caused by a transient die swell effect, which continuously appears right at the capillary exit. In practical applications, the capillary-to-orifice distance H is set below its minimum value, and the pull-out instability occurs [25, 26].

It should be pointed out that the flow focusing effect produced by the air current substantially differs from that caused by outer gaseous streams in Gas-Jet/Electrospinning [27, 28] and Gas-Assisted Melt Electrospinning [29]. In gaseous flow focusing, the high-speed air stream is the only source of energy powering the liquid ejection. The critical element of this technique is that the gas is injected in front of the discharge orifice. In this way, the liquid uses not only the viscous drag force exerted by the gas stream but also the pressure drop that drives that stream in the ejection region. For this reason, the applied pressure drop is much smaller than that necessary to emit the jet in Solution Blow Spinning [30] and Gas Jet Process [31].

11.3.1 Applications

The flow focusing principle has been applied to form viscoelastic threads by purely hydrodynamic means [24, 32]. These threads can be used for smooth printing (Fig. 11.8) and bioplotting [25]. Threads of low-molecular-weight Polyvinylpirrolidone (PVP) and poly(3,4-ethylenedioxythiophene):poly(styrene sulfonate) (PEDOT:PSS) have been produced with gaseous flow focusing and gently deposited on solid substrates.

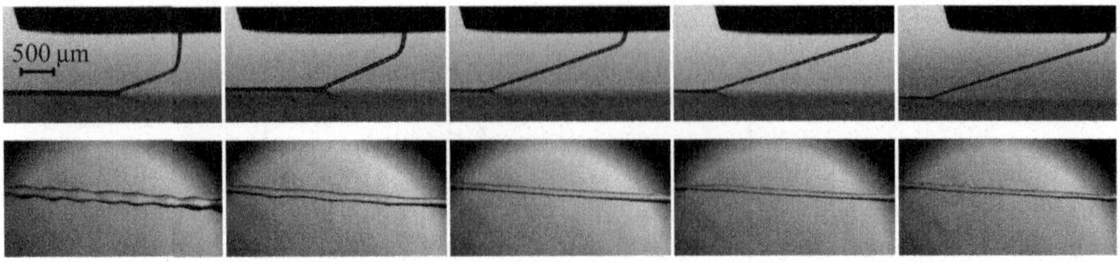

Fig. 11.8 Lateral (upper images) and bottom (lower images) views of the deposition process for different values of the substrate velocity v_s. From left to right, the images correspond to v_s =0.22, 0.44, 0.65, 0.85, and 0.95 m/s. The results were obtained for $H = 2.5$ mm, $Q_i = 5$ ml/h, $\Delta p = 35$ mbar, and the ejector-to-surface distance $Z = 1$ mm. The liquid was a PAA aqueous solution 1000 ppm in concentration (reprinted from Ref. [25] with permission of Elsevier)

These results make viscoelastic flow focusing a promising technique in several fields. PVP is a synthetic polymer widely used in pharmacy, cosmetics, biology, or medicine [33]. In particular, PVP has been used to produce scaffolds for tissue engineering due to its biocompatibility and low toxicity [34], to fabricate water-soluble filters owing to its high solubility in water and organic solvents [35], or to form hydrogels by modifying its molecular structure [34]. PVP has also been utilized as a blend polymer with different substances for applications ranging from drug delivery to functional materials [33, 36]. On the other hand, PEDOT:PSS is probably the best polymer for applications in printed and flexible electronics due to its high electrical conductivity, transparency, physical and chemical stability in air, and capability to form films.

The above application of the flow focusing principle constitutes the hydrodynamic version of near-field electrospinning and can also be applied to other fields. For instance, it is well-known that the continuous jet impingement of a liquid onto a surface can be used as a high-capacity cooling method. The addition of small amounts (on the order of 100 ppm) of polymers to water-based jets leads to a noticeable increase in the cooling rate compared to that of pure water [37]. The water-based impingement suspension of polymers can also reduce the friction in rails. In this case, the liquid is usually applied onto the rail top surface with sprayers mounted on maintenance vehicles [38].

The spinning of nanofibers made of low-molecular-weight polymers is a challenging task because of the small viscoelasticity of the polymeric solution [33]. Fibers of low-molecular-weight Polyvinylpirrolidone (PVP) and other polymers have been produced with gaseous flow focusing as well (Fig. 11.9) [26, 39]. The fiber diameter ranges from a few microns down to hundreds of nanometers.

11.4 Weakly Viscoelastic Transonic Flow Focusing

Transonic flow focusing exhibits significant differences with respect to the original incompressible version of this technique [5]. In almost all the practical realizations

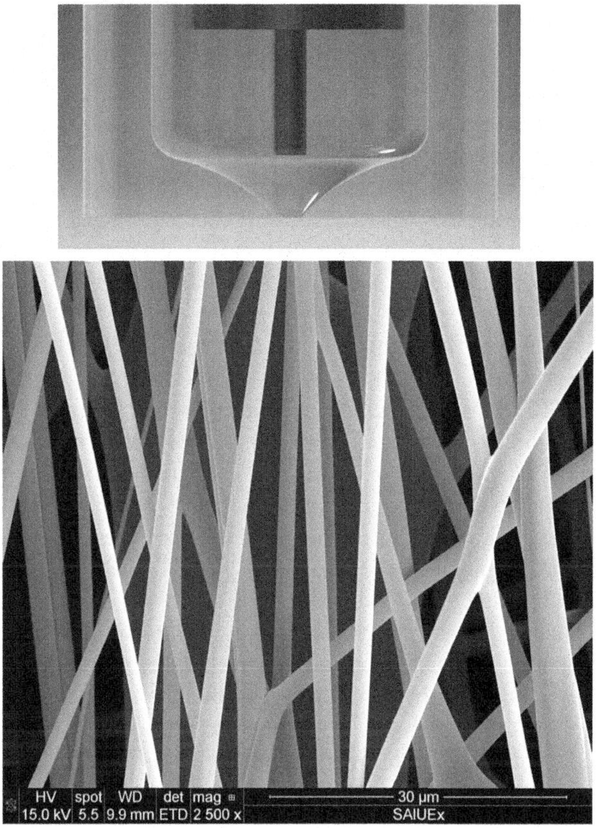

Fig. 11.9 SEM images of fibers produced with gaseous flow focusing for a flow rate $Q_i = 5$ ml/h and an applied pressure drop $\Delta p = 115$ mbar. The ejector is shown in the upper image. The liquid was a solution of 10 wt % of PVP in ethanol. The average diameter is 1.87 μm, and the standard deviation is 0.7 μm (reprinted from Ref. [26] with permission of Elsevier)

of transonic flow focusing, the microjetting mode instability originates in the tapering meniscus. The convective-to-absolute instability (microjetting-to-microdripping) transition [40] is not observed due to the large values of the jet's Weber number [12]. The whipping (bending) instability [41] is suppressed because the jet is discharged into a low-pressure chamber, which considerably reduces the destabilizing hydrostatic pressure force exerted by the coflowing current. We devote this section to analyze the behavior of this configuration when the focused fluid is a weakly-viscoelastic (quasi-Newtonian) liquid.

11.4.1 Viscoelastic Transition

In ideal dilute polymer solutions, electrically neutral chains adopt a random coil configuration at equilibrium and stretch when the polymer solution is subjected to a straining flow. If the straining flow is not sufficiently intense, stretched polymers relax to the coil configuration. The relaxation from the stretched to coil conformations consists of multiple processes. Suppose the time characterizing the slowest process is much larger than the rest of the spectrum. In that case, the polymer solution elasticity can approximately be quantified by a single polymer characteristic time λ_p related to that process.

It has been recognized for a long time that sufficiently high-velocity gradients can cause the transition from a coil to a nearly complete stretched conformation of the polymer chains [42]. Several flows can produce these gradients. A typical example is the uniaxial (simple) extensional flow taking place during the capillary breakup of a liquid bridge [43] (see Chap. 4). According to the Oldroyd-B model [44], the coil-stretch transition occurs in a uniaxial extensional flow if the extensional rate $\dot{\varepsilon}$ exceeds the threshold $\dot{\varepsilon} = 1/(2\lambda_e)$, where λ_e is the polymer stress relaxation time [45, 46]. In this case, the axial polymer stress grows exponentially over time. The polymer stress relaxation time λ_e, defined in the context of the continuum theory, is related to the microscopic polymer relaxation time λ_p mentioned above.

In the microjetting mode of transonic flow focusing, the outer gas stream causes extremely high-velocity gradients in the tip of the tapering meniscus. Rubio et al. [47] have found that these velocity gradients can induce the coil-stretch transition of polymer molecules even with low molecular weights (low relaxation times λ_p). In that case, the polymer stress built up in the meniscus tip leads to a substantial stabilization of the microjetting regime.

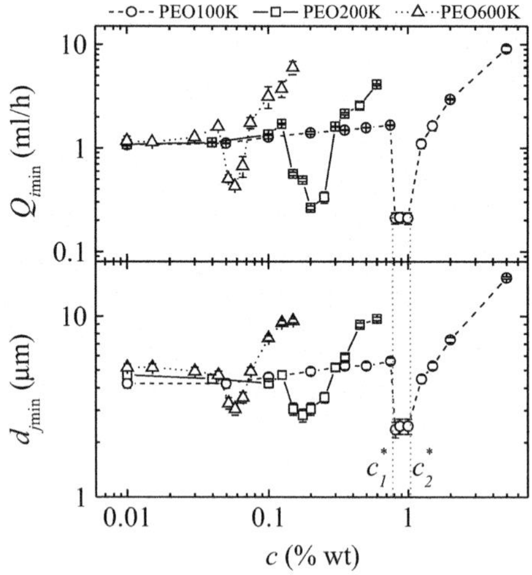

Fig. 11.10 Minimum flow rate $Q_{i\min}$ leading to the microjetting regime and the corresponding jet diameter $d_{j\min}$ for aqueous solutions of PEO with different molecular weights (reprinted with permission from Ref. [47]. Copyright (2022) by the American Physical Society)

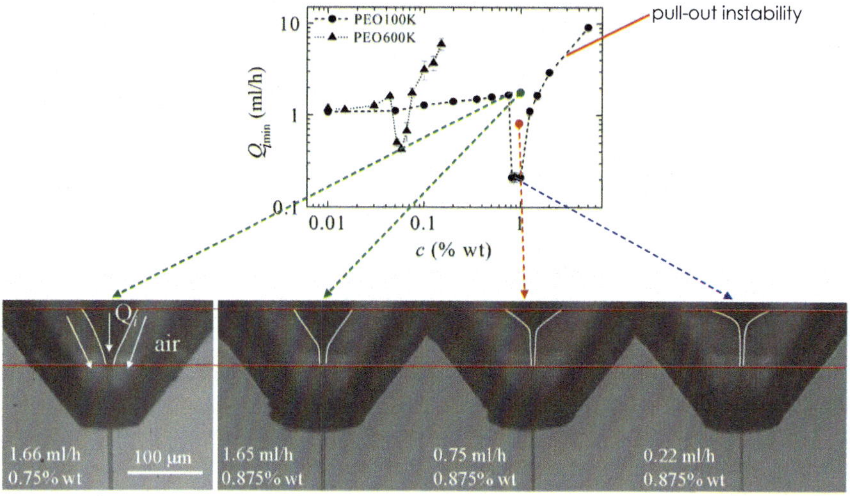

Fig. 11.11 Viscoelastic meniscus shrinkage. The images show the shape of the tapering meniscus for a subcritical concentration of PEO100K (left-hand image) and for a supercritical concentration as the flow rate decreases (right-hand images) (reprinted with permission from Ref. [47]. Copyright (2022) by the American Physical Society)

Figure 11.10 shows the minimum flow rate $Q_{i\min}$ and the corresponding jet diameter $d_{j\min}$ for viscoelastic solutions of polyethylene oxide (PEO) with molecular weights $M_w = 100 \times 10^3, 200\times 10^3$, and 600×10^3 g/mol (the acronym PEOX stands for a PEO aqueous solution with a molecular weight of X g/mol). The minimum flow rate $Q_{i\min}$ plunges at a critical concentration c_1^*, which increases as the molecular weight decreases. The critical flow rate remains almost constant in a small interval of the polymer concentration. Finally, $Q_{i\min}$ sharply increases beyond a second critical concentration c_2^* owing to the pull-out instability mentioned in the previous section.

Figure 11.11 shows the viscoelastic effect on the tapering meniscus when the PEO100K concentration is slightly increased above c_1^*, from 0.75 to 0.875 % wt. The meniscus for $c = 0.875$ % wt considerably shrinks as Q_i decreases. In fact, the meniscus shape for $c = 0.875$ % wt and $Q_i = 0.22$ ml/h has not been observed in any Newtonian flow focusing realization [12, 48], regardless of the liquid viscosity and gas speed.

Fluid particles undergo an intense extensional flow next to the meniscus interface dragged by the outer gas stream. The liquid velocity increases from values ~ 1 mm/s at the capillary exit up to the jet speed $v_j \sim 10$ m/s along distances $s \sim 100$ μm. The resulting strain rate, $\dot{\varepsilon} \simeq v_j/s \sim 10^5$ s^{-1}, is commensurate with the inverse of the extensional relaxation time λ_e for PEO100K (see Chap. 4). Under the above conditions, the coil-stretch transition of the dissolved polymers occurs. The elastic stresses produced by that transition collaborate with those exerted by the outer flow to push the liquid throughout the meniscus-jet region. The resulting extra acceleration explains the meniscus shrinkage. When the flow rate is decreased, the jet speed increases, and so does the strain rate in the meniscus. This effect enhances the building-up of the elastic stress, which explains why the meniscus shrinks as Q_i decreases.

11.4.2 Scaling of the Viscoelastic Transition

As mentioned above, the viscoelastic transition takes place when the strain rate in the tapering meniscus exceeds the critical value leading to the coil-stretch transition. For this reason, we define the Weissenberg number $\mathrm{Wi} = \lambda_e \langle \dot{\varepsilon} \rangle$, where $\langle \dot{\varepsilon} \rangle = v_\ell/\ell_m$ is the mean strain rate along the meniscus axis. Here, v_ℓ is the liquid velocity at the end of the meniscus and ℓ_m the meniscus length. The viscoelastic transition is expected to occur at a fixed value of Wi.

For $c < c_1^*$, the tapering meniscus can be regarded as a Newtonian low-viscosity meniscus. In this case, the minimum flow rate scales as $Q_D = D_o\mu_i/\rho_i$ (see Chap. 10) [49, 50]. For $c > c_1^*$, the stretching of the polymer chains drastically increases the solution extensional viscosity μ_e. The minimum flow rate is expected to scale as the viscous-capillary flow rate $Q_v^* = (2R_i)^2\sigma/\mu_e$ (see Chap. 10) [48]. Following the work of Rubio et al. [47], we use the capillary diameter $2R_i$ as the characteristic length.

Because of the extensional character of the flow in the meniscus, the characteristic flow rate Q_v^* must involve the extensional viscosity μ_e instead of the shear viscosity μ_i considered in the Newtonian (isotropic) case. When the coil-stretch transition takes place, the scale of the extensional viscosity can be calculated as $\mu_e = 3\lambda_e\sigma/d_\ell$ [45], where d_ℓ is the diameter at the meniscus end. The above scaling leads to the following definition of the dimensionless minimum flow rate:

$$\Phi = \begin{cases} \mathrm{Re} = Q_{i\min}/Q_D & \text{for } c < c_1^* \\ \mathrm{Ca} = Q_{i\min}/Q_v^* & \text{for } c > c_1^* \end{cases} . \tag{11.7}$$

The dimensionless number Φ reduces to the Reynolds number based on the shear viscosity for $c < c_1^*$ and the capillary number based on the extensional viscosity μ_e beyond the viscoelastic transition. Figure 11.12 shows how the curves $\Phi(\mathrm{Wi})$ for two polymers with considerably different molecular weights practically overlap. The viscoelastic transition takes place around Wi = 1/2, the critical value for the coil-stretch transition in a uniaxial extensional flow [45, 51].

11.4.3 The Jet Stability

The solution of the Oldroyd-B model [52] shows that a relaxed (i.e., in the absence of polymeric axial stress) viscoelastic jet is more unstable than its Newtonian counterpart (i.e., that with the same Ohnesorge number) due to the apparent shear thinning caused by viscoelasticity [53, 54]. The capillary instability in a viscoelastic jet is followed by a nonlinear phase in which a thin cylindrical thread connects two consecutive drops (beads), giving rise to the so-called Beads-on-a-String (BOAS) morphology [53, 55–57]. The surface tension drives the thread thinning against the growing axial stress caused by the elasticity of the dissolved polymer molecules. In this elasto-capillary regime, the thread diameter decreases following an exponential function, which delays the jet breakup a few extensional relaxation times [23, 58].

Experiments show that the viscoelastic transition in transonic flow focusing produces jets that do not exhibit the expected behavior described above. They are much more stable jets than their Newtonian counterparts. The capillary instability is delayed, and the jet breakup occurs at distances from the ejector of the order of tens of thousands times the jet diameter (Fig. 11.13) [59].

Axial viscoelastic stress is built up by the intense converging extensional flow in the ejection point of the tapering meniscus. It is well-known that this stress drastically increases the extensional viscosity, suppressing the capillary instability [60, 61]. However, the viscoelastic stress should vanish at distances from the ejector much

Fig. 11.12 Φ(Wi) for PEO100K and PEO600K (reprinted with permission from Ref. [47]. Copyright (2022) by the American Physical Society)

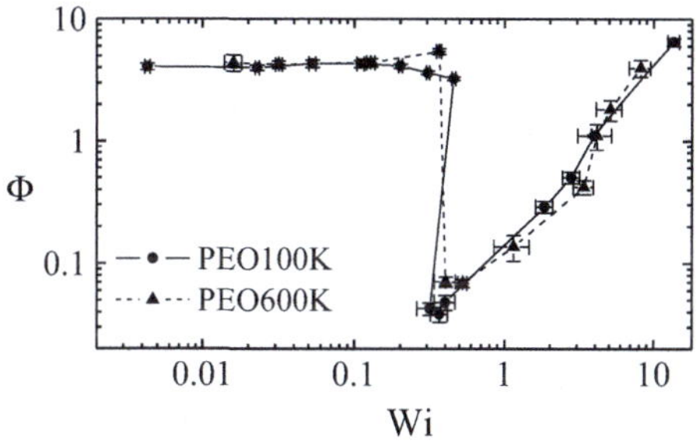

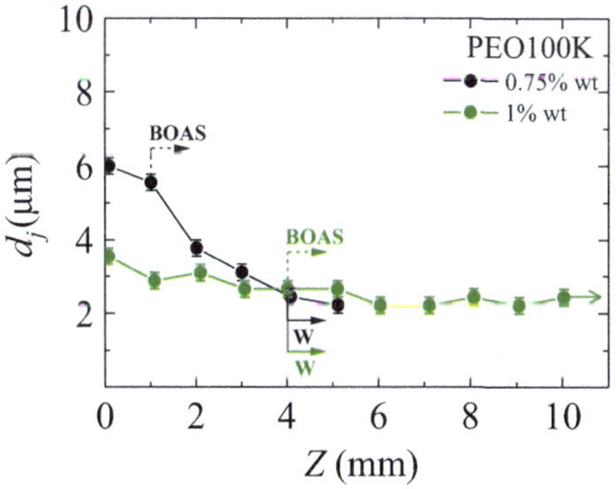

Fig. 11.13 Jet diameter d_j versus the distance Z from the ejector exit for the subcritical ($c = 0.75\%$ wt, $Q = 1.6$ ml/h) and supercritical ($c = 1\%$ wt, $Q_i = 0.4$ ml/h) PEO100K jets. The vertical solid and dashed lines indicate the beginning of whipping (W) and BOAS instability, respectively. The chevron-right green arrow indicates that the supercritical jet broke up at distances from the ejector larger than 10 mm (reprinted from Ref. [59] with permission of AIP Publishing)

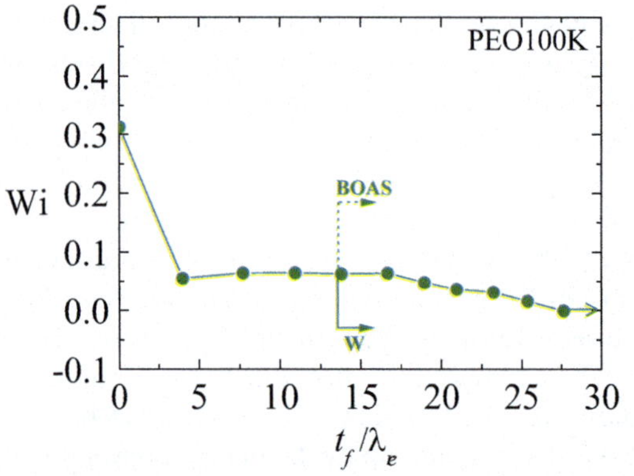

Fig. 11.14 Weissenberg number Wi versus the jet flight time t_f/λ_e for the supercritical jet ($c = 1\%$ wt, $Q_i = 0.4$ ml/h). The vertical solid and dashed lines indicate the beginning of whipping (W) and BOAS instabilities, respectively. The chevron-right green arrow indicates that the supercritical jet of PEO100K broke up at dimensionless times larger than 30 (reprinted from Ref. [59] with permission of AIP Publishing)

smaller than the jet length, as explained below. Therefore, the jet super-stability cannot be simply explained in terms of the unrelaxed viscoelastic stress.

The combination of the upper-convected Maxwell model and the 1D approximation [62] leads to the following equation for the axial deviatoric stress τ_{zz} in an aerodynamically-driven viscoelastic jet [63]:

$$\tau_{zz} = \tau_{zz}^0 \exp\left[-\left(1 - 2\text{Wi}_j\right)(t^* - t_0^*)\right], \tag{11.8}$$

where $t^* = t/\lambda_e$, $\text{Wi}_j = \lambda_e \, dv_j/dz$ is the jet Weissenberg number, and we have assumed that $dv_j/dz \simeq$ const.. The jet Weissenberg number takes values much smaller than 0.5 (Fig. 11.14). Therefore, and according to Eq. (11.8), the viscoelastic stress should relax exponentially at times of the order of λ_e. However, the jet survives over flight times t_f much longer than that characteristic time (Fig. 11.14).

The prolonged survival of the supercritical jet cannot be attributed to the breakup delay caused by the elasto-capillary regime, which occurs on times of the order of λ_e. It cannot be caused by solvent evaporation either. The Hertz-Knudsen-Langmuir evaporation speed in a vacuum [64, 65] is given by

$$v_{\text{ev}} = \frac{\alpha_c P_s(T)}{\sqrt{2\pi}\,\rho(\hat{R}_{gv}T)^{1/2}}, \tag{11.9}$$

where α_c is the evaporation coefficient ($\alpha_c \simeq 0.2$ in our case), $P_s(T)$ is the vapor pressure at the temperature T, and $\hat{R}_{gv}$ is the gas constant of the vapor. The non-dimensional number $G = v_{\text{ev}}\pi d_j L_j/(v_j \pi d_j^2/4)$ measures the relative rate of evaporation along a jet of diameter d_j and length L. The velocity of a low-viscosity jet

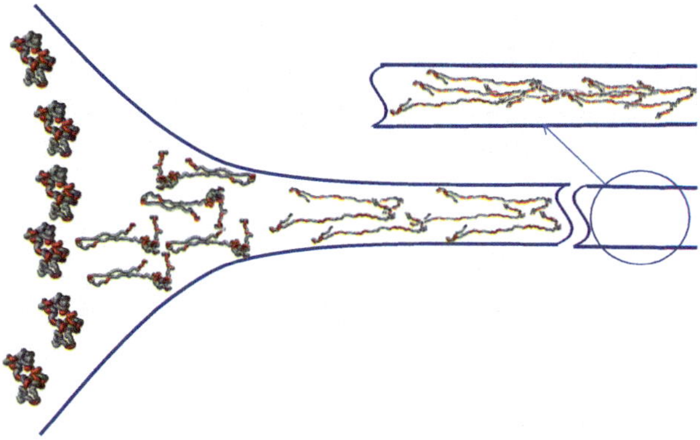

Fig. 11.15 Sketch of the coil-stretch polymer transition and entanglement or self-assembly of the stretched polymers

flow-focused by gas can be estimated as $v_j \simeq (2p_0/\rho)^{1/2}$, which leads to

$$G \simeq \frac{\alpha_c P_s(T)}{2\sqrt{\pi}(\rho p_0 \hat{R}_{gv} T)^{1/2}} \frac{4L_j}{d_j}. \quad (11.10)$$

For the supercritical jet of PEO100K, G is of the order of 10^{-4}. Therefore, evaporation does not significantly increase the polymer concentration.

Rubio et al. [59] have ruled out other possible explanations for the enormous stability exhibited by the supercritical PEO100K jet. They have hypothesized that this phenomenon can be explained as follows. The large acceleration (intense extensional flow) undergone by the liquid in the supercritical meniscus triggers the coil-stretch polymer transition. This transition drastically increases the extensional viscosity at the emission point. The strong converging flow in the meniscus tip causes some entanglement or self-assembly of the stretched polymers (Fig. 11.15). This effect hinders the extensional viscosity relaxation and delays the capillary instability. The viscoelastic stress does not relax in the filament formed between the beads, making the jet breakup at distances larger than 10 mm.

References

1. Chapman HN et al (2011) Femtosecond X-ray protein nanocrystallography. Nature 470:73–79
2. Gañán-Calvo AM (2019) Scaling laws of an exploding liquid column under an intense ultrashort X-ray pulse. Phys Rev Lett 123(064):501
3. Stan CA et al (2016) Liquid explosions induced by X-ray laser pulses. Nat Phys 12:966–971
4. Wiedorn MO et al (2018) Megahertz serial crystallography. Nat Commun 9:4025
5. Gañán-Calvo AM (1998) Generation of steady liquid microthreads and micron-sized monodisperse sprays in gas streams. Phys Rev Lett 80:285–288

6. DePonte DP, Weierstall U, Schmidt K, Warner J, Starodub D, Spence JCH, Doak RB (2008) Gas dynamic virtual nozzle for generation of microscopic droplet streams. J Phys D Appl Phys 41(195):505
7. Gañán-Calvo AM, DePonte DP, Herrada MA, Spence JCH, Weierstall U, Doak RB (2010) Liquid capillary micro/nanojets in free-jet expansion. Small 6:822–824
8. Beyerlein KR, Adriano L, Heymann M, Kirian R, Knoska J, Wilde F, Chapman HN, Bajt S (2015) Ceramic micro-injection molded nozzles for serial femtosecond crystallography sample delivery. Rev Sci Instrum 86(125):104
9. Piotter V, Klein A, Plewa K, Beyerlein KR, Chapman HN, Bajt S (2018) Development of a ceramic injection molding process for liquid jet nozzles to be applied for X-ray free-electron lasers. Microsyst Technol 24:1247–1252
10. Nazari R, Zaare S, Alvarez RC, Karpos K, Engelman T, Madsen C, Nelson G, Spence JCH, Weierstall U, Adrian RJ, Kirian RA (2020) 3D printing of gas-dynamic virtual nozzles and optical characterization of high-speed microjets. Opt Express 28:21,749–21,765
11. Acero AJ, Rebollo-Muñoz N, Montanero JM, Gañán-Calvo AM, Vega EJ (2013) A new flow focusing technique to produce very thin jets. J Micromech Microeng 23(065):009
12. Rubio M, Rubio A, Cabezas MG, Herrada MA, Gañán-Calvo AM, Montanero JM (2021) Transonic flow focusing: stability analysis and jet diameter. Int J Multiphase Flow 142(103):720
13. Shapiro AH (1953) Compressible fluid flow. Wiley, New York
14. Flynn GP, Hanks RV, Lemaire NA, Ross J (1963) Viscosity of nitrogen, helium, neon, and argon from $-78.5°$ to $100°C$ below 200 atmospheres. J Chem Phys 38:154–162
15. Zahoor R, Bajt S, Sarler B (2018) Influence of gas dynamic virtual nozzle geometry on micro-jet characteristics. Int J Multiphase Flow 104:152–165
16. Zahoor R, Belsak G, Bajt S, Sarler B (2018) Simulation of liquid micro-jet in free expanding high-speed co-flowing gas streams. Microfluidics and Nanofluidics 22:87
17. Hirt CW, Nichols BD (1981) Volume of Fluid (VOF) method for the dynamics of free boundaries. J Comput Phys 39:201–225
18. Vega EJ, Montanero JM, Herrada MA, Gañán-Calvo AM (2010) Global and local instability of flow focusing: the influence of the geometry. Phys Fluids 22(064):105
19. Sarler B, Zahoor R, Bajt S (2021) Alternative geometric arrangements of the nozzle outlet orifice for liquid micro-jet focusing in gas dynamic virtual nozzles. Materials 14:1572
20. Zahoor R, Bajt S, Sarler B (2018) Numerical investigation on influence of focusing gas type on liquid micro-jet characteristics. Int J Hydromechatronics 1:222–237
21. Zahoor R, Regvar R, Bajt S, Sarler B (2020) Simulation of liquid micro-jet in free expanding high-speed co-flowing gas streams. Microfluid Nanofluid 20:71–83
22. McKinley GH, Tripathi A (2000) How to extract the Newtonian viscosity from capillary breakup measurements in a filament rheometer. J Rheol 44:653–670
23. Anna SL, McKinley GH (2001) Elasto-capillary thinning and breakup of model elastic liquids. J Rheol 45:115–138
24. Ponce-Torres A, Montanero JM, Vega EJ, Gañán-Calvo AM (2016) The production of viscoelastic capillary jets with gaseous flow focusing. J Non-Newtonian Fluid Mech 229:8–15
25. Ponce-Torres A, Vega EJ, Castrejón-Pita AA, Montanero JM (2017) Smooth printing of viscoelastic microfilms with a flow focusing ejector. J Non-Newtonian Fluid Mech 249:1–7
26. Ponce-Torres A, Ortega E, Rubio M, Rubio A, Vega EJ, Montanero JM (2019) Gaseous flow focusing for spinning micro and nanofibers. Polymer 178(121):623
27. Lin Y, Yao Y, Yang X, Wei N, Li X, Gong P, Li R, Wu D (2007) Preparation of poly(ether sulfone) nanofibers by gas-jet/electrospinning. J Appl Polym Sci 107:909–917
28. Wang B, Yao Y, Peng J, Lin Y, Liu W, Luo Y, Xiang R, Li R, Wu D (2009) Preparation of poly(ester imide) ultrafine fibers by gas-jet/electrospinning. J Appl Polym Sci 114:883–891
29. Zhmayev E, Cho D, Joo YL (2010) Nanofibers from gas-assisted polymer melt electrospinning. Polymer 51:4140–4144
30. Oliveira JE, Moraes EA, Costa RGF, Afonso AS, Mattoso LHC, Orts WJ, Medeiros ES (2011) Nano and submicrometric fibers of poly(d, l-lactide) obtained by solution blow spinning: process and solution variables. J Appl Polym Sci 122:3396–3405

31. Benavides RE, Jana SC, Reneker DH (2012) Nanofibers from scalable gas jet process. ACS Macro Lett 1:1032–1036
32. Hofmann E, Krüger K, Haynl C, Scheibel T, Trebbin M, Förster S (2018) Microfluidic nozzle device for ultrafine fiber solution blow spinning with precise diameter control. Lab Chip 18:2225–2234
33. Cena CR, Silva MJ, Malmonge LF, Malmonge JA (2018) Poly(vinyl pyrrolidone) submicrofibers produced by solution blow spinning. J Polym Res 25:238
34. de Lima GG, Lyons S, Devine DM, Nugent MJD (2018) Hydrogels. Gels horizons: from science to smart materials. Springer, Singapore, pp 219–258 (chap Electrospinning of Hydrogels for Biomedical Applications)
35. Mikheev AY, Kanev IL, Morozova TY, Morozov VN (2013) Water-soluble filters from ultrathin polyvinylpirrolidone nanofibers. J Membr Sci 448:151–159
36. Chuangchote S, Sagawa T, Yoshikawa S (2009) Electrospinning of poly(vinyl pyrrolidone): effects of solvents on electrospinnability for the fabrication of poly(p-phenylene vinylene) and tio2 nanofibers. J Appl Polym Sci 114:2777–2791
37. Sarkar I, Behera DK, Jha JM, Pal SK, Chakraborty S (2016) Effect of polymer additive on the cooling rate of a hot steel plate by using water jet. Exp Therm Fluid Sci 70:105–114
38. Keshavarz B, Green SI, Eadie DT (2012) Elastic liquid jet impaction on a high-speed moving surface. AIChE J 58:3568–3577
39. Vasireddi R, Kruse J, Vakili M, Kulkarni S, Keller TF, Monteiro DCF, Trebbi M (2019) Solution blow spinning of polymer/nanocomposite micro-/nanofibers with tunable diameters and morphologies using a gas dynamic virtual nozzle. Sci Rep 9(14):297
40. Huerre P, Monkewitz PA (1990) Local and global instabilites in spatially developing flows. Annu Rev Fluid Mech 22:473–537
41. Acero AJ, Ferrera C, Montanero JM, Gañán-Calvo AM (2012) Focusing liquid microjets with nozzles. J Micromech Microeng 22(065):011
42. Peterlin A (1966) Hydrodynamics of linear macromolecules. Pure Appl Chem 12:563–586
43. Sousa PC, Vega EJ, Sousa RG, Montanero JM, Alves MA (2017) Measurement of relaxation times in extensional flow of weakly viscoelastic polymer solutions. Rheol Acta 56:11–20
44. Bird RB, C R, Armstrong, Hassager O (1987) Dynamics of polymeric liquids volume I: fluid mechanics; volume II: kinetic theory. Wiley, New York
45. Eggers J, Villermaux E (2008) Physics of liquid jets. Rep Prog Phys 71(036):601
46. Snoeijer JH, Pandey A, Herrada MA, Eggers J (2020) The relationship between viscoelasticity and elasticity. Proc R Soc A 476(20200):419
47. Rubio A, Galindo F, Vega EJ, Montanero JM, Cabezas MG (2022) Viscoelastic transition in transonic flow focusing. Phys Rev Fluids 7(074):201
48. Montanero JM, Rebollo-Muñoz N, Herrada MA, Gañán-Calvo AM (2011) Global stability of the focusing effect of fluid jet flows. Phys Rev E 83(036):309
49. Herrada MA, Gañán-Calvo AM, Ojeda-Monge A, Bluth B, Riesco-Chueca P (2008) Liquid flow focused by a gas: jetting, dripping, and recirculation. Phys Rev E 78(036):323
50. Si T, Li F, Yin XY, Yin XZ (2009) Modes in flow focusing and instability of coaxial liquid-gas jets. J Fluid Mech 629:1–23
51. Entov VM, Hinch EJ (1997) Effect of a spectrum of relaxation times on the capillary thinning of a filament of elastic liquid. J Non-Newtonian Fluid Mech 72:31–53
52. Oldroyd JG (1950) On the formulation of rheological equations of state. Proc Roy Soc Lond 200:523–541
53. Goldin M, Yerushalmi J, Pfeffer R, Shinnar R (1969) Breakup of a laminar capillary jet of a viscoelastic fluid. J Fluid Mech 38:689–711
54. Liu Z, Liu Z (2006) Linear analysis of three-dimensional instability of non-Newtonian liquid jets. J Fluid Mech 559:451–459
55. Chang HC, Demekhin EA, Kalaidin E (1999) Iterated stretching of viscoelastic jets. Phys Fluids 11:1717–1737
56. Oliveira MSN, McKinley GH (2005) Iterated stretching and multiple beads-on-a-string phenomena in dilute solutions of highly-extensible flexible polymers. Phys Fluids 17(071):704

57. Clasen C, Eggers J, Fontelos MA, Li J, McKinley GH (2006) The beads-on-string structure of viscoelastic threads. J Fluid Mech 556:283–308
58. Bazilevsky AV, Voronkov SI, Entov VM, Rozhkov AN (1981) Orientational effects in capillary breakup of jets and threads of dilute polymer solutions (English version in, vol 26). Doklady Akademii Nauk SSSR 257:336–339
59. Rubio A, Vega EJ, Gañán-Calvo AM, Montanero JM (2022) Unexpected stability of micrometer weakly viscoelastic jets. Phys Fluids 34(062):014
60. Goren S, Gottlieb M (1982) Surface-tension-driven breakup of viscoelastic liquid threads. J Fluid Mech 120:245–266
61. Mohamed AS, Herrada MA, Gañán-Calvo AM, Montanero JM (2015) Convective-to-absolute instability transition in a viscoelastic capillary jet subject to unrelaxed axial elastic tension. Phys Rev E 92(023):006
62. Eggers J (1997) Nonlinear dynamics and breakup of free-surface flows. Rev Mod Phys 69:865–929
63. Yarin AL, Pourdeyhimi B, Ramakrishna S (2014) Fundamentals and applications of micro- and nanofibers. Cambridge University Press, Cambridge
64. Alty T (1933) The maximum rate of evaporation of water. Lond Edinb Dublin Philos Mag J Sci 15:82–103
65. Persad AH, Ward CA (2016) Expressions for the evaporation and condensation coefficients in the Hertz-Knudsen relation. Chem Rev 116:7727–7767

The manufacturer's authorised representative in the EU is Springer Nature Customer Service Centre GmbH, Europaplatz 3, 69115 Heidelberg, Germany. If you have any concerns regarding our products, please contact ProductSafety@springernature.com

Printed and bound by CPI Group (UK) Ltd, Croydon, CR0 4YY
01/10/2026
02225907-0007